Pathogens, Vectors, and Plant Diseases: Approaches to Control

Pathogens, Vectors, and Plant Diseases: Approaches to Control

EDITED BY

KERRY F. HARRIS

Department of Entomology
Texas A&M University
College Station, Texas

KARL MARAMOROSCH

Waksman Institute of Microbiology
Rutgers University
New Brunswick, New Jersey

ACADEMIC PRESS 1982

A Subsidiary of Harcourt Brace Jovanovich, Publishers

New York London

Paris San Diego San Francisco São Paulo Sydney Tokyo Toronto

ACADEMIC PRESS, INC.
111 Fifth Avenue, New York, New York 10003

United Kingdom Edition published by
ACADEMIC PRESS, INC. (LONDON) LTD.
24/28 Oval Road, London NW1 7DX

Library of Congress Cataloging in Publication Data
Main entry under title:

Pathogens, vectors, and plant diseases.

Includes bibliographical references and index.
1. Pest control. 2. Vector control. 3. Plant diseases. 4. Insects as carriers of disease.
I. Harris, Kerry F. II. Maramorosch, Karl.
SB950.P32 632'.3 81-4893
ISBN 0-12-326440-5 AACR2

PRINTED IN THE UNITED STATES OF AMERICA

82 83 84 85 9 8 7 6 5 4 3 2 1

CONTENTS

CONTRIBUTORS

Numbers in parentheses indicate the pages on which the authors' contributions begin.

Jean B. Adams (221), Agriculture Canada Research Station, P. O. Box 20280, Fredericton, New Brunswick, Canada E3B-427

M. Basile (57), Istituto di Nematologia Agraria del C.N.R., 70126 Bari, Italy

Sven Bingefors (187), Department of Plant and Forest Protection, Swedish University of Agricultural Sciences, Uppsala, Sweden

Shlomo Cohen (45), The Volcani Institute, Bet-Dagan, Israel

J. A. Foster (151), USDA–APHIS–PPQ, U.S. Plant Introduction Station, Glenn Dale, Maryland

Isaac Harpaz (1), Department of Entomology, Hebrew University of Jerusalem, Rehovot, Israel

Robert P. Kahn (123), Plant Protection and Quarantine Programs, Animal and Plant Health Inspection Service, U.S. Department of Agriculture, Hyattsville, Maryland

G. LaBonne (95), I.N.R.A., Pathologie Vegetale, Petit-Bourg, Guadeloupe, French West Indies

F. Lamberti (57), Istituto di Nematologia Agraria del C.N.R., 70126 Bari, Italy

Karl Maramorosch (265), Waksman Institute of Microbiology, Rutgers University, New Brunswick, New Jersey 08854

J. Marrou (95), I.N.R.A., 149 Rue de Grenelle, Paris 7e, France

J. B. Quiot (95), I.N.R.A., Pathologie Vegetale, Petit-Bourg, Guadeloupe, French West Indies

Raoul A. Robinson[1] (245), 2 Balmoral Terrace, Trinity Hill, St. Helier, Jersey, United Kingdom

John M. Simons (71), JMS Flower Farms, Inc., 1105 25th Avenue, Vero Beach, Florida 32960

[1]Present address: Department of Biological Sciences, Simon Fraser University, Vancouver, British Columbia V5A1S6, Canada

T. Richard Tarn (221), Agriculture Canada Research Station, P. O. Box 20280, Fredericton, New Brunswick, Canada E3B–427

J. A. Tomlinson (23), National Vegetable Research Station, Wellesbourne, Warwick, England

PREFACE

This is the fifth and final volume in a series of books on the general topic of vectors of plant pathogens. The first three volumes, *Aphids as Virus Vectors, Leafhopper Vectors and Plant Disease Agents,* and *Vectors of Plant Pathogens,* are up-to-date treatises on pathogen–vector–host interactions and how such interactions define vector-dependent transmission systems. The fourth volume, *Plant Diseases and Vectors: Ecology and Epidemiology,* covers timely topics that illustrate some of the incipient overriding principles relating to transmission ecology—the study of how various biotic and abiotic components of an ecosystem influence pathogen–vector–host compatibility and, hence, the efficiency of pathogen transmission and disease epidemiology.

Our intent in this volume is to illustrate how knowledge of pathogen–vector–host interactions, vector ecology, and disease epidemiology is being applied to disease prevention and control. Some of the more commonplace control strategies, such as breeding plants for resistance to particular pathogens, are not discussed per se, since these approaches to control, especially the latter, have been the subject of innumerable review articles, including many current ones. Instead, our objective has been to treat less well-known but innovative strategies on the frontier of control-oriented research that have proven applicability in the field or show considerable promise in this regard. To accomplish our objective, we enlisted the talents of sixteen outstanding scientists from six countries.

Chapter 1 discusses nonpesticidal control of vector-borne viruses (evasion measures, vector repellants, sticky traps, barriers and baffles, biological and integrated control of vectors, etc.), whereas Chapter 2 discusses chemotherapy (pyrimidines and purines, antibiotics, hormones, fungicides, and herbicides) in controlling plant viruses and virus diseases. Chapter 3 is a state-of-the-art report on the use of color mulches to manipulate control vectors, especially whiteflies. Chapters 4 and 5 are detailed accounts of chemical (fumigant and nonfumigant nematicides) control of nematode vectors and the use of oil sprays and reflective surfaces in controlling aphid-borne plant viruses, respectively. Chapters 6, 7, and 8 focus on the roles that man, host plant, and nature can play in spreading or preventing the spread of vectors and vector-borne pathogens. Chapter 9 is an in-depth study of inherited nematode resistance in plants. Chapter 10 describes systems for electronically monitoring aphid probing and feeding, and their use in plant breeding programs to screen plants for resistance to aphids. An interesting and provocative discussion of evolutionary stable strategy of an aphid pathosystem is presented in Chapter 11. And, finally, Chapter 12 provides information on the latest research developments in biocontrol and chemical control of mycoplasmas and the vectors that transmit them.

The editors thank the authors in this and the previous volumes for their scholarly contributions and the staff of Academic Press for their encouragement and support.

Chapter 1

NONPESTICIDAL CONTROL OF VECTOR-BORNE VIRUSES

Isaac Harpaz

Department of Entomology
Hebrew University of Jerusalem
Rehovot, Israel

1.1 INTRODUCTION

Efforts to control viral infections in plants by direct application of chemical virucides, similar to the treatment with fungicides against phytopathogenic fungi, have so far met with minimal, if any, practical success. In order to avoid misunderstanding, the term "chemical virucides," mentioned above, is obviously distinct from "physical virucides" such as heat therapy, which for quite a long time has been successfully employed as a direct virus-killing measure in the production of virus-free stocks for commercial plant propagation (Nyland and Goheen, 1969).

ISBN 0-12-326440-5

Indeed, a rather impressive amount of advanced research has already been reported on the occurrence, identity, and mode of action of a variety of antiviral agents in plants [see, for instance, Gupta (1977), Misra (1977) and Sela (1981)]. However, not one of these agents, whether of natural or synthetic origin, has yet been developed to a stage whereby it can be applied in the field for the control of infection or spread of a virus disease in a crop. Hence, with plant virus chemotherapy being still in its infancy, the only available means of controlling plant viruses are, at least for the time being, in the area of indirect control approaches. These can be divided into the following categories according to their nature:

(A) *Cultural measures*, such as (1) genetic manipulation aimed at producing plant varieties that are resistant to infection and/or to the pathogen's vectors, (2) culturing of plant tissue fragments, viz. meristem tips, for obtaining virus-free propagation stocks, (3) elimination of inoculum sources, whether by preventive legislative measures or by actual eradication of infected and suspected material, (4) cultivation practices aimed at breaking the infection cycle by introducing sufficiently wide gaps in the availability of susceptible host plants to the virus, its vector, or both, e.g., bare fallowing, rotation of crops, and the like, and (5) evasive measures based on taking advantage of the epidemiological pattern of a virus disease in order to adjust planting and harvesting dates to evade infection.

(B) *Technical measures* devised to (1) reduce, to a maximum possible extent, the number of inoculative vector individuals that are active in the field or (2) interfere with the transmission process at any of its phases, thereby arresting the spread of the virus in the field. The former can be achieved by killing vectors out rightly (via pesticides or natural enemies), by luring them away from crop plants, or by repelling them from and thus barring their access to plants.

For obvious reasons, the measure which is most commonly applied in control approach "B1" is still conventional treatment with pesticides in a variety of formulations. However, in cases where nonpersistent viruses are involved, spraying with aphicides is sometimes likely to increase rather than decrease the spread of the disease in the crop. Situations like this were reported, for instance, by Broadbent *et al.* (1963) with respect to narcissus yellow stripe virus following spraying with DDT, or by Münster and Murbach (1952) with respect to potato viruses. The reasons for these economically negative effects of aphicides may be found in the studies of Lehmann *et al.* (1975, 1976, 1977) on the probing behavior of aphids under the influence of various insecticidal treatments.

With increasing concern throughout the world over the environmental effects of continued reliance on toxic chemicals for pest control, let alone the rising costs of these products, far greater attention should be devoted to the development of alternative, less toxic, and less disruptive methods for controlling pests in general and vector-borne viruses in particular.

A number of the aforementioned indirect control approaches have been recently reviewed in the literature. Thus, for instance, certain aspects of genetic manipulation (category "A1") were the subject of a comprehensive review by Gibson and Plub (1977), whereas the control of nonpersistent virus transmission by oils and other inhibitors, which falls under category "B2," was reviewed by Vanderveken (1977) and further updated by the studies of Simons and Beasley (1977), Simons *et al.* (1977) and Loebenstein and Raccah (1980). Chapters written by Mellor and Stace-Smith (1977) and Quak (1977) deal with meristem tip culture (category "A2").

The present chapter attempts to summarize some further research efforts in the area of less conventional control of vector transmission of plant viruses. It will mainly cover topics that fall under categories A5 and B1 of the scheme of control approaches outlined above. Their potential field applicability was the major reason for including the research works referred to in the present review.

1.2 EVASIVE MEASURES

Maize rough dwarf virus (MRDV), in years of outbreak, causes heavy losses to maize crops in Italy, Israel and other parts of the Old World where the disease occurs. The vectors of this virus are a number of delphacid planthopper species, *Laodelphax striatella* (Fallén) being the principal one. The virus propagates in its vector and can be observed in such a large variety of vector tissues that it might even be regarded as an insect pathogen, besides its causing a lethal disease in maize plants (Harpaz, 1972). Epidemiological studies carried out in Israel revealed that spread of the disease in the field practically ceases after early June. This led Harpaz (1961) to conclude, rather prematurely, that the sharp drop in the level of the planthopper population, caused by the oncoming hot and dry Eastern Mediterranean summer, is responsible for the arrest of spread of the virus by its vector. It should perhaps be explained that maize, being an exotic species in the Old World, is not a host plant of the vector and that transmission takes place quite accidentally when inoculative planthoppers probe newly emerged maize seedlings while in search of their natural, graminaceous summer host plants.

The same epidemiological pattern pertains to the Po Valley of northern Italy, where very few new infections occur in maize from mid-July onwards. However, at the very same time of midsummer heat, the population of *L. striatella* in the Po Valley rises to its annual peak owing to the rapid development on one of its preferred, natural host plants—irrigated rice. Thus, susceptible maize seedlings that are growing there adjacent to rice fields, which are spilling over their excess *L. striatella* adults, still remain virtually free from new infection throughout the planthopper peak season. Obviously, the reason for this lack of transmission during the height of summer does not lie in the paucity of vector individuals, as suggested above with respect to the MRDV situation in Israel. Incidentally, this erroneous attribution of lack of MRDV spread to the heat-induced decline of the vector population was also quoted in Broadbent's (1969) review on vector control.

A detailed study, therefore, had to be conducted on the effect of heat on each phase of the vector transmission cycle: beginning with a wheat experimental source plant, continuing through the latent period and inoculation feed of the vector, and ending in the postinoculation period of symptom development in a maize test plant. The study revealed that the only phase in the cycle that is sensitive to heat is the latent period in the vector, during which the virus multiplies in the planthopper body and accumulates in the salivary glands to attain an inoculative titer. No cessation of virus synthesis, or at least masking of visible symptoms, could be noticed either in the source plant or in the test plant following an exposure of 14 days or more to a constant temperature of 36°C as long as the plants remained alive. On the other hand, however, the same heat treatment virtually supresses the vector's transmission potential by differentially inhibiting propagation of the virus within the insect body system, but not in the plant medium as explained above. The very fact that heat sensitivity of the same virus may vary from one host species to another is not altogether surprising, considering the basic difference between a plant and an animal medium. Moreover, it has already been shown that such variance can exist between two taxonomically related host plant species; namely, the same isolate of sugarcane mosaic virus in sorghum plants will not undergo masking at a temperature which completely suppresses it in maize plants (Klein *et al.*, 1973).

Returning to the control of MRDV, it should be further explained that for epidemiological purposes the effect on the vector's inoculativity of an experimental exposure to 36°C, as described earlier, is empirically equivalent in microclimatic terms (relative to the insect's immediate ambience) to a macroclimatic daily or monthly mean of 24°C. Based on these findings, a recommendation regarding cultural control of MRDV has been developed whereby sowing of maize should begin only at such date when the emergent seedlings will be exposed to an expected daily mean temperature of not lower than 24°C. In fact, implementation of this recommendation has indeed resulted in effective reduction of MRDV incidence in the field. Table I provides the relevant meteorological data for some of the areas in the world where the virus occurs. And the epidemiology of the disease, as known from these localities, actually bears out the above assumption (Harpaz, 1972). Thus in Israel, for instance, by merely postponing the sowing of maize from the customary date of mid-April to the last week of May, incidence of MRDV among plants, emerging early in June, dropped from a countrywide average of 45% to no more than 3%, which is below any economic threshold.

TABLE I

Monthly mean temperatures (°C) of various localities where MRDV occurs. The bold figures fall within the epidemiological range of 17-24°C at which the disease actually spreads in the field. (From Harpaz, 1972).

	JAN.	FEB.	MAR.	APR.	MAY	JUN.	JUL.	AUG.	SEP.	OCT.	NOV.	DEC.
Milan	2.4	3.6	8.3	12.6	**17.6**	**22.0**	24.6	23.7	19.6	13.1	7.6	2.7
Zaragoza	5.5	8.3	10.6	13.3	**17.6**	**21.4**	24.6	24.7	20.8	14.6	9.4	5.7
Tel Aviv	13.9	15.0	16.3	**19.7**	**21.9**	24.5	26.7	27.6	25.9	23.0	18.5	14.5
Prague	–1.4	0.6	4.1	8.6	13.9	**16.9**	**18.9**	**18.3**	14.7	9.4	3.3	–0.3

This may serve as another example whereby complete practical control of a major vector-borne disease of a crop can be accomplished without recourse to any pesticidal treatment against the vector. At the same time there is an important lesson to be learned. An empirical experience can be pursued with highly successful practical results in one country, even if the reasons attributed to the success are scientifically unfounded. It becomes, however, very dangerous when these scientifically wrong interpretations serve as a basis for extrapolating to similar situations in other countries.

1.3 REPELLENCE BY REFLECTIVE SURFACES

The state of the art regarding this new approach to the control of vector-borne viruses, as it stood in the late sixties, was written up by Smith and Webb (1969). At that stage, aluminum (whether as a reflective foliar spray or as mulch) and white plastic mulches appeared to be the most promising, practical repellents of migrating winged aphids. Subsequent work in this particular approach to vector control, which was published during the seventies and cited by Wyman *et al.* (1979), further widened the range of virus diseases and vector species against which the method has been successfully tried. Thus, Loebenstein *et al.* (1975) reported from Israel that, besides aluminum foil, grey plastic mulch was slightly more effective than white plastic on black backing in the protection of peppers (*Capsicum annuum* L.) against cucumber mosaic virus and potato virus Y, both transmitted by *Myzus persicae*. Infection rates at harvest time were 4.5-6% in the mulched plots versus 45% in untreated controls, i.e., an 87-90% reduction in disease incidence. Strikingly similar results were obtained by Wyman *et al.* (1979) in the Imperial Valley of California, who succeeded in reducing the incidence of watermelon mosaic virus in summer squash (*Cucurbita pepo* L.) by 94 and 77% using, respectively, aluminum foil and white plastic mulches. In the same experiment, weekly applications of methyl demeton (a systemic, organophosphorus insecticide) against *M. persicae*, the predominant aphid in this case, had no effect on the incidence of this nonpersistently transmitted virus.

Recent developments in the use of repellent surfaces to control whitefly-borne viruses are discussed in Chapter 3 of the present volume.

Yet with all the promise shown by this noninsecticidal control approach, it still contains a number of limitations which apparently hamper its wider use in practice. Even when only 69% of the soil surface has to be mulched, as in the summer squash experiments of Wyman *et al.* (1979), it is still a costly operation. It can be economical only for high-value crops, taking into consideration the additional beneficial effects of mulching on yields, e.g. soil temperature adjustment, weed control and water conservation. Loebenstein *et al.* (1975), however, observed an occasional drop in pepper yields, presumably due to a too low soil-temperature underneath the aluminum mulch. Another limitation is that, with the growth of the crop plants, their developing foliage progressively shades the repellent reflection of the mulch surfaces. This is particularly noteworthy in regard to recumbent plants like watermelon and cucumber (Smith and Webb, 1969). However, Wyman *et al.* (1979) pointed out that repellency by aluminum mulch was significant even during harvest time, when the majority of the reflective surface was covered by the squash growth, whereas plastic mulch was no longer repellent at that stage.

Not in all instances has aluminum-foil mulch lived up to expectations. Shands and Simpson (1972), for example, tried it on potatoes in northeastern Maine and reported no reduction in numbers of *M. persicae*, even when 95% of the soil surface was mulched. A probably more practicable approach will be discussed in the following section.

1.4 STICKY YELLOW TRAPS

The fact that winged aphids are attracted to yellow has been recognized for quite a long time (Moericke, 1950). An early utilization of this phenomenon in the practice of controlling aphid-borne virus diseases was the construction and operation of the yellow water pan trap. This was used in many parts of the world as an important instrument in sampling and monitoring of aphid vector populations in the field (Moericke, 1951; Gonzalez and Rawlins, 1968).

A significant step forward in the direction of utilizing yellow traps not merely as an ancillary, indicatory tool, but rather as a primary device for direct vector control, was made by Cohen and Marco (1973) in Israel. They used yellow polyethylene sheets (dominant light reflection at a wavelength of 562 nm) covered with a layer of transparent glue that remains sticky for about three weeks, after which it has to be renewed. The sheets were erected 70 cm above ground at a distance of 6 m from the windward border of a pepper (*Capsicum annuum* L.) field, four days before germination. At the end of a 114-day period, these sticky sheets managed to lure away from the plants and trap such a large number of inflying viruliferous aphids that the cumulative percentage of infection of potato virus Y and cucumber mosaic virus (both nonpersistently transmitted viruses) in the protected plots reached only 26.2%, compared to 52% in the unprotected check plots–a reduction of 51% in disease incidence without the use of conventional chemical insecticides. Even better results were obtained by Zimmerman-Gries (1979) in her trials to protect seed potatoes against a persistently transmitted virus, potato leafroll, by surrounding the trial plots with identical sticky sheets. Here, the percentage infection in the harvested tubers in one trial year was 17.2 in unprotected plots versus no more than 2% in the protected plots, and 29 versus 6%, respectively, in the subsequent year. An infection rate of 6% in the progeny is still a little too high for certified seed potatoes. Hence additional measures, such as roguing out diseased plants in the field, will probably have to be applied.

Based upon the aforementioned work of Cohen and Marco (1973), sticky yellow plastic sheets have since become a standard, widely applied practice in Israel for the control of spread of PVY and CMV in peppers. Both these viruses (PVY being the predominant one) are a major problem in pepper growing there. The relevant recommendation of Israel's Agricultural Extension Service (Shoham, 1977) advises the erection of a yellow "protective belt" around the pepper field, or at least along the windward side, at a distance of 7 m from the terminal row. The belt should consist of 2- to 3-m-long polyethylene sheets with 2- to 3-m spaces in between. These standard, 120 cm-wide sheets are folded over and attached to 130-cm high trelising stakes in a matter that their open lower edge is 70 cm above ground (Fig. 1). The outward surface (60 cm in width) is painted with a thin layer of transparent glue. A belt like this will provide protection to 200 m downwind, as long as the drying glue is renewed in time.

One should, however, be warned against attempts to extrapolate from these quite successful examples to other situations comprising different combinations of crops, viruses and vectors. The following is a case in point. In a series of comparative tests, conducted in an aphid flight chamber, Weiss (1976) found significant differences in color sensitivity between *Myzus persicae* and *Aphis gossypii*, a matter which could actually be expected as already mentioned by Moericke (1969). Both an increase in wavelength (i.e., modifying the hue of the color) within the yellow-orange range of the spectrum, and a decrease in the color's intensity (i.e., the percentage reflectivity of the peak of the curve, compared with a white standard) render the color more attractive to *M. persicae*, whereas no such correlation could be found with respect to *A. gossypii*. The tint of the color (i.e., the amount of white added to the hue) did not seem to have any noticeable effect in this respect on either of the two aphid species within the tested range of 557-577 nm. Hence, the standard yellow color which was used in the above-described trials of Cohen and Marco (1973) and of Zimmerman-Gries (1979) proved inferior in its attractiveness to *M. persicae* and, therefore, should be replaced by a different color (namely, #404, Tambour-Askar Paints Co., Akko, Israel, having a dominant wavelength of 577 nm) wherever protection against *M. persicae* is required. When tested outdoors, this #404 color trapped

FIG. 1. Section of a yellow "protective belt" of sticky polyethylene sheets erected along the windward side of a pepper field. (Photo by C. Shoham.)

over 10% more *M. persicae* aphids than the standard color, whereas the reverse was recorded in regard to *A. gossypii*.

Furthermore, the reflection of the painted surface, though in itself not a component of the color substance, was also found to significantly influence aphid attraction. This feature has to do more with the surface onto which the color is applied and with the nature of the microcrystals formed upon the drying of the color substance. Hence, comparison between colors, regarding attractiveness to aphids, should be done under uniform reflection conditions. This is a matter which has apparently been overlooked in previous studies and could perhaps account for inconsistency in the relevant results.

Much more surprising, however, were the results of a trial by Weiss *et al.* (1977) to control the spread of vector-borne viruses in a squash field (cv. Bushy White) by the above-described yellow polyethylene sticky sheets. In a thrice-repeated experiment, the percentage of plants that became infected in plots surrounded by polyethylene sheets was significantly higher than in unprotected ones (Fig. 2). Sixty-five percent of the infected plants in the protected plots had watermelon mosaic virus (WMV), while the remaining 35% were dually infected with WMV and cucumber vein yellowing virus. The latter is a semipersistent whitefly-borne virus (Harpaz and Cohen, 1965). This negative control, which stands in complete contrast to the case of peppers (see above), has occurred consistently, in spite of a very drastic reduction in alate aphid counts in the polyethylene-protected plots throughout the entire duration of the trial.

Two different explanations were offered for this unexpected result:

(A) The aphid primarily responsible for the spread of WMV in this case was not the prevalent *A. gossypii*, but rather a much less abundant species, namely

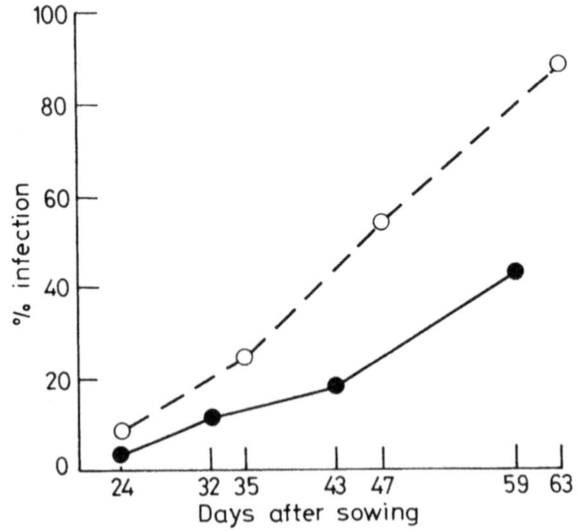

FIG. 2. Spread of watermelon mosaic virus in squash plots surrounded by sticky yellow sheets (○), compared to unprotected plots (•). (From Weiss, 1976.)

A. fabae Scopoli, which is also a recognized vector of WMV (Molnar and Schmelzer, 1964). This aphid was indeed found in protected plots in greater numbers than in unprotected ones. *M. persicae* did not show up in any of the plots. The conclusion to be derived from this explanation is that *A. fabae* is less attracted to the yellow traps than *A. gossypii*, hence its higher numbers on the plants. However, no direct evidence has yet been provided to support this claim of difference in yellow sensitivity between these two congeneric aphid species, although the phenomenon in principle has already been demonstrated, as mentioned above. At the same time, some support for the suggestion that "itinerant" rather than "colonist" aphids were involved in this case, may be found in a somewhat similar situation described by Leclant (1978, p. 286) in the south of France. There, spread of CMV in melon plots treated with dimethoate (a systemic aphicide which prevents colonization but not short-time probing by alate aphids) was significantly faster than in plots that were left untreated.

(B) Strange as it may sound, the high number of aphids that were feeding on the squash plants in the untreated plots (Fig. 3) actually protected these plants against viral infection by means of their excessive saliva injection. This may seem somewhat speculative, but similar cases have already been described (Nishi, 1969). Moreover, in a laboratory trial designed to check this assumption, half-leaves of

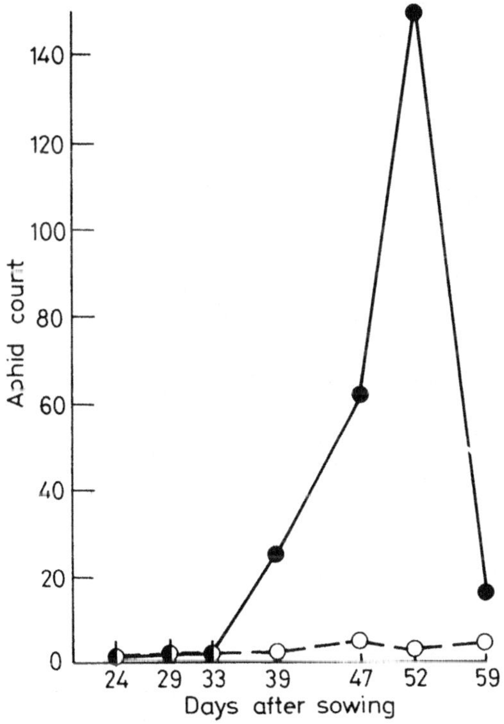

FIG. 3. Counts of alate *Aphis gossypii* on squash plants in plots surrounded by sticky yellow sheets (○), compared to unprotected plots (•). (From Weiss *et al.*, 1977.)

Chenopodium amaranticolor test plants were each caged for 24 hr with 100 virus-free individuals of *M. persicae* and then inoculated mechanically with CMV. The total number of local lesions that developed on the aphid-fed half-leaves was 66 for the five replicates, compared to 660 lesions in the aphid-free half-leaves. When the same experiment was repeated with cucumber seedlings, which are systemically susceptible to CMV, 85 out of 100 seedlings, inoculated with a 1:10-diluted sap, became infected in the aphid-free series, compared to only 44% among seedlings that were previously caged for 24 hr with 100 aphids each.

Regardless of whether either or neither of the above two explanations will eventually prove correct, the practical fact remains that the sticky trap control measure which is highly effective for one crop-virus-vector combination can be equally ineffective, or even harmful, for another combination.

1.5 BARRIERS AND BAFFLERS

The idea of using barriers to prevent vectors from reaching their target plants is almost as old as the science of plant virology. Over 50 years ago, Kunkel (1929) already reported that surrounding small aster plots with 18-mesh wire screens, 1.2-2.4 m high, reduced the incidence of aster yellows (a leafhopper-borne plant pathogen) from 80 to 20%. In 1969 Broadbent reviewed the work that was done in the 40 intervening years on the subject, including plant and soil barriers. With respect to barrier crops, Broadbent pointed out that these are likely to be more effective in protecting plants from nonpersistent viruses than from persistently transmitted ones, because aphids will often lose the nonpersistent virus when they probe the barrier plants. An observation relevant to this point was made by Cohen and Marco (1973) regarding sticky yellow sheets (see preceding section). They recorded a significant decrease in virus spread also in plots which were surrounded by nonsticky yellow plastic sheets, though not to such extent as in those surrounded by sticky sheets. The question, however, still remains open as to whether the glueless yellow sheets act merely as mechanical flight barriers or as preferential attractants to winged viruliferous aphids, thus causing the latter to lose their nonpersistent virus charge while attempting to probe the plastic material.

Undoubtedly, the most effective way to apply the concept of barring access would be to grow crops in vector-proof screenhouses. For obvious reasons, this would very rarely prove practical, except for most valuable crops, such as virus-free propagation stocks. Nevertheless, some new developments have indeed taken place in this particular area, which will be discussed below.

The observation of Weiss (1976), made in an aphid flight chamber, suggested that from a distance of 50 cm both *M. persicae* and *A. gossypii* failed to "see" and be attracted to yellow disks measuring 55 mm in diameter. The latter is about the size of the horizontal silhouette of an emergent vegetable seedling. Proper detection of such alightment targets could be done only from distances of 20 cm or less. Earlier indications that aphids are indeed so "nearsighted" can be inferred from Kring's (1966) flight chamber observations, where direct flights towards a yellow target (in his case a 125 X 175 mm card) began only from a distance of some 65 cm. This has duly led Cohen *et al.* (1978) to assume that alate aphids could be prevented from alighting on plants by means of inexpensive, wide-mesh nets instead of the costly dense screens used in conventional insect-proofing. In a series of field trials, involving synthetic-fiber nets of different colors and mesh sizes, they were able to show that incidence of PVY and CMV caused by *M. persicae* in peppers 105 days after sowing can be reduced from 90% to no more than 2% merely by stretching a white-colored 8-mesh net over the plants at a height of 70 cm or more above ground (Fig. 4). The supremacy of white over grey in this respect is somewhat surprising, since Loebenstein *et al.* (1975) found grey plastic mulch to be slightly more repellent than white against the same aphid species in the same crop. However, one should not equate the repellency of a continuous, uniform surface of a mulch layer to that of a reticular texture of a net. Furthermore, it may very well be that the mode of action of these white nets is more in the way of baffling the aphids by camouflaging the contrast between the plant and its soil-surface background rather than mere repellency by the light reflected from the net's fibers. This contrast was shown by a number of workers to be instrumental in the aphid's ability to

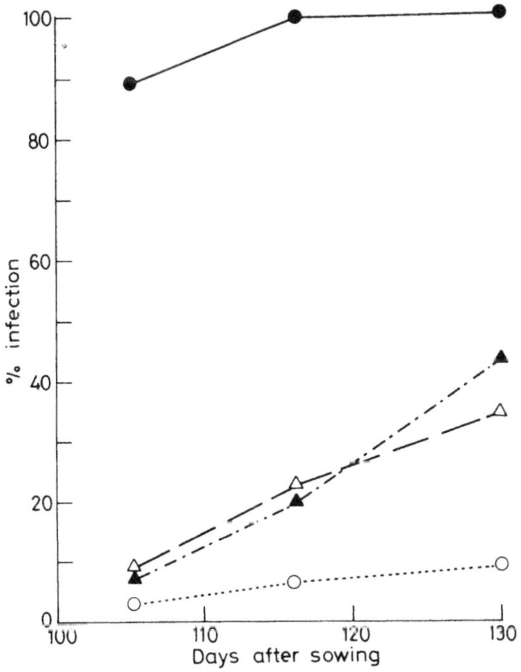

FIG. 4. Spread of both PVY and CMV in pepper plots protected by white (○), yellow (▲), and grey (△) large-mesh nets, in comparison to unprotected (•) plots. (Cohen,1981).

locate its target plant for alignment (see Smith, 1976). The matter is apparently even more complex than that, considering the observation of Cohen *et al.* (1978) that many alate aphids were seen wandering on the outside of the yellow net which obviously attracted them, but most of them failed to find their way to the plants through the holes, although these were wide enough (about 3 mm in diameter) for an aphid's body size.

Results as satisfactory as those represented in Fig. 4 were obtained with a knitted, white, high-density polyethylene net (2 × 8 mesh, corresponding to a quadrangular hole size averaging 12 × 3 mm), a type widely used nowadays as bagging material and also for protecting crops against hail. This flexible net is considerably cheaper than the 8-mesh molded nets referred to in Fig. 4. Its holes are large enough to allow normal development of the plants, since the amount of shading it casts does not exceed 18%. The aphid exclusion capacity of the net was demonstrated by the scarcity of aphid individuals that managed to penetrate it in comparison to the number trapped in the unnetted plots, namely 115 versus 4019, a reduction of more than 40 times (Cohen, 1981).

With the rapidly developing technology of growing out-of-season protected crops under plastic-covered structures of various designs, the application of the above-described new technique of control of aphid-borne diseases will most probably become economically feasible in the very near future.

A note of caution should, however, be registered here again against generalizing from this particular experience. In fact, when the same knitted net, mentioned above, was tested on squash in the south of France, where aphid populations in summer reach much greater proportions than in Israel, the results were rather disappointing (Cohen and Marco, 1979). Only a slight delay in CMV spread could be achieved. Later on, the 3% or so of aphids that manage to infiltrate through the net (see above) are still numerous enough, in absolute terms, for spreading disease to an impermissible extent. Moreover, once these "infiltrators" start colonizing the plants under the partial shade of the net cover, they develop much larger populations than in uncovered plots. A similar difficulty was encountered by Cohen and Marco (1979) in their attempt to protect potato plants with the same net against spread of potato leaf roll virus (PLRV). Populations of *M. persicae*, the main vector of this persistently transmitted virus, could indeed be maintained at a low level by aid of the net. However, spread of PLRV was prevented (0-3% in netted test plots versus 12-48% in unnetted controls) only in plots where virus-free seed was used. Contrarily, in plots contaminated with seed-borne infection, the rate of PLRV spread under the net was even greater than in unprotected controls. Most of this increased spread was perpetrated by apterous aphids, the numbers of which were considerably higher in the net-covered plots thanks to shade and presumable concealment form natural enemies.

A novel method of preventing alate aphids from colonizing their host plants was preliminarily tried by Bar-Joseph and Frankel (1979). By spraying lime (*Citrus aurantifolia* Swingle) seedlings with a clay-mineral suspension containing a mixture of kaolinite and montmorillonite, they managed to reduce the number of winged individuals of the spirea aphid (*Aphis citricola* van der Goot) from an average of 13.5 per plant in the untreated series to no more than 2-3 in the sprayed plants. A similar reduction was observed in a comparative treatment where the seedlings were covered with a white 2X8-mesh net of the type described above. The assumption is that the white coating which is formed on the leaves by the clay-mineral spray hinders the flying aphids from settling, in a manner somewhat similar to that of the white net. However, more studies are still required to determine the effect of these mineral coats on the development of the plant, as well as their persistence under rainy or overhead-irrigation conditions.

1.6 SOIL SOLARIZATION

The principal vectors of soil-inhabiting plant viruses are ectoparasitic nematode species of the Dorylainid order, belonging to the genera *Xiphinema, Longidorus, Trichodorus, Paratrichodorus* and probably also to the genus *Paralongidorus* (Hooper, 1975). Control of these vectors in the open field is done primarily by chemical nematicides applied to the soil in either liquid or fumigant formulation. For small-scale purposes such as greenhouse benches, soil pasteurization by electric heating, hot water or steam is a common practice which takes care of nematodes as well as fungi, bacteria and weed seeds. However, all these control practices are considerably expensive, in many instances ecologically disruptive, and not always sufficiently efficacious.

A novel method for the control of noxious organisms in the soil has been recently developed in Israel, utilizing solar radiation, which is abundant there, for raising the soil temperature to partially sterilizing levels. The technique is based on mulching* the irrigated, moist soil with transparent 0.03 mm-thick polyethylene sheets (Fig. 5) during the months of July and August. This will heat the soil to maximum temperatures of 49-52 and 42°C at depths of 5 and 15 cm, respectively (Katan, 1980).

In a field trial devised to test, inter alia, the efficacy of this solarization (or pasteurization) technique against nematodes (Grinstein *et al.*, 1979), the soil of a potato field naturally infested with the nematode *Pratylenchus thornei* Sher and Allen was irrigated and mulched with transparent 0.04 mm-thick polyethylene sheets for a period of 31 days before planting. Maximum temperatures in the mulched soil were 46-54 and 40-44°C at depths of 5 and 20 cm, respectively. In the unmulched soil the temperatures were 8-12°C lower. The elevated temperatures in the solarized plots have effectively suppressed the nematode infestation, maintaining it at a very low level both in the soil and on the plant roots, and eventually reducing it to zero by the 140th day after planting. These control results proved even better than those obtained in comparative plots treated with a 4:1 mixture of ethylene dibromide and chloropicrin at a rate of 350 liters/ha.

*Pullman and DeVay (1977) used the term "tarping" (abbreviated from tarpaulin) instead of mulching for this operation.

FIG. 5. Mechanized soil "tarping" with transparent polyethylene sheets. (Photo by J. Katan.)

The latter is a standard soil fumigant against nematodes in the country. It seems pertinent to mention in this connection that the soil heating efficiency of this fuelless method is comparable to flaming, as can be derived from Miyamoto's (1961) trials to control soil-borne wheat mosaic (a fungus-transmitted virus) by means of soil treatment. He also used, inter alia, a kerosene flame thrower, which in several minutes of flaming did not raise the soil temperature to more than 40-60°C at a depth of 15 cm.

Although the nematode genus involved in the above trial (*Pratylenchus*) is not one of the four genera of virus vectors mentioned earlier, one may still safely assume that the same soil solarization technique will be equally, if not more, effective against vector nematodes. The assumption is based upon the studies of Cohn (1969) on the vertical distribution in the soil of various species of *Xiphinema* and *Longidorus*. He reported maximum population levels in the 0-30 cm layer, with a decrease in numbers with increasing depths. As shown above in the work of Grinstein *et al.* (1979), this is the soil layer into which the nematicidal solar heat penetrates in the mulched soil. Furthermore, all virus-transmitting nematodes, being ectoparasites, are more susceptible than endoparasites (like the *Pratylenchus* species) to environmental stress factors such as high temperature. In fact, with respect to *Paratrichodurus minor* Colbran, for instance, which is a vector of tobacco rattle virus, Schilt and Cohn (1975) found that a soil temperature of 30°C is already supraoptimal. Hence a temperature of 42°C, prevailing in the solarized soil at a depth of 20 cm (see above), will probably be lethal to this nematode as well as to its ecologically equivalent, allied species.

It should perhaps be added in this connection that high soil temperature is likely to affect the soil-borne virus even more severly than its vector. Thus, for instance, Debrot (1964) reported that raising the soil temperature to 30°C was enough to reduce to zero the transmission rate of a strain of raspberry ringspot virus to cucumber test plants by the nematode *Longidorus macrosoma* Hooper. This temperature is definitely not lethal to the nematode directly, but apparently only suppresses feeding, thereby preventing spread of the virus. This situation is quite reminiscent of the one discussed in Section 1.2 regarding differential heat sensitivity of maize rough dwarf virus, its planthopper vector and its cereal host plants.

1.7 BIOLOGICAL AND INTEGRATED CONTROL OF VECTORS

Biological control of crop pests can be defined as the utilization of natural enemies (predators, parasites and pathogens) of a noxious organism to prevent it from causing economic damage. A distinction should first be made, in relation to the term "economic damage," between the minimal population density of a pest that is likely to cause direct economic damage to a crop (also termed the "economic threshold") on the one hand and, on the other, the minimal population density of a vector that is likely to spread a virus in a crop beyond an economically tolerable level. Although the spreading of a disease agent by a vector is regarded as *indirect* economic damage, its significance in monetary terms may often greatly exceed the *direct* losses inflicted by the mere feeding of the vector on the crop plants. A small number of virus-carrying aphids in a crop will hardly cause by their feeding any noticeable, direct damage to plants that would economically warrant the application of a control measure. Yet the same sparse, far-below-threshold aphid population is capable of inoculating a very large percentage of the plants with virus. This is particularly, though not exclusively, true in regard to nonpersistent viruses, whereby alate, noncolonizer aphids, while briefly probing plants and taking off to alight soon on others, transmit such viruses with much greater efficiency than resident colonist aphids [see Harris (1977) and Section 1.4]. An instance in point is the recently published observation by Roitberg *et al.* (1979) that even apterous pea aphids, *Acyrthosiphon pisum* (Harris), readily dispersed to new host plants in the presence of adult predators of *Coccinella californica* Mannerheim.

However, where vector spread of persistent viruses is concerned, the picture is quite different. Here, a much closer relationship exists between the size of the "resident" vector population and the rate of virus spread in the crop. Hence, any measure which will significantly decrease the number of vector individuals present in the field, including the activity of biological control agents, is likely to result in a corresponding reduction in virus incidence.

It should, however, be remembered that biocontrol efforts have thus far been successful in such situations where the economic threshold, density of the target pest population has been rather high, being considerably higher, in fact, than the threshold levels pertaining to virus-spreading vector populations, even when persistent viruses are concerned. This somewhat discouraging state of affairs has duly led Remaudière (1971) to state, while discussing the possibility of controlling aphid-borne beet viruses by entomophagous fungi, that "it seems illusory, today, to hope to prevent the appearance of viruses transmitted through aphids in beet fields, by means of *Entomophthora*." However, a considerable decrease in the rate of virus spread can yet be expected if the aphid population in the field is adequately reduced by the action of a microbial or any other biological control agent.

Notwithstanding the fact that Remaudière's above statement is virtually as relevant today as it was 10 years ago, there could still be certain special circumstances in which biological control might serve as the main factor in curbing the spread of a vector-borne virus disease. A case in point is the exceptionally severe outbreak of the aphids *Metopolophium dirhodum* (Walker) and *Sitobion avenae* (Fabricius) which has been going on over vast areas of cereal crops in Latin America for the past ten years (Zúniga and Suzuki, 19676). Barley yellow dwarf virus, which is persistently transmitted by these two aphids (Rochow, 1970), has since become coextensively widespread. The disease now presents a major problem to cereal cultivation in the affected countries, attaining unparalleled crisis proportions.

The assumption in this case is that the two aphid species have only recently arrived in South America, leaving behind in their native lands a number of important aphidophagous species that would normally keep them under effective natural control. As is often the case, insect species invading a new territory, devoid of some of the ecological factors which restrain them in their area of origin, will "break loose" and spread like wildfire. This usually goes on for a number of years until local agents of the disrupted, invaded ecosystem eventually put an end to the outbreak and restore balance. However, the urgent need to protect crops against such ravages does not allow us to wait for nature to take its course. Immediate human intervention is doubtlessly required in such situations.

In the Mediterranean area, for example, where the two aphid species are indigenous, their populations have never been reported to reach even anywhere near such explosive levels as encountered by the present author during a recent visit to Chile (Harpaz, unpublished). Nor does BYDV, for the same reasons, constitute any real problem in Israel for instance, although its presence there has been documented (Harpaz and Klein, 1965). An internationally cooperative project has therefore been initiated to import a number of the missing natural enemies of *M. dirhodum* and *S. avenae* from the Mediterranean and other areas into South America (van den Bosch, 1976). One such beneficial insect, *Aphidius ervi* Haliday (Hymenoptera: Aphidiidae), has already become established in Chile. Its effect is expected to be shortly noticed in containing the enormous vector populations and the resultant spread of BYDV.

Past experience has, however, taught us that a problem of such magnitude as that of cereal aphids and BYDV spread in the South American continent, is not likely to be solved merely by importation and release of one or more species of exotic parasites and predators. The fuller answer lies in the combination of judicious use of selective pesticides together with intensive development and application of biological, cultural and other control measures, some of which were discussed in the preceding sections. Such a combined approach to the solution of the problem, better known as integrated pest control, is currently regarded as the most reasonable, most promising, and most feasible alternative to exclusively chemical warfare against harmful insects, including vectors of plant pathogens.

One such integrated control program has already been implemented to control virus yellows of sugarbeets in 21 cooperating European countries. Although this complex system has met with a considerable degree of success, yellows can still spread widely in some years in certain regions, which means that the battle against these viruses is not yet won (Heathcote, 1976).

1.8 SOME CONCLUDING REMARKS

Energy conservation and increased food production to meet the requirements of a rapidly growing world population are undoubtedly going to be the key words in matters relating to human welfare in this ninth decade of the century. Minimizing crop losses by the application of nonpesticidal, i.e., energy-saving, control strategies is simultaneously, therefore, a direct contribution toward both of the aforementioned objectives. Some of the advances made in this general direction during the past decade have been discussed in preceding sections of this treatise. Although these refer to one particular area of crop protection, namely the prevention of losses caused by vector-borne plant viruses, their importance in economic terms is paramount.

However, not all these newer, noninsecticidal measures have yet been developed to the stage when they can be recommended to the respective growers as an alternative to conventional pesticide application. Further intensive research efforts are still required to achieve that goal, but the potential and promise of these approaches have been demonstrated. The inescapable need to reduce the use of pesticides in agriculture will duly expedite the development of noninsecticidal measures, as long as efforts to combat plant viruses remain within the domain of "indirect control approaches" (see Section 1.1).

Even the latest technique of soil solarization (Section 1.6), which is an outstanding example of harnessing solar energy for crop protection in place of "fossil," energy-consuming, soil sterilants, still has a number of limitations that impede its wider application: (a) its use is restricted to those parts of the world where solar radiation is available in ample amounts; (b) the cost of the plastic mulching sheets, which are usually used only once, is not negligible; (c) treated fields must be left unworked for at least 30 days, a matter likely to create some farm-management problems, and (d), as far as virus control is concerned, solarization is limited to soil-borne viruses, the economic importance of which is rather minor in the overall global picture.

The science of insect control, in general, has already given rise to the development of a number of novel approaches (e.g., pheromones, male sterilization by radiation or chemosterilants, and genetic control) which are not treated in this chapter. The reason for their omission is given in the introduction where it is stated that the major criterion for inclusion of control approaches in the chapter was their potential for field applicability in controlling virus spread by vectors. To my knowledge, none of these novel methods has yet been successfully tried for the above purpose. In at least one reported case, the alarm pheromone of aphid cornicles (trans-β-farnesene) was indeed tried. According to Yang and Zettler (1975), the pheromone failed to significantly reduce the transmission rate of blackeye cowpea mosaic virus by *A. craccivora* and *M. persicae*. Nevertheless, Nault and Montgomery (1977), while discussing this experiment in a wider context, still imply optimism for the future use of pheromones in reducing spread of aphid-borne viruses. Yet, in a later-reported trial, where synthetic trans-β-farnesene was incorporated into polyvinylchloride bars to regulate release of the pheromone, still no significant reduction could be measured in alate aphid-caused spread of potato virus Y in seed potato plots following repeated applications of the repellent in two different years (Hille Ris Lambers and Schepers, 1978).

It is indeed not altogether surprising that serious difficulties are encountered in attempts to apply these modern, more sophisticated approaches for the control of plant virus vectors. The reasons for that have already been discussed in Section 1.7 in connection with the different economic thresholds of insect population densities that apply to virus inoculators as distinct from mere feeders. Nevertheless, one may quite safely foresee that in a chapter like the present one, to be written five or ten years hence, many practicable examples of these and even newer methods will be cited with respect to control of vector-borne viruses of plants.

1.9 ACKNOWLEDGMENTS

I am obliged to Mr. C. Shoham, Extension Service, Israel Ministry of Agriculture, Tel Aviv, and Prof. J. Katan, Department of Plant Pathology, Hebrew University, Rehovot, for kindly lending unpublished photographs for reproduction in this chapter, as acknowledged in the relevant captions of Figs. 1 and 5. The courtesy of Keter Publishing House, Jerusalem, in providing Table I is likewise acknowledged.

1.10 REFERENCES

Bar-Joseph, M., and Frankel, H. (1979). Spraying lime plants with clay suspensions prevents their colonization by aphids. *Hassadeh (Tel Aviv)* **60**:65-67 (in Hebrew with an abstract in English).

Broadbent, L. (1969). Disease control through vector control. *In* "Viruses, Vectors, and Vegetation" (K. Maramorosch, ed.), pp. 593-630. Wiley (Interscience), New York.

Broadbent, L., Green, D. E., and Walker, P. (1963). Narcissus virus diseases. *Daffodil Tulip Yearbook* **28**:154-160.

Cohen, S. (1981). Reducing the spread of aphid-transmitted viruses in peppers by coarse-net cover. *Phytoparasitica* **9**:69-76.

Cohen, S., and Marco, S. (1973). Reducing the spread of aphid-transmitted viruses in peppers by trapping the aphids on sticky yellow polyethylene sheets. *Phytopathology* **63**: 1207-1209.

Cohen, S., and Marco, S. (1979). Reducing virus spread in vegetables and potatoes by net cover. *Phytoparasitica* **7**:40-41.

Cohen, S., Yoles, K., and Ben-Yossef, R. (1978). Prevention of virus spread in sweet pepper by coverage with large-mesh nets. *Hassadeh (Tel Aviv)* **58**:1775-1779 (in Hebrew).

Cohn, E. (1969). The occurrence and distribution of species of *Xiphinema* and *Longidorus* in Israel. *Nematologica* **15**:179-192.

Debrot, E. A. (1964). Studies on a strain of raspberry ringspot virus ocurring in England. *Ann. Appl. Biol.* **54**:183-191.

Gibson, R. W., and Plumb, R. T. (1977). Breeding plants for resistance to aphid infestation. *In* "Aphids as Virus Vectors" (K. F. Harris and K. Maramorosch, eds.), pp. 473-500. Academic Press, New York.

Gonzalez, D., and Rawlins, W. A. (1968). Aphid sampling efficiency of Moericke traps affected by height and background. *J. Econ. Entomol.* **61**:109-114.

Grinstein, A., Orion, D., Greenberger, A., and Katan, J. (1979). Solar heating of the soil for the control of *Verticillium dahliae* and *Pratylenchus thornei* in potatoes. *In* "Soil-Borne Plant Pathogens" (B. Schipper and W. Gams, eds.), pp. 431-438. Academic Press, London.

Gupta, B. M. (1977). Inhibition of plant virus infections by antiviral agents. *In* "Aphids as Virus Vectors" (K. F. Harris and K. Maramorosch, eds.), pp. 455-471. Academic Press, New York.

Harpaz, I. (1961). *Calligypona marginata,* the vector of maize rough dwarf virus. *FAO Plant Protect. Bull.* **9**:144-147.

Harpaz, I. (1972). "Maize Rough Dwarf, a Planthopper Virus Disease affecting Maize, Rice, Small Grains and Grasses." Israel Universities Press, Jerusalem.

Harpaz, I., and Cohen, S. (1965). Semipersistent relationship between cucumber vein yellowing virus (CVYV) and its vector, the tobacco whitefly *(Bemisia tabaci* Gannadius). *Phytopathol. Z.* **54**:240-248.

Harpaz, I., and Klein, M. (1965). Occurrence of barley yellow dwarf virus (BYDV) in Israel. *Plant Dis. Reptr.* **49**:34-35.

Harris, K. F. (1977). An ingestion-egestion hypothesis of noncirculative virus transmission. *In* "Aphids as Virus Vectors" (K. F. Harris and K. Maramorosch, eds.), pp. 165-220. Academic Press, New York.

Heathcote, G. D. (1976). Insects as vectors of plant viruses. *Z. Angew. Entomol.* **82**:72-80.

Hille Ris Lambers, D., and Schepers, A. (1978). The effect of trans-B-farnesene, used as a repellent against landing aphid alatae in seed potato growing. *Potato Res.* **21**:23-26.

Hooper, D. J. (1975). Virus vector nematodes–taxonomy and general introduction. *In* "Nematode Vectors of Plant Viruses" (F. Lamberti, C. E. Taylor, and J. W. Seinhorst, eds.), pp. 1-14. Plenum Press, London and New York.

Katan, J. (1980). Solar pasteurization of soils for disease control: Status and future prospects. *Plant Dis.* **64**:450-454.

Klein, M., Harpaz, I., Greenberger, A., and Sela, I. (1973). A mosaic disease of maize and sorghum in Israel. *Plant Dis. Reptr.* **57**:125-128.

Kring, J. B. (1966). An aphid flight chamber: construction and operation. *J. Econ. Entomol.* **59**:1518-1520.

Kunkel, L. O. (1929). Wire-screen fences for the control of aster yellows. *Phytopathology* **19**:100.

Leclant, F. (1978). Étude bioécologique des aphides de la région méditerranéenne. Implications agronomiques. Thèse du Doctorat d'État, Université des Sciences et Techniques du Languedoc, Montpellier, France.

Lehmann, W., Claus, S., and Karl, E. (1975). Das Verhalten von Aphiden unter dem Einfluss verschiedener Pflanzenarten und ihrer chemischen Behandlung, bezogen auf den Anteil einstechender Tiere und kurzfristigen Probesaugstiche. *Arch. Phytopathol. Pflanzenschutz* **11**:273-282.

Lehmann, W. Claus, S., and Karl E. (1976). Der Einfluss verschiedener Pflanzenarten und ihrer chemischen Behandlung auf die Anzahl der Probesaugstiche und die Dauer des ersten Einstichs von Aphiden. *Arch. Phytopathol. Pflanzenschultz* **12**:345-354.

Lehmann, W., Claus, S., and Karl, E. (1977). Der Einfluss verschiedener Pflanzenarten und ihrer chemischen Behandlung auf die Laufaktivität von Aphiden. *Arch. Phytopathol. Pflanzenschutz* **13**:137-144.

Loebenstein, G., Alper, M., Levy, S., Palevitch, D., and Menagem, E. (1975). Protecting peppers from aphid-borne viruses with aluminum foil and plastic mulch. *Phytoparasitica* **3**: 43-53.

Loebenstein, G., and Raccah, B. (1980). Control of nonpersistently transmitted aphid-borne viruses. *Phytoparasitica* **8**:221-235.

Mellor, F. C., and Stace-Smith, R. (1977). Virus-free potatoes by tissue culture. *In* "Applied and Fundamental Aspects of Plant Cell, Tissue and Organ Culture" (J. Reinert and Y. P. S. Bajaj, eds.), pp. 616-635. Springer-Verlag, Berlin, Heidelberg, and New York.

Misra, A. (1977). The use of antibiotics for the control of plant virus diseases. *Z. Pflanzenkr. Pflanzenschutz* **84**:244-252.

Miyamoto, Y. (1961). Studies on soil-borne, cereal mosaics. VII. Controlling soil-borne cereal mosaics with special reference to the effect of soil treatment with pyroligneous acid (wood vinegar). *Ann. Phytopathol. Soc. Japan* **26**:90-97. (in Japanese with a summary in English).

Moericke, V. (1950). Über das Farbsehen der Pfirsichblattlaus (*Myzodes persicae* Sulz.). *Z. Tierpsychol.* **7**:265-274.

Moericke, V. (1951). Eine Farbfalle zur Kontrolle des Fluges von Blattläusen in besonders der Pfirsichblattlaus *Myzodes persicae* (Sulz.). *Nachrichtenbl. Dtsch. Pflanzenschutzdienstes (Braunschweig)* **3**:23-24.

Moericke, V. (1969). Host plant specific colour behaviour by *Hyalopterus pruni (Aphididae). Entomol. Exp. Appl.* **12**:524-534.

Molnar, A., and Schmelzer, K. (1964). Beiträge zur Kenntnis der Wassermelonenmosaik-Virus. *Phytopathol. Z.* **51**:361-384.

Münster, J., and Murbach, R. (1952). L'application d'insecticides contre les pucerons vecteurs des viroses de la pomme de terre peut-elle garantir la production de plants de qualité? *Rev. Rom. Agric.* 8:41-43.

Nault, L. R., and Montgomery, M. E. (1977). Aphid pheromones. *In* "Aphids as Virus Vectors" (K. F. Harris and K. Maramorosch, eds.), pp. 527-545. Academic Press, New York.

Nishi, Y. (1969). Inhibition of viruses by vector saliva. *In* "Viruses, Vectors, and Vegetation" (K. Maramorosch, ed.), pp. 579-591. Wiley (Interscience), New York.

Nyland, G., and Goheen, A. C. (1969). Heat therapy of virus diseases of perennial plants. *Annu. Rev. Phytopathol.* **7**:331-354.

Pullman, G. S., and DeVay, J. E. (1977). Control of *Verticillium dahliae* by plastic tarping. *Proc. Am. Phytopathol. Soc.* **4**:210.

Quak, F. (1977). Meristem culture and virus-free plants. *In* "Applied and Fundamental Aspects of Plant Cell, Tissue and Organ Culture" (J. Reinert and Y. P. S. Bajaj, eds.), pp. 598-615. Springer-Verlag, Berlin, Heidelberg, and New York.

Remaudière, G. (1971). Vers l'utilisation pratique des Entomophthorales parasites de pucerons. *Parasitica* **27**:115-126.

Rochow, W. F. (1970). Barley yellow dwarf virus. CMI/AAB Descriptions of Plant Viruses, No. 32, Commonwealth Agricultural Bureaux, Slough, England.

Roitberg, B., Myers, J. H., and Frazer, B. D. (1979). The influence of predators on the movement of apterous pea aphids between plants. *J. Anim. Ecol.* **48**:111-122.

Schilt, H. G., and Cohn, E. (1975). Pathogenicity and population increase of *Paratrichodorus minor* as influenced by some environmental factors. *Nematologica* **21**:71-80.

Sela, I. (1981). Plant-virus interactions related to resistance and localization of viral infections. *Adv. Virus Res.* **26**:201-237.

Shands, W. A., and Simpson, G. W. (1972). Effects of aluminum foil mulches upon abundance of aphids on, and yield of potatoes in Northeastern Maine. *J. Econ. Entomol.* **65**: 507-510.

Shoham, C. (1977). Recommendations for the control of pests in vegetable crops 1977/78. Israel Ministry of Agriculture, Extension Service, Tel Aviv (in Hebrew).

Simons, J. N., and Beasley, C. A. (1977). Visualization of oil on leaf surfaces—a technique. *J. Econ. Entomol.* **70**:307-308.

Simons, J. N., McLean, D. L., and Kinsey, M. C. (1977). Effects of mineral oil on probing behavior and transmission of stylet-borne viruses by *Myzus persicae*. *J. Econ. Entomol.* **70**:309-315.

Smith, F. F., and Webb, R. E. (1969). Repelling aphids by reflective surfaces, a new approach to the control of insect-transmitted viruses. *In* "Viruses, Vectors, and Vegetation" (K. Maramorosch, ed.), pp. 631-639. Wiley (Interscience), New York.

Smith, J. G. (1976). Influence of crop background on aphids and other phytophagous insects on Brussels sprouts. *Ann. Appl. Biol.* **83**:1-13.

van den Bosch, R. (1976). Report on a second visit to Chile as a consultant to the program of biological and integrated control of cereal aphids. Unpublished.

Vanderveken, J. J. (1977). Oils and other inhibitors of nonpersistent virus transmission. *In* "Aphids as Virus Vectors" (K. F. Harris and K. Maramorosch, eds.), pp. 435-454. Academic Press, New York.

Weiss, M. (1976). Trials in controlling virus diseases in vegetable crops by luring aphids to sticky color traps. M.Sc. Thesis, Hebrew University, Faculty of Agriculture, Rehovot (in Hebrew with a summary in English).

Weiss, M., Cohen, S., Marco, S., and Harpaz, I. (1977). Failure of sticky yellow traps to reduce aphid-borne virus incidence in squash. *Hassadeh (Tel Aviv)* **58**:75-78 (in Hebrew).

Wyman, J. A., Toscano, N. C., Kido, K., Johnson, H., and Mayberry, K. S. (1979). Effects of mulching on the spread of aphid-transmitted watermelon mosaic virus to summer squash. *J. Econ. Entomol.* **72**:139-143.

Yang, S. L., and Zettler, F. W. (1975). Effects of alarm pheromone on aphid probing behavior and virus transmission efficiency. *Plant Dis. Reptr.* **59**:902-905.

Zimmerman-Gries, S. (1979). Reducing the spread of potato leaf roll virus, alfalfa mosaic virus and potato virus Y in seed potatoes by trapping aphids on sticky yellow polyethylene sheets. *Potato Res.* **22**:123-131.

Zúniga, E., and Suzuki, H. (1976). Ecological and economic problems created by aphids in Latin America. *Outlook Agric.* **8**:311-319.

Chapter 2

CHEMOTHERAPY OF PLANT VIRUSES AND VIRUS DISEASES

J. A. Tomlinson

National Vegetable Research Station
Wellesbourne, Warwick
England

2.1 INTRODUCTION

Most crop plants can become infected with a variety of different pathogens including fungi, bacteria, mycoplasmas and viruses resulting in varying reductions of yield or quality. With fungus diseases, direct methods of control have been possible for many years using protectant fungicides (Woodcock, 1968; Bent, 1969) and lately further dramatic advances have been made following the dis-

ISBN 0-12-326440-5

covery and use of systemic fungicides (Evans, 1971; Wain and Carter, 1977). Direct control with antibiotics is also possible for diseases caused by bacteria such as fire blight (Schroth *et al.*, 1974) and halo-blight (Taylor and Dudley, 1977) and tetracyclines have been used for the control of pear decline, caused by a mycoplasma (Nyland and Moller, 1973).

Few chemicals are known, however, that will eliminate or substantially reduce the multiplication of viruses in infected plants, although certain purine and pyrimidine analogues including 2-thiouracil and 8-azaguanine will decrease virus multiplication without eradicating the virus (Matthews and Smith, 1955). Because of the narrow safety margin between antiviral effectiveness and phytotoxicity such substances have found no practical application (Matthews, 1970).

The importance of plant viruses derives from their ability to cause systemic infections and to persist in the vegetative parts of plants and sometimes in their seeds. Once infected, the virus persists in perennial plants indefinitely, so that the progeny of vegetatively propagated plants are also usually infected.

The symptoms and the concentration of virus in infected plants frequently change with plant age and changes in environment until infected plants are sometimes indistinguishable from healthy plants. However, there is no true recovery as can occur with animal virus diseases in which the infecting virus is neutralized by interferon and antibodies. Ideally, the simplest method of controlling plant viruses has been by breeding immune, resistant or tolerant cultivars. Where such sources of resistance cannot be found, however, control depends on measures to protect plants by killing virus vectors, removing sources of infection or by using healthy seed and virus-free vegetative stock derived by heat-therapy or meristem-tip culture (Hollings, 1965).

The objective of plant virus chemotherapy is to find a chemical that will provide an effective method of control or, failing that, will augment existing measures. A successful chemotherapeutant might conceivably supplant the time-consuming method of meristem-tip culture to establish healthy clones, eradicate virus from infected seeds or inhibit symptoms in field infected crops and prevent loss of yield. Although the purist might consider a successful chemotherapeutant within the strict confines of the interference or inhibition of virus replication, I prefer a broader definition which includes any substance with protective or therapeutic effects which significantly enhances resistance, tolerance, recovery or suppression of symptoms.

2.2 EARLIER SEARCHES FOR CHEMOTHERAPEUTANTS

During the 1940's and 1950's many compounds were found to interfere or inhibit infection by plant viruses. They included a wide variety of plant and fungal growth products, enzymes and other synthetic products. When mixed with virus preparations or if applied as leaf sprays *before* inoculation, they reduced or totally prevented infection. One of the best known examples occurs in pokeweed (*Phytolacca americana*) which contains a glycoprotein with a molecular weight of 13,000 (Wyatt and Shepherd, 1969). Recent studies showed that, as well as inhibiting infection of plant cells, the inhibitor also prevented infection of animal cells by influenza virus (Tomlinson *et al.*, 1974). Studies on the sources and modes of action of these substances have been reviewed (Bawden, 1954; Gupta, 1977).

In addition to inhibitors, numerous substances, including dyes (Takahashi, 1948), were tested which might prevent or interfere with virus multiplication *after* inoculation. The rationale behind the selection of these compounds and how they were applied if briefly described.

By 1949, it had been established that RNA of turnip yellow mosaic virus (TYMV) was essential for infectivity (Markham and Smith, 1949); that RNA was the genetic material of most plant viruses and, in virus replication RNA served both as messenger and template. Also, despite the intimate relationship between virus and host cell, the RNA and protein of a given virus are virus specific and distinct from those of the host cell. Thus, it might be possible to inhibit selectively the synthesis of virus-induced RNA and protein without simultaneously inhibiting host RNA and protein. Since RNA was the most important part of the virus and the bases the most important part of RNA, early investigations were made by testing for virus-inhibitory compounds among synthetic analogues of the natural bases.

For these reasons, Matthews (1953a) used the purine 8-azaguanine (AZ) and found that, when sprayed onto plants, it had a marked effect on the spread of virus within the plant. It was most effective against alfalfa and cucumber mosaic viruses in tobacco plants. Later, Matthews (1955) demonstrated that the incorporation of AZ into TYMV RNA rendered a proportion of the virus particles non-infective. While the use of AZ and other purines and pyrimidines are discussed below, the studies of Matthews demonstrated that selective inhibition of a virus in a living plant was possible.

2.3 TEST METHODS

Various methods of applying compounds to plant tissues prior to inoculation with virus have been used but those most commonly practiced have been (i) detached leaves or leaf discs floated or placed in a solution of the compound (Commoner and Mercer, 1951; Schlegel and Rawlins, 1954), (ii) vacuum leaf impregnation (Kirkpatrick and Lindner, 1954), (iii) spraying leaves with a solution of the compound (Matthews, 1953a) and (iv) watering the compound onto plant roots growing in soil (Matthews, 1953b; Tomlinson *et al.*, 1976). The limitations of these methods have been discussed by Matthews and Smith (1955). Since viruses that cause systemic infections are themselves translocated within the plant, a logical approach to the discovery of new systemic chemotherapeutants is the examination of compounds, also known to be systemically translocated, for their ability to prevent or reduce virus multiplication in all parts of the plant. Accordingly, Ramaswamy (1975) recommended the simultaneous application of aqueous solutions of candidate chemicals to leaves and roots of test seedlings.

2.4 PYRIMIDINES AND PURINES

Commoner and Mercer (1951, 1952) found that the pyrimidine, 2-thiouracil (TU) almost completely inhibited synthesis of tobacco mosaic virus (TMV) in tobacco leaf discs floated on solutions of TU containing 4.3×10^{-5} M. Bawden and Kassanis (1954) found that TU also reduced the multiplication of potato viruses X (PVX) and Y (PVY).

TU is incorporated into TMV RNA (Jeener and Rosseels, 1953; Matthews, 1956), the reduction in infectivity being correlated with the degree of replacement of uracil (Francki and Matthews, 1962). The reduced virus synthesis is thought to be caused by incorporation of TU into the negative strand of the replicative structure which fails to function in producing RNA strands.

TU also reduced the multiplication of TYMV in chinese cabbage (*Brassica pekinensis*) discs. It was not incorporated into TYMV RNA but induced an increase in the production of virus particles containing no RNA (Francki and Matthews, 1962), an effect thought to be caused by the inhibition of uridylic acid by inhibiting orotidylic acid decarboxylase (Ralph and Wojcik, 1976).

When sprayed on leaves of intact plants TU caused marked phytotoxic effects and attempts to use it as a therapeutic chemical have so far been unsuccessful.

Of the substituted purines, 8-azaguanine (AZ) has been most studied. Its effect on alfalfa mosaic (AMV) and cucumber mosaic virus (CMV) are one of the nearest approaches found in the protection of plants from virus infection. Compared with many other purines, AZ had the greatest latitude between antiviral effectiveness and phytotoxicity (Lindner *et al.*, 1960).

When solutions of AZ containing *c.* 0.005 M were sprayed on leaves of tobacco (*Nicotiana tabacum*) or *N. glutinosa* it reduced the number of lesions and delayed systemic spread of AMV. It had no effect on tomato spotted wilt virus (TSWV) in tomato, PVX and PVY in potato or tobacco and pea mosaic virus (PMV) in peas (Matthews, 1953a). Others have since reported its effectiveness in preventing or delaying the systemic development of brome mosaic virus in wheat (Chiu and Sill, 1962) and tomato leaf curl virus in tomato (Varma, 1977).

Analyses of TMV and TYMV from AZ-treated plants showed that the analogue was incorporated into the RNA of virus particles, replacing guanine to a few per-cent of the residues (Matthews, 1953b, 1955). AZ is thought to interfere with the functioning of the replicative RNA or with the efficiency of the messenger function (Matthews, 1970). In marked contrast to TU, AZ had no significant effect on the proportion of virus particles containing no RNA.

2.5 ANTIBIOTICS

Many of the well-known antibiotics developed for use against human bacterial diseases, were ineffective when tested against TMV. They included penicillin, streptomycin, aureomycin, chloromycetin, terramycin and subtilin (Manil, 1947; Beale and Jones, 1951; Kutsky, 1952).

Schlegel and Rawlins (1954), using inoculated floated tobacco leaf discs, found that an antibiotic produced by the actinomycete *Nocardia formica* and known to prolong the life of mice infected with swine influenze, reduced the multiplication of TMV. Gray (1955) found that when the same antibiotic (noformycin) in solutions containing 1000 parts per million (ppm) was sprayed on to leaves, it inhibited both local lesions and systemic infection caused by southern bean mosaic virus (SBMV) and TMV in intact plants.

Cytovirin, a metabolite of a *Streptomyces* sp., prevented local lesions production in SBMV in bean and TMV in *Nicotiana rustica* (Gray, 1957). Cytovirin sprayed on to test plants had no effect on the aphid transmission of several viruses including PVY, cauliflower mosaic virus (CaMV) and CMV (Shanks and Chapman, 1965).

Other antibiotics with antiviral activity isolated respectively from *Streptomyces griseus, S. griseochromogenes* and *S. lavendulae* include cycloheximide (actidione), blasticidin S and laurusin.

Hirai and Shimomura (1960) studying naramycin (an analogue of cycloheximide) showed that it inhibited the multiplication of TMV in tobacco leaf discs. As cycloheximide is a well-known inhibitor of cellular protein synthesis it is possible that virus specific proteins were not synthesized. However, because the concentrations required for virus inhibition were highly phytotoxic, cycloheximide appears unlikely to have practical applications.

Blasticidin S at 0.05 ppm inhibited the production of TMV in tobacco leaf discs by 50% and was considerably more effective than TU (Hirai and Shimomura, 1965). When applied to tobacco leaves soon after inoculation with TMV, the antibiotic inhibited TMV RNA synthesis without affecting host cell RNA synthesis (Hirai *et al.*, 1966).

Laurusin (formycin B) strongly inhibited TMV multiplication and symptom development in tobacco and, in comparative tests, was less toxic to plant tissue than cycloheximide and blasticidin S (Huang *et al.*, 1966).

Recent reports of other antibiotics which reduced or prevented the multiplication of viruses *in vivo* include aabomycin A (Yamaguchi *et al.*, 1969), the inhibition of TMV in tobacco protoplasts with gentamycin (Kassanis *et al.*, 1975) and the inhibition of the tobacco tumor virus in tobacco hybrids (*N. clevelandii* × *N. glutinosa*) with chloramphenicol, daunomycin, mithramycin and tubercidin (Misra and Nienhaus, 1977).

Because antibiotics are readily biodegradable and are being used increasingly in agriculture (Misato, 1973), their possible role as antiviral chemotherapeutants appears to merit further study.

2.6 HORMONES

Because systemic virus infection frequently results in growth abnormalities (reviewed by Holmes, 1964), many workers have considered the possibility that certain plant virus symptoms are the result of virus-mediated alterations in endogenous hormone levels. Supporting evidence has been sought by comparing levels of certain hormones in virus-infected and healthy plants and by attempts to reverse symptoms or reduce virus multiplication by applying appropriate hormones and related compounds.

Growth and development of healthy plants is known to be partly controlled by internal gradients of at least five different hormones or growth regulatory substances. These are (i) the *auxins*: the main representative being indoleacetic acid (IAA), (ii) the *gibberellins*: a large group of related compounds. At present 52 different compounds have been identified and all of them have a terpenoid structure. Usually several are present in a single plant, (iii) the *cytokinins*: all are N^6–(substituted) adenines. Seven naturally occurring compounds are known (Hall, 1973), typified by zeatin. The role of cytokinins in healthy plants and plants infected with fungi and bacteria have been reviewed by Dekhuijzen (1976), (iv) *inhibitors*: the best known of which is abscisic acid (ABA) and (v) *ethylene*: the hormonal role of which has been reviewed (Abeles, 1972).

The physiological action of these compounds is not fully known and their functions may be similar, synergistic or antagonistic. In the intact plant, each group may induce many growth and physiological effects (Letham, 1967; Evans, 1974) and there is evidence that specific binding sites are involved (Kende and Gardner, 1976). In this complex situation, growth is governed by the integrated effects of the hormones. The study of hormones in virus-infected plants may be thus useful in understanding their role in healthy plants. For example, the gibberellins were discovered by directly studying abnormal growth of rice plants affected by the fungus *Gibberella fujikuroi* (Kurosawa, 1926).

2.6.1 Auxins

Stunting of plant growth, often associated with plant virus diseases, has been correlated with a lowered auxin content. Thus, with potato degeneration, caused by potato leaf roll virus (PLRV), diseased tubers contained 53% less auxin than healthy tubers (Jahnel, 1939). These findings were confirmed by others (Lucas, 1939; Söding and Funke, 1941) who demonstrated a decline in the auxin-content of virus-infected plants, particularly in the stem tips and young foliage. Reduced auxin activity in virus-infected leaf extracts has also been reported following infection of tomato plants with TSWV (Grieve, 1943) and TMV (Pavillard, 1952) and of various hosts by sugar beet curly top virus (Smith *et al.*, 1968). However, the subject is not entirely clear because, with the solanaceous hosts, there is frequently a concomitant accumulation of an inhibitor (scopoletin) which can either inhibit or stimulate the photo-oxidation of IAA depending on conditions (Sequeira, 1963).

Several investigators have reported significant effects of auxins on disease symptom suppression or masking and on virus concentration after applying them to diseased plants. Naphthalene acetic acid (NAA) applied to tobacco tissue cultures reduced the amount of TMV (Kutsky and Rawlins, 1950) and Augier de Montgremier and Morel (1948) suggested that less TMV developed in tobacco callus cultures with high NAA content. Auxin content is less in TMV-infected tobacco plants than in healthy control plants (Pavillard, 1954). Conversely studies on the multiplications of TMV in cotton (*Gossypium hirsutum*) normally highly resistant to TMV showed that IAA and 2,4-dichlorophenoxyacetic acid (2,4-D) enhanced virus multiplication by stimulating ethylene production (Cheo, 1971). He concluded that auxin could act in two ways, by stimulating TMV synthesis and by repressing metabolic resistance with the production of ethylene in a resistant host.

More recently Singh (1972) reported that IAA and indolebutryic acid at 50 ppm reduced the symptoms of papaya (*Carica papaya*) plants infected with papaya leaf reduction virus, a result similar to the partial inhibition of symptoms in chili (*Capsicum annum*) caused by CMV after treatment of infected plants with IAA and NAA (Joshi and Dubey, 1975).

2.6.2 Gibberellins

Several attempts have been made to reverse the growth stunting effects of virus infection by applying gibberellic acid (GA) to infected plants. Stunting was partly prevented in tobacco infected with tobacco etch virus (Chessin, 1957) and in clover infected with wound tumor virus (Maramorosch, 1957). GA sprayed three times at weekly intervals at 100 ppm overcame the stunting of clover but other symptoms were not prevented and the plants retained the virus. Stein (1962) applied GA to tobacco plants infected with severe etch virus and greatly stimulated plant growth. The treated, infected plants were 2.5 times taller than untreated infected plants and 0.8 times as tall as treated healthy plants.

Similar responses of tomato plants infected with CMV to foliar sprays of GA (1 and 5 ppm) were reported by Selman and Arulpragasam (1970) and of barley (*Hordeum vulgare*) infected with barley yellow dwarf virus (Russell and Kimmins, 1971). GA reversed the suppression of stem elongation in citrus plants infected with the tristeza virus without preventing other characteristic symptoms (Singh, 1971). Early infection of cucumber (*Cucumis sativus*) plants by CMV, resulting in pronounced root stunting and reduced stem and leaf growth, was associated with a reduction in the concentration of endogenous gibberellins (Bailiss, 1974).

From these results, there is considerable evidence that the reduced or stunted growth associated with some virus diseases is correlated with a reduced GA content.

2.6.3 Cytokinins

Most studies have been with kinetin (6-furfurylamino purine) and N^6-benzyladenine (BA), a synthetic compound structurally related to kinetin. The results have shown that virus multiplication in systemically infected plants, leaf discs and excised leaves can be altered by the application of cytokinins prior to or after inoculation (Table I).

Kinetin reduced the numbers of local lesions and inhibited virus multiplication of TSWV in tomato and petunia leaf (Selman, 1964), TMV in tobacco leaf discs (Kiraly and Szirmai, 1964), TYMV in Chinese cabbage (Chalcroft and Matthews, 1967; Berridge and Ralph, 1969) and tobacco ringspot virus (TRSV) in cowpea (Kuriger and Agrios, 1977). The same cytokinins also apparently stimulated virus multiplication in treated leaves after inoculation (Daft, 1965; Aldwinckle and Selman, 1967). An explanation of these contrasting results has now been offered from recent studies on the multiplication of TMV in BA treated *N. glutinosa* leaves. Thus, Aldwinckle (1975) showed that virus multiplication could be reduced or enhanced depending on the time lag between inoculation and the application of BA. BA treatment 1 min after inoculation reduced virus

TABLE I. Effect of Kinetin (K) and Benzyladenine (BA) on Virus Multiplication

Virus[a]	Host	Type of tissue	Cytokinin and concentration (ppm)	Virus increase (+) or decrease (–)	Reference
TMV	*N. glutinosa*	leaf disc	K (50)	–	Kiraly and Szirmai, 1964
	N. tabacum	excised leaf	K (10)	+	Daft, 1965
	N. glutinosa	excised leaf	K (20)	+	Milo and Srivastra, 1969
	N. tabacum			+	
	N. rustica			+	
	Tomato	leaf disc	K (0.002)	+	Bailiss *et al.*, 1977b
			K (20.0)	–	
	N. glutinosa	intact plant	BA (17)[b]	– and +	Aldwinckle, 1975
TSWV	Tomato	excised leaf	K (50)	–	Selman, 1964
	N. rustica	excised leaf	BA (10^{-6} M)	–[c]	Aldwinckle and Selman, 1967
		intact plant	BA (10^{-5} M)	–[d]	
		intact plant	BA (10^{-4}M)	+[e]	
TYMV	*Brassica pekinensis*	leaf disc	K (5)	– (slight)	Berridge and Ralph, 1969
TRSV	*Vigna sinensis*	intact plant	K (10^{-4} M)	–	Kuriger and Agrios, 1977

[a]TMV (tobacco mosaic), TSWV (tomato spotted wilt), TYMV (turnip yellow mosaic), TRSV (tobacco ringspot); [b]BA applied 1 min after inoculation was inhibitory; applied 5-120 min after inoculation it was stimulatory; [c]BA applied before or after inoculation; [d]BA applied before inoculation; [e]BA applied 1 day after inoculation and twice daily for 6 days.

multiplication but if BA was applied 5-120 min after inoculation, virus multiplication was enhanced.

Cytokinin levels and kinetin-virus interactions in the roots of TRSV infected cowpea (*Vigna sinensis*) plants have been studied by Kuriger and Agrios (1977). These studies seem particularly pertinent as cytokinins are thought to be produced mainly in root apical meristems (Weiss and Vaadia, 1965; Short and Torrey, 1972). The authors showed that TRSV could be detected in cowpea root tips 3 days after leaf inoculation, and the virus reached its highest concentration 2 days later. The cytokinin levels of root extracts were lower 7 days after infection than those of healthy roots, and virus production in roots of plants, after treating their leaves with kinetin, was also lower and delayed compared with untreated plants. Kinetin-treated inoculated leaves, in which virus synthesis was apparently lower, were slower to develop yellow senescing symptoms than leaves of untreated inoculated plants. They (Kuriger and Agrios, 1977) suggested that senescence of the inoculated non-kinetin treated plants could be attributed to reduced cytokinin production by the roots and transport to the shoots and leaves.

Beneficial effects of applying cytokinin to virus-infected plants were also shown by Berridge and Ralph (1969). They applied kinetin to floated Chinese cabbage discs infected with TYMV and studied the changes in cellular components by analytical ultracentrifugation. The results showed that the main effect of applying kinetin at 5 ppm was to maintain ribosome levels. Additional effects were an increase in the weight of leaf discs by 100% over 5 days, stimulation of the production of DNA by 25%, slight reduction in the production of whole virus and an increase in the production of empty virus protein shells.

2.6.4 Abscisic Acid and Ethylene

The precise roles of abscisic acid (ABA) and ethylene in plant virus infections have stimulated various investigations, particularly those associated with the phenomenon of acquired systemic resistance. This resistance can be demonstrated in certain plants after inoculation and following the formation of local lesions in the lower leaves. When the upper leaves of the same plants are subsequently inoculated, fewer and smaller lesions are formed compared with the lesions in the leaves of control plants not previously inoculated (Ross, 1966; Loebenstein, 1972). Such effects have been extensively described. They are considered here because acquired resistance has been attributable to a translocatable antiviral substance and efforts have been made to isolate and characterize it (Loebenstein *et al.*, 1966). Clearly, the study of such host defense mechanisms could offer a valuable contribution to disease control if the number of infectible sites were reduced. Recent studies, however, have not supported the presence of an antiviral substance which confers resistance. Thus Balazs *et al.* (1976) and Balazs *et al.* (1977) suggested that the induced resistance in tobacco var. Xanthi-nc simply reflected a failure of lesions to turn necrotic without inhibition of virus multiplication. Also studies by Fraser (1979) with TMV in tobacco plants which reacted hypersensitively to certain strains of TMV but which did not contain the N gene for hypersensitivity, showed that lesion size in the upper (challenged) leaves could be reduced without a concomitant reduction in virus multiplication. Studies with the hypersensitive host *Nicotiana glutinosa* suggested that in 'resistant' leaves, the alteration in lesion size and number were separate effects which could vary independently (Fraser *et al.*, 1979). Mechanisms currently suggested to explain the alteration in lesion number involve altered water stress in the challenged leaves (Cassells *et al.*, 1978) associated with changes in endogenous abscisic acid content (Fraser, 1979; Fraser *et al.*, 1979).

The association of ethylene with plant virus infections commenced with the study by Ross and Williamson (1951) who indicated that the symptoms of *Physalis floridana* plants infected with PVY (leaf epinasty and abscission) were related to the evolution of ethylene and that the amount of gas produced was connected with the degree of necrosis of leaf tissue and not the concentration of virus in the leaf. Nakagaki *et al.* (1970), studying the production of ethylene by detached leaves infected with TMV, showed that the amount of gas produced by infected tissue was proportional to the number and size of lesions, i.e., the degree of necrosis. Similar results with TMV and CMV in tobacco cv. Xanthi were also reported by Balazs *et al* (1969) and with TMV by Pritchard and Ross (1975).

Treatment of plants with ethrel (2-chloroethylphosphonic acid), which releases ethylene, accelerates tissue senescence and causes apparent increases in virus multiplication of TMV in cotton and cucumber but not in *Physalis floridana* (Cheo, 1971) and increases TMV production in tobacco (Balazs and Gaborjanyi, 1974). Studies by Bailiss *et al.* (1977a) showed that tobacco cv. Xanthi-nc leaves pre-treated with ethrel or ABA subsequently exhibited senescence and that the treatment caused an increased number of TMV local lesions not associated with a similar increase in recoverable, infective TMV.

2.7 FUNGICIDES

Benzimidazoles have recently received attention as potential virus chemotherapeutants. While probably best known as fungicides, they possess an extraordinary range of other biological properties. For example, benzimidazole derivatives have been shown to have mite ovicidal (Delp and Klöpping, 1968), anti-senescent (Person *et al.*, 1957) anti-helminthic (Staron and Allard, 1964), fungicidal (Delp and Klöpping, 1968), aphicidal (Tomlinson and Ward, 1976) and anti-tumor (Davidse and Flack, 1977) properties.

The possibility that benzimidazoles might be active against plant viruses was implied from the knowledge that (i) the compound 2-(hydroxybenzyl)-benzimidazole inhibited animal picorna viruses (Tamm and Eggers, 1963) and (ii) the fungicide benomyl, methyl (1-butylcarbamoyl)-benzimidazol-2yl-carbamate, had cytokinin-like activity (Schruft, 1970; Skene, 1972; Thomas, 1974). However, other properties favored their exploration. Thus, they were known to be systemic and remain active after root absorption and translocation to leaves (Marsh, 1977) and large amounts could be applied to plants without causing toxicity.

The first experiments were done with methyl benzimidazole-2yl-carbamate (carbendazim), the water decomposition product and fungitoxic principle of benomyl (Clemons and Sisler, 1969). Test solutions were made of Bavistin (BASF 3460F), a wettable powder containing 50% (w/w) carbendazim, because the compound was stable and soluble at pH 7.0 in aqueous preparations of the commercial formulation, whereas the pure carbendazim is water insoluble at pH 7.0.

Although the stage was set, early investigations by the author using tobacco leaf discs infected with TMV gave no indication of a reduction in virus content when the discs were floated on aqueous solutions of carbendazim. However, the application of carbendazim to the *roots* of tobacco plants infected with the Rothamsted strain of TMV caused an almost total suppression of leaf symptoms (Tomlinson *et al.*, 1976). In particular, chloroplast breakdown and consequent leaf yellowing and senescence were greatly reduced. Tests of treated plants by infectivity and electron microscopy showed no reduction in virus content which tended to be higher overall presumably because of the lack of leaf senescence. It appeared, therefore, that carbendazim acted mainly in the alleviation of the effects of infection. One hundred days after inoculation, when plants were 1 m in height, the leaves of untreated, infected plants were yellowish and brittle, whereas those of carbendazim-treated, infected plants were flexible, uniformly green and devoid of symptoms. The amounts of virus present in the leaves of treated and untreated plants were similar, but the symptomless leaves of the carbendazim-treated plants contained approximately three times as much chlorophyll-*a* as the control leaves. Later studies (Tomlinson *et al.*, 1977) showed that when carbendazim (1 g/plant) was applied to the roots of tobacco plants infected with other TMV strains (designated: cherry, grapevine, Dahlemense and Ohio), the compound suppressed symptom development irrespective of whether the dominant symptom was a mosaic (cherry strain) or a severe, debilitating leaf malformation (Dahlemense strain).

Recent studies by Fraser and Whenham (1978a, b) confirmed the effectiveness of carbendazim as a prophylactic chemical in TMV-infected tobacco plants. They also demonstrated that carbendazim reduced the accumulation of TMV (grapevine strain) by 20-90%, depending on the age of the plant after the application of the chemical, and that much lower doses of carbendazim were required to inhibit virus multiplication than to inhibit the production of symptoms.

The effect of carbendazim on the beet western yellows virus disease (BWYV) of lettuce, a disease which causes chloroplast breakdown and an intense interveinal leaf yellowing and premature senescence was also investigated. The virus was cultured in *Montia (Claytonia) perfoliata* and transferred to lettuce using *Myzus persicae* fed on infected *M. perfoliata* plants for two days and then transferred to 21-day old lettuce plants grown in pots of soil in the glasshouse. Half of the inoculated and half of the uninoculated plants were treated with 100 ml of a 1 g/l suspension of Bavistin applied to each plant 7 days before and 5 days after infestation with viruliferous aphids. Each treated plant thus received 0.1 g carbendazim. There were 120 plants in each treatment and 5 days after the second treatment, each group of 120 plants was planted in the field into four, randomized, 30-plant plots. Disease symptoms were recorded on the plants 50 days after inoculation, at which time they were marketable (Table II).

Many of the inoculated plants not receiving carbendazim (treatment 1) showed marked yellowing symptoms, but most of those receiving carbendazim were symptomless. Of equal interest was the fact that whereas almost one quarter of the untreated uninoculated plants showed severe symptoms, only one of the treated uninoculated plants showed symptoms, suggesting that carbendazim was also effective against disease caused by natural spread. Return inoculations to *M. perfoliata* using aphids taken from each of the thirty plants in a plot of treatment 2, however, showed that, although symptomless, every plant was infected with BWYV.

These results have been confirmed in other trials in which it was shown that Bavistin (50% w/w carbendazim) and Benlate (50% w/w benomyl) were equally effective in controlling BWYV symptoms when applied at 0.1 g per plant to young lettuce as root drenches, but neither compound was effective as a leaf spray. Other chemicals with cytokinin activity not effective when applied as leaf sprays included benzimidazole, benzyladenine, NC5392 (Fisons) and SD 8339 (Shell Chemicals).

The amount of carbendazim in the leaves of lettuce plants following a pre-transplant treatment with a root drench of 0.1 g carbendazim was determined by bioassay (Tomlinson *et al.*, 1977). The results showed that two weeks after transplanting this was equivalent to 30-80 ppm but, after a further three weeks when the plants were almost mature, it had decreased to a level too low for detection by bioassay (<3 ppm).

TABLE II. Effect of Carbendazim on Beet Western Yellows Virus (BWYV) Disease Symptoms in Lettuce 50 Days After Inoculation

Treatment	Inoculation	Carbendazim (0.1 g per plant)	No. plants	Plants with symptoms No.	Plants with symptoms %
1	+[a]	–	118	84	71.2
2	+	+[b]	119	4	3.4
3	–	–	117	26[c]	22.2
4	–	+	112	1[c]	0.8

[a] 21-day old lettuce plants were infested with green peach aphids, *Myzus persicae*, previously given a 2-day acquisition-access feeding period on BWYV-infected *Montia perfoliata* plants.

[b] Plants were treated with 100 ml of a 1 g/l suspension of Bavistin (50% w/w carbendazim) applied as a root drench to each plant 7 days before and 5 days after infestation with viruliferous aphids.

[c] Symptoms shown by these plants resulted from natural spread of the disease after test plants had been transplanted to field plots.

The mechanism of carbendazim suppression of BWYV symptoms in lettuce is unknown but studies indicate that the effect is mediated through the host plant. In BWYV-infected lettuce, the effect is associated with preventing or delaying the degeneration of chloroplasts; events that can be followed by electron microscopy of sections of chloroplasts of carbendazim-treated and untreated lettuce. Because of the close similarity between the degenerative changes in chloroplasts of BWYV-infected lettuce and those in naturally senescing lettuce leaves, it is possible that the symptom suppressant effects of carbendazim are due to its cytokinin, anti-senescent properties (Tomlinson and Webb, 1978).

Tests of the effect of carbendazim on other virus diseases suggest that its symptom suppressant effects are specific phenomena and depend on the host/virus combination. It is known, for example, that when applied as a root drench carbendazim does not alleviate the symptoms caused by CMV in tobacco, lettuce mosaic virus (LMV) in lettuce, BWYV in *M. perfoliata* (Tomlinson and Webb, 1978) or beet mild yellowing virus in sugar beet (Tomlinson *et al.*, 1976). Carbendazim has now been shown to prevent the symptoms (leaf chlorosis, stunting and lack of tillering) in rice plants caused by infection with the rice tungro virus (Thomas and John, 1980). Clearly, many other host/virus combinations remain to be tested.

Recently, Bailiss *et al.* (1977b) showed that when TMV-inoculated tomato leaf discs were floated on a solution containing benomyl there was an increase in virus multiplication above certain concentrations of the chemical. Similar effects were also obtained with CMV in treated cucumber cotyledon discs and in cucumber seedlings.

The chemotherapeutic effect of a systemic triazole fungicide was reported by Joshi and Joshi in 1974. They found that when banana rhizomes infected with banana mosaic virus were soaked for 90 min in a 0.25% aqueous solution of 4-n-Butyl-1,2,4,triazole, 60% of the plants failed to develop symptoms. Whether symptom suppression was associated with virus inhibition was not stated.

2.8 HERBICIDES

Treatment of virus-infected plants with herbicides, notably 2-methyl-4-chlorophenoxyacetic acid (MCPA) and 2,4-dichlorophenoxyacetic acid (2,4-D), were found to cause a variety of effects on symptoms and rates of virus multiplication.

MCPA reduced multiplication of PVX and PVY in tobacco plants (Limasset *et al.*, 1948). When sprayed on lettuce plants infected with lettuce mosaic virus, the symptoms caused by MCPA were additional to those caused by the virus (Way, 1968).

2,4-D suppressed the symptoms of potato leaf roll virus (Locke, 1948) and increased 15-fold the area of lesions caused by TMV in tobacco and *Datura stramonium* (Simons and Israel, 1972). When applied after inoculation, 2,4-D stimulated the production of TMV in *Physalis floridana* and in cotton (Cheo, 1971).

The growth retardant 2-chloroethyltrimethylammonium chloride (chlormequat) reduced TMV multiplication by 80% in inoculated tobacco leaf discs, without phytotoxicity (Rawlins, 1962).

Other herbicides reported to suppress virus symptoms include 2,4,5-trichlorophenoxyacetic acid (2,4,5-T) shown by Singh (1969) to retard watermelon mosaic virus in *Cucurbita pepo* cv. Caseria, and maleic hydrazide, di-allate and chloroxuron shown by Schuster (1972) to suppress PVX and PVY symptoms in tobacco plants. Horvath and Hunyadi (1973) demonstrated increased resistance of the primary leaves of bean (*Phaseolus vulgaris* cv. Pinto) to inoculation with viruses after the plants had grown in soil treated 17 days previously with trifluralin. They showed that when the leaves were inoculated with alfalfa mosaic virus or TMV, the soil treatment reduced the numbers of local lesions by 76 and 65% respectively.

2.9 CHEMOTHERAPY OF ANIMAL VIRUSES

Since Matthews and Smith (1955) concluded that "there is no well-established case of the effective use of chemotherapy for any virus disease," various chemicals have been developed with antiviral properties (Maugh, 1976). Five compounds that have been extensively tested during this period against human virus diseases include 5-iodo-2-deoxyuridine (Idoxuridine) for use against herpes keratitis eye infection, N-methyl isatin 3-thiosemicarbazone (Marboran) effective against smallpox and vaccinia virus, amantadine hydrochloride (Symmetrel) used prophylactically against influenza, 9-β-D-arabinofuranosyladenine (vidarabine) effective against herpes keratitis and 1,-β-D-ribofuranosyl-1,2,4,-triazole-3-carboxamide or ribavirin (Virazole). The latter chemical appears to have the widest spectrum of activity and is effective *in vitro* against RNA and DNA viruses including herpes, hepatitis and influenza (Sidwell *et al.*, 1972). The mode of action lies in the inhibition of inosine monophosphate dehydrogenase, a key enzyme of guanine nucleotide synthesis in cells infected by viruses such as herpes. By inhibiting this enzyme, the ribavirin interferes with the synthesis of guanine nucleotides and hence the synthesis of viral DNA.

Ribavirin was recently shown to inhibit PVX. In studies on the regeneration of plants from protoplasts of PVX-infected tobacco leaves, Shepard (1977) showed that it would eradicate the virus in shoots derived from protoplasts. It has also been demonstrated by Lerch (1977) that when roots of the tobacco plants were grown in solutions containing 5 μg/ml and the lower leaves of the plants inoculated with PVX, multiplication of the virus in the systemically-infected upper leaves was reduced.

A most promising recent discovery is 9-(2-hydroxyethoxymethyl) guanine (acylcoguanosine) which is effective against herpes simplex and varicella zoster (Shaeffer *et al.*, 1978). Its high therapeutic index (the ratio between the effective dose and the toxic dose) is largely because the drug is phosphorylated into the active compound by the herpes-specified thymidine kinase and is much less affected by mammalian thymidine kinases.

Extensive studies have also been made with other compounds and these have provided information on the mechanisms of virus replication. One such class of substances has been the benzimidazoles (Tamm and Eggers, 1963; Tamm and Sehgal, 1978). Compounds of this type, e.g., 2-(α-hydroxybenzyl) benzimidazole (HBB) and related derivatives, were found to act specifically on RNA synthesis of picorna viruses by inhibiting the activity of RNA polymerase. HBB, however, has not proved useful in animals infected with picorna viruses because virus strains resistant to its action rapidly arose and other virus strains also developed that could not multiply unless HBB was present.

The types of compounds studied by Tamm and co-workers have provided an important clue for studies on the effect of the fungicide, methyl benzimidazole-2-yl carbamate (carbendazim) on plant viruses which resulted in a method of preventing the yellowing leaf symptoms in lettuce caused by beet western yellows virus as described above. Although the mechanism of preventing chloroplast breakdown with carbendazim is still unsolved (Tomlinson and Webb, 1978), it is an example of the merging of common interests of workers in plant and animal virus chemotherapy.

2.10 DISCUSSION

By combining a number of different control measures, numerous crops have been protected from virus diseases and the incidence of some virus diseases has decreased (Broadbent, 1964). Nevertheless, many virus diseases continue to present problems with no immediate prospect of effective control.

A chemical method of control would be highly desirable particularly for the numerous annual crops grown on a large scale. A foliar spray which reduced virus multiplication or prevented chloroplast breakdown in infected leaves would be of great value in the control of insect-transmitted diseases such as those caused by the 'yellowing' viruses of sugar beet or infection of cereals. Likewise, a compound which eliminated virus in the embryos of seeds would be of immense value in the treatment of various seed-transmitted viruses (Shepherd, 1972). Chemotherapy of important virus diseases of trees and other perennials, however, is likely to be much more difficult as these are grown for several years and are subject to continual re-infection.

Knowledge on plant virus chemotherapy is still fragmentary and many uncertainties remain in interpreting the results reviewed above; but it is clear that efforts to find chemotherapeutants have continued unabated since the subject was reviewed by Matthews and Smith (1955). This is perhaps somewhat remarkable because, for practical purposes, very few chemicals have yet been found to be of any practical significance.

These difficulties arise from the intrinsic nature of viruses and are different from those encountered with bacteria and other microorganisms. As viruses have no independent metabolism and multiply by nucleic acid replication, one basic question that emerges is how do viruses alter the metabolism of the cell resulting in disease symptoms. Why, for example, do leaves of tomato plants infected with cucumber mosaic virus become 'fern-like,' with greatly reduced lamina, whereas those of chickweed (*Stellaria media*) remain symptomless when infected with the same virus? These questions are so far unanswerable, beyond the fact that the symptoms result from the addition of viral genetic elements to the host cell.

It is clear that antiviral effectiveness is one of several properties required of a successful chemotherapeutant. The chemical will need to possess a broad spectrum of antiviral activity to justify development costs for agricultural use, and it must be non-phytotoxic and safe for consumers if applied to edible plants. With these constraints what *modus operandi* should now be adopted that might lead to future success in plant virus chemotherapy? I shall conclude by venturing certain generalizations.

Screening tests should be designed to test the full potential of a selected candidate chemical. Plant virus infection frequently causes premature plant aging, and the consequent derangement of infected tissues may result from such premature senescence. Parallels exist between the effects of infection and those associated with senescence in whole plants involving growth regulatory substances. In these circumstances a leaf-disc method of assay, while yielding information on the inhibitory properties of a chemical on virus multiplication, would not demonstrate its potential symptom suppressant properties and is therefore of limited relevance to the control of viruses or virus symptoms in whole plants. Leaf disc tests, for example, detected no effect of carbendazim on the multiplication of TMV in tobacco or *N. glutinosa* leaf discs; however, when the chemical was applied to the roots of whole plants, it caused an almost complete suppression of symptoms of TMV in tobacco and of BWYV in lettuce (Tomlinson *et al.*, 1976).

It is obvious from the many synthetic and naturally occurring substances tested that there is no easy guide in the search for viral chemotherapeutants and at present it appears that their discovery will be largely fortuitous. This is because, although the objectives of virus chemotherapy and the specification of a successful therapeutant are easily stated, the chemical and biological properties of hypothetical compounds are not. In listing a specification, we can enunciate desirable properties such as non-phytotoxicity and ability to be absorbed and translocated, but we have very little information on the molecular configuration necessary to block some key replicative pathway of virus synthesis. Accordingly, no rational approach exists. The recent review by Bachrach (1979) underlines the highly specialized and complex nature of virus replicative mechanisms of different virus groups. Although the basic strategies of replication are now understood, ways of designing a chemical that will, for example, prevent uncoating, block transcription or translocation, prevent replication of the genome or interfere with viral assembly are not known.

For these reasons few investigators appear ready to abandon empiricism completely and, although otherwise undesirable, empiricism continues to be an essential feature of biological studies in virus chemotherapy. Studies of natural resistance or tolerance in plants, in which virus multiplication is greatly restricted, may eventually provide clues from which strategies for chemotherapy can be developed on a rational basis.

In the past decade, the early stages of research and development of the systemic fungicides inexorably followed the development of systemic insecticides and now a great number of novel systemically active chemicals with fungicidal activity have been developed. As some of these chemicals interfere with nucleic acid and protein synthesis and seemingly pose no great mammalian hazard, many of them appear to be ideally suited for testing against viruses and virus diseases.

2.11 REFERENCES

Abeles, F. B. (1972). Biosynthesis and mechanism of action of ethylene. *Ann. Rev. Plant Physiol.* **23**:259-292.

Aldwinckle, H. S. (1975). Stimulation and inhibition of plant virus replication *in vivo* by 6-benzylaminopurine. *Virology* **66**:341-343.

Aldwinckle, H. A., and Selman, I. W. (1967). Some effects of supplying benzyladenine to leaves and plants inoculated with viruses. *Ann. Appl. Biol.* **60**:49-58.

Augier de Montgremier, H., and Morel, G. (1948). On the diminution of the virus (*Marmor tabaci*, Holmes) content in tobacco tissues cultivated *in vitro*. *Compt. Rend.* **227**:688.

Bachrach, H. L. (1978). Comparative strategies of animal virus replication. *Adv. Virus Res.* **22**:163-186.

Bailiss, K. W. (1974). The relationship of gibberellin content to cucumber mosaic virus infection of cucumber. *Physiol. Plant Pathol.* **4**:73-79.

Bailiss, K. W., Balazs, E., and Kiraly, Z. (1977a). The role of ethylene and abscisic acid in TMV-induced symptoms in tobacco. *Acta Phytopath. Acad. Sci. Hung.* **12**:133-140.

Bailiss, K. W., Cocker, F. M., and Cassells, A. C. (1977b). The effect of benlate and cytokinins on the content of tobacco mosaic virus in tomato leaf disks and of cucumber mosaic virus in cucumber cotyledon disks and seedlings. *Ann. Appl. Biol.* **87**:383-392.

Balazs, E., and Gaborjanyi, R. (1974). Ethrel induced leaf senescence and increased TMV susceptibility in tobacco. *Z. Pflanzenkr. Pflanzenschutz.* **81**:389-393.

Balazs, E., Gaborjanyi, R., Toth, A., and Kiraly, Z. (1969). Ethylene production in Xanthi tobacco after systemic and local virus infections. *Acta Phytopathol. Acad. Sci. Hung.* **4**:355-358.

Balazs, E., Barna, B. and Kiraly, Z. (1976). Effect of kinetin on lesion development and infection sites in Xanthi-nc tobacco infected by TMV: Single-cell local lesions. *Acta Phytopathol. Acad. Sci. Hung.* **11**:1-9.

Balazs, E., Sziraki, I., and Kiraly, Z. (1977). The role of cytokinins in the systemic acquired resistance of tobacco hypersensitive to TMV. *Physiol. Plant Pathol.* **11**:29-37.

Bawden, F. C. (1954). Inhibitors and plant viruses. *Adv. Virus Res.* **2**:32-55.

Bawden, F. C., and Kassanis, B. (1954). Some effects of thiouracil on virus infected plants. *J. Gen. Microbiol.* **10**:160-173.

Beale, H. P., and Jones, C. R. (1951). Virus diseases of tobacco mosaic and potato yellow dwarf not controlled by certain purified antibiotics. *Contrib. Boyce Thompson Inst.* **16**:395-407.

Bent, K. J. (1969). Fungicides in perspective. *Endeavour* **27**:129-134.

Berridge, M. V., and Ralph, R. K. (1969). Some effects of kinetin on floated chinese cabbage leaf disks. *Biochem. Biophys. Acta* **182**:266-269.

Broadbent, L. (1964). Control of plant virus diseases. *In* "Plant Virology" (M. K. Corbett and H. D. Sisler, eds.), pp. 330-364. University of Florida Press, Gainesville.

Cassels, A. C. , Barnett, A., and Barlass, M. (1978). The effect of polyacrylic acid treatment on the susceptibility of *Nicotiana Tabacum* cv. Xanthi-nc to TMV. *Physiol. Plant Pathol.* **13**:13-21.

Chalcroft, J. P., and Matthews, R. E. F. (1967). Virus strains and leaf ontogeny as factors in the production of leaf mosaic patterns by turnip yellow mosaic virus. *Virology* **33**: 167-171.

Cheo, P. C. (1971). Effect of plant hormones on virus-replicating capacity of cotton infected with tobacco mosaic virus. *Phytopathology* **61**:869-872.

Chessin, M. (1957). Growth substances and stunting in virus infected plants. *Proc. 3rd Conf. Potato Virus Diseases*, pp. 80-84. Lisse-Wageningen, Netherlands.

Chiu, R. J., and Sill, W. H., Jr. (1962). Chemotherapeutic effects of some substituted purines and pyrimidines and other compounds against three cereal viruses and tobacco mosaic virus. *Phytopathology* **52**:432-438.

Clemons, G. P., and Sisler, H. D. (1969). Formation of a fungitoxic derivative from Benlate. *Phytopathology* **59**:705-706.

Commoner, B., and Mercer, F. L. (1951). Inhibition of the biosynthesis of tobacco mosaic. *Nature* **168**:113-114.

Commoner, B., and Mercer, F. L. (1952). The effect of thiouracil on the rate of tobacco mosaic virus biosynthesis. *Arch. Biochem. Biophys.* **35**:278-288.

Daft, M. J. (1965). Some interactions of kinetin and temperature on tobacco leaves infected with tomato aucuba mosaic virus. *Ann. Appl. Biol.* **55**:51-56.

Davidse, L. C., and Flach, W. (1977). Differential binding of methyl benzimidazol-2-yl carbamate to fungal tubulin as a mechanism of resistance to this antimitotic agent in mutant strains of *Aspergillus nidulans. J. Cell Biol.* **72**:174-193.

Delp, C. J., and Klopping, H. L. (1968). Performance attributes of a new fungicide and mite ovicide candidate. *Plant Dis. Reptr.* **52**:95-99.

Evans, E. (1971). Systemic fungicides in practice. *Pest. Sci.* **2**:192-196.

Evans, M. L. (1974). Rapid responses to plant hormones. *Annu. Rev. Plant Physiol.* **25**: 195-223.

Francki, R. I. B., and Matthews, R. E. F. (1962). Relation between incorporation of 2-thiouracil into tobacco mosaic nucleic acid and virus inhibition. *Virology* **17**:22-29.

Fraser, R. S. S. (1979). Systemic consequences of the local lesion reaction to tobacco mosaic virus in a tobacco variety lacking the N gene for hypersensitivity. *Physiol. Plant Pathol.* **14**:383-394.

Fraser, R. S. S., and Whenham, R. J. (1978a). Inhibition of the multiplication of toabcco mosaic virus by methyl benzimidazol-2-yl-carbamate. *J. Gen. Virol.* **39**:191-194.

Fraser, R. S. S., and Whenham, R. J. (1978b). Chemotherapy of plant virus disease with methyl benzimidazol-2-yl carbamate: Effects on plant growth and multiplication of tobacco mosaic virus. *Physiol. Plant Pathol.* **13**:51-64.

Fraser, R. S. S., Loughlin, S. A. R. , and Whenham, R. J. (1979). Acquired systemic susceptibility to infection by tobacco mosaic virus in *Nicotiana glutinosa* L. *J. Gen. Virol.*, **43**:131-141.

Gray, R. A. (1955). Activity of an antiviral agent from *Nocardia* on two viruses in intact plants. *Phytopathology* **45**:281-285.

Gray, R. A. (1957). Inhibition of local lesion and systemic plant virus infection with a new antiviral agent cytovirin. *Phytopathology* **47**:522.

Grieve, B. J. (1943). Studies in the physiology of host-parasite relations. 4. Some effects of tomato spotted wilt virus on growth. *Austral. J. Exp. Biol. Med. Sci.* **21**:89-101.

Gupta, B. M. (1977). Inhibition of plant virus infections by antiviral agents. *In* "Aphids as Virus Vectors" (K. F. Harris and K. Maramorosch, eds.), pp 455-471. Academic Press, New York.

Hall, R. H. (1973). Cytokinins as a probe of developmental processes. *Annu. Rev. Plant Physiol.* **24**:415-444.

Hirai, T., and Shimomura, T. (1960). The mode of action of some antibiotics in their inhibitory effect on tobacco mosaic virus multiplication. *Phytopathol. Z.* **40**:35-44.

Hirai, R., and Shimomura, T. (1965). Blasticidin S, and effective antibiotic against plant virus multiplication. *Phytopathology* **55**:291-295.

Hirai, T., Hirashima, A., Itoh, T., Takahashi, T., Shimomura, T., and Hayashy, Y. (1966). Inhibitory effect of blastocidin S on tobacco mosaic virus multiplication. *Phytopathology* **56**:236-240.

Hollings, M. (1965). Disease control through virus-free stock. *Annu. Rev. Phytopathol.* **3**:367-396.

Holmes, F. O. (1964). Symptomatology of viral diseases in plants. *In* "Plant Virology" (M. K. Corbett and H. D. Sisler, eds.), pp 17-38. Univ. of Florida Press, Gainesville.

Horvath, J., and Hunyadi, K. (1973). Studies on the effects of herbicides on virus multiplication. I. Effects of trifluralin on alfalfa mosaic and tobacco mosaic viruses in *Phaseolus vulgaris* L. cv. Pinto. *Acta Phytopathol. Acad. Sci. Hung.* **8**:347-350.

Huang, K. T. , Katagiri, M., and Misato, T. (1966). Effects of laurusin (Formycin B) on TMV multiplication and its symptom development. *J. Antibiotics, Ser A.* **19**:75-77.

Jahnel, H. (1939). Wuchsstoffuntersuchungen an abbaukranken Kartoffein. II. *Phytopath Z.* **12**:312-317.

Jeener, R., and Rosseels, J. (1953). Incorporation of 2-thiouracil^{35}S in the ribose-nucleic acid of tomato mosaic virus. *Biochem. Biophys. Acta* **11**:438.

Joshi, D. M., and Joshi, H. U. (1974). The control of banana mosaic. *Ind. J. Hort.* **31**: 382-385.

Joshi, R. D., and Dubey, L. N. (1975). Studies on the inhibition of cucumber mosaic virus in chili (*Capsicum annum* L.) III Effect of growth regulators. *Sci. Cult.* **41**:351-353.

Kassanis, B., White, R. F., and Woods, R. D. (1975). Inhibition of multiplication of tobacco mosaic virus in protoplasts by antibiotics and its prevention by divalent cations. *J. Gen. Virol.* **28**:185-191.

Kende, H., and Garnder, G. (1976). Hormone binding in plants. *Annu. Rev. Plant Physiol.* **27**:267-290.

Kiraly, Z., and Szirmai, J. (1964). The influence of kinetin on tobacco mosaic virus production in *Nicotiana glutinosa* leaf discs. *Virology* **23**:286-288.

Kirkpatrick. H. C., and Lindner, R. C. (1954). Studies concerning chemotherapy of two plant viruses. *Phytopathology* **44**:529-533.

Kuriger, W. E., and Agrios, G. N. (1977). Cytokinin levels and kinetin-virus interactions in tobacco ringspot virus-infected cowpea plants. *Phytopathology* **67**:604-609.

Kurosawa, E. (1926). Experimental studies on the secretion of the "bakanae" fungus on rice plants. *Trans. Nat. Hist. Soc. Formosa* **16**:213-217.

Kutsky, R. J. (1952). Effects of indolebutryric acid and other compounds on virus concentration in plant tissue cultures. *Science* **115**:19-20.

Kutsky, R. J., and Rawlins, T. E. (1950). Inhibition of virus multiplication by naphthalene acetic acid in tobacco tissue cultures, as revealed by a spectrophotometric method. *J. Bacteriol.* **60**:763-766.

Lerch, B. (1977). Inhibition of the biosynthesis of potato virus X by ribavirin. *Phytopathol. Z.* **89**:44-49.

Letham, D. S. (1967). Chemistry and physiology of kinetin-like compounds. *Annu. Rev. Plant Physiol.* **18**:349–364.

Limasset, P., Levieil, F., and Sechet, M. (1948). Influence d'une phytohormone de synthese sur le developpement des virus X et Y de la pomme de terre chez le tabac. *Compt. Rend.* **227**:643-645.

Lindner, R. C., Cheo, P. C., Kirkpatrick, H. C., and Govindu, H. C. (1960). Some effects of 8-azaguanine on tobacco mosaic virus replication. *Phytopathology* **50**:884-889.

Locke, S. B. (1948). Studies on the chemotherapy of potato virus diseases. *Phytopathology* **38**:916.

Loebenstein, G. (1972). Localization and induced resistance in virus-infected plants. *Annu. Rev. Phytopathol.* **10**:177-206.
Loebenstein, G., Rabina, S., and van Praagh, T. (1966). Induced interference phenomena in virus infections. *In* "Viruses of Plants" (A. B. R. Beemster and J. Dijkstra, eds.), pp 151-157. North Holland Publ., Amsterdam.
Lucas, H. (1939). Weitere Untersuchungen an abbaukranken Kartoffeln. II. *Phytopathol. Z.* **12**:312-317.
Manil, P. (1947). Action negative, sur le virus de la mosaique du tabac, de la penicilline et pronotosil. *Ann. Inst. Pasteur.* **73**:294.
Maramorosch, K. (1957). Reversal of virus caused stunting in plants by gibberellic acid. *Science* **126**:651-652.
Markham, R. and Smith, K. M' (1949). Studies on the virus of turnip yellow virus. *Parasitology* **39**:330-342.
Marsh, R. W. , Ed. (1977). "Systemic Fungicides." Longman London.
Matthews, R. E. F. (1953a). Chemotherapy and plant viruses. *J. Gen. Microbiol.* **8**:277-288.
Matthews, R. E. F. (1953b). Incorporation of 8-azaguanine into nucleic acid of tobacco mosaic virus. *Nature* **171**:1065-1066.
Matthews, R. E. F. (1955). Infectivity of turnip yellow mosaic virus containing 8-azaguanine. *Virology* **1**:165-175.
Matthews, R. E. F. (1956). Thiouracil in tobacco mosaic virus. *Biochem. Biophys. Acta* **19**: 559.
Matthews, R. E. F. (1970). "Plant Virology." Academic Press, New York.
Matthews, R. E. F., and Smith, J. D. (1955). The chemotherapy of viruses. *Adv. Virus Res.* **3**:49-148.
Maugh, T. H. (1976). Chemotherapy: antiviral agents come of age. *Science* **192**:128-132.
Milo, G. E., and Strivastava, B. I. S. (1969). Effects of cytokinins on tobacco mosaic virus production in local lesion and systemic hosts. *Virology* **38**:26-31.
Misato, T. (1974). The development of agricultural antibiotics in Japan. *Proc. 3rd Int. Symp. Chemical and Toxicological Aspects of Environmental Quality.* Tokyo, 19-22 November, 1973.
Misra, A. and Nienhaus, F. (1977). Inhibition of virus-tumour formation in tobacco by antibiotics. *Phytopathol. Z.* **89**:76-81.
Nakagaki, Y., Hirai, T., and Stahmann, M. A. (1970). Ethylene production by detached leaves infected with tobacco mosaic virus. *Virology* **40**:1-9.
Nyland, G., and Moller, W. J. (1973). Control of pear decline with a tetracycline. *Plant Dis. Reptr.* **57**:634-637.
Pavillard, J. (1952). Recherches sur la croissance des plantes virosees; virus et auxines. *Compt. Rend.* **235**:87-88.
Pavillard, J. (1954). Contribution a l'etude de la croissance des plantes virosees: virus et auxines. *Ann. Biol.* **30**:169-181.
Person, C., Samborski, D. J., and Forsyth, F. R. (1957). Effect of benzimidazole on detached leaves. *Nature* **180**:1294-1295.
Pritchard, D. W., and Ross, A. F. (1975). The relationship of ethylene to formation of tobacco mosaic virus lesions in hypersensitive responding tobacco leaves with and without induced resistance. *Virology* **64**:295-307.
Ralph, R. K., and Wojcik, S. J. (1976). Inhibition of turnip yellow mosaic virus synthesis by pyrimidine analogues. *Biochem. Biophys. Acta* **444**:261-268.
Ramaswamy, S. (1975). Techniques and problems in research on chemotherapy of cucumber mosaic virus. *In* "Cucumber Mosaic Virus" (W. Welvaert, ed.), pp 197-202. Gent, Belgium.
Rawlins, T. E. (1962). The inhibitory effect of 2-chloroethyltrimethyl-ammonium chloride treatment on tobacco mosaic virus (TMV) multiplication. *Plant Dis. Reptr.* **46**:170.
Ross, A. F. (1966). Systemic effects of local lesion formation. *In* "Viruses of plants"

(A. B. R. Beemster and J. Dijkstra, eds.), pp 127-150. North Holland Publ., Amsterdam.

Ross, A. F., and Williamson, C. E. (1951). Physiologically active emanations from virus infected plants. *Phytopathology* **41**:431-438.

Russell, S. L., and Kimmins, W. C. (1971). Growth regulators and the effect of barley yellow dwarf virus on barley. *Ann. Bot.* **35**:1037-1043.

Schroth, M. N., Thomson, S. V., and Hildebrand, D. C. (1974). Epidemiology and control of fire blight. *Annu. Rev. Phytopathol.* **12**:389-412.

Schruft, G. (1970). Die Wirkung des Botrytis-Praparates Benomyl in Amaranthus-Cytokinin test. *Wein-Wissenschaft* **25**:329-333.

Schuster, G. (1972). Die Beeinflussung von Virussymptomen durch Herbizide und andere Pflanzenschutzmittel. *Ber. Inst. Tabakforsch.* **19**:14-21.

Selman, I. W. (1964). The effect of kinetin on infection of petunia and tomato leaves with tomato spotted wilt virus. *Ann. Appl. Biol.* **53**:67-76.

Selman, I. W., and Arulpragasam, P. V. (1970). Response of tomato plants infected with cucumber mosaic virus to foliar sprays of gibberellic acid. *Ann. Bot.* **34**:1107-1114.

Sequira, L. (1963). Growth regulators in plant disease. *Annu. Rev. Phytopathol.* **1**:5-30.

Shanks, C. H., and Chapman, R. K. (1965). The use of antiviral chemicals against some viruses transmitted by aphids. *Virology* **25**:83-87.

Shepard, J. F. (1977). Regeneration of plants from protoplasts of potato virus X-infected tobacco leaves. II. Influence of virazole on the frequency of infection. *Virology* **78**: 261-266.

Shepherd, R. J. (1972). Transmission of viruses through seed and pollen. *In* "Principles and Techniques in Plant Virology" (C. I. Kado and H. O. Agrawal, eds.), pp 267-292. Van Nostrand-Reinhold, Englewood Cliffs, New Jersey.

Short, K. C., and Torrey, J. G. (1972). Cytokinins in seedling roots of pea. *Plant Physiol.* **49**:155-160.

Sidwell, R. W., Huffman, J. H., Khare, G. P., Allen, L. B., Witkowski, J. T., and Robins, R. K. (1972). Broad spectrum antiviral activity of Virazole: 1,-β-D-ribofuranosyl-1,2,4, -triazole-3-carboxamide. *Science* **177**:705-706.

Simons, T. J., Israel, H. W., and Ross, A. F. (1972). Effect of 2,4-dichlorophenoxyacetic acid on tobacco mosaic virus lesions in tobacco and on the fine structure of adjacent cells. *Virology* **48**:502-515.

Singh, A. B. (1976). Retarding effects of some plant growth regulators on papaya plants infected with papaya leaf reduction virus. *Phytopathol. Medit.* **11**:197-199.

Singh, B. P. (1971). Influence of gibberellic acid on the physiology of tristeza infected citrus. *Indian Phytopathol.* **24**:769-776.

Singh, R. (1969). The retarding effect of 2,4,5-trichlorophenoxy acetic acid on watermelon mosaic virus symptoms. *Acta Phytopathol. Sci. Hung.* **4**:317-319.

Skene, K. G. M. (1972). Cytokinin-like properties of the systemic fungicide benomyl. *J. Hort. Sci.* **47**:179-182.

Smith, S. H., McCall, S. R., and Harris, J. H. (1968). Alterations in the auxin levels of resistant and susceptible hosts induced by the curly top virus. *Phytopathology* **58**:575-577.

Soding, H., and Funke, H. (1941). Uber den Wuchsstoffhaushalt Abbaukrander Kartoffeln. *Phytopathol. Z.* **12**:334-350.

Staron, T., and Allard, C. (1964). Proprietes antifongiques du 2-(4-thiozolyl) benzimidazole ou thiabendazole. *Phytiat. Phytopharm.* **13**:163-168.

Stein, D. B. (1962). The developmental morphology of *Nicotiana tabacum* "White Burley" as influenced by virus infection and gibberellic acid. *Am. J. Bot.* **49**:437-443.

Takahashi, W. N. (1948). The inhibition of virus increase by malachite green. *Science* **107**:226.

Tamm, I., and Eggers, H. J. (1963). Specific inhibition of replication of animal viruses. *Science* **142**:24-33.

Tamm, I., and Sehgal, P. B. (1978). Halobenzimidazole ribosides and RNA synthesis of cells and viruses. *Adv. Virus Res.* **22**:188-258.

Taylor, J. D. and Dudley, C. L. (1977). Seed treatment for the control of halo blight of beans (*Pseudomonas phaseolicola*). *Ann. Appl. Biol.* **85**:223-232.

Thomas, J., and John, V. T. (1980). Suppression of symptoms of rice tungro virus disease by carbendazim. *Plant Dis.* **64**:402-403.

Thomas, T. H. (1974). Investigations into the cytokinin-like properties of benzimidazole derived fungicides. *Ann. Appl. Biol.* **76**:237-241.

Tomlinson, J. A. (1978). Chemotherapy of plant virus diseases. *Proc. 1977 Br. Crop. Protect. Conf. Pests and Diseases*, pp 807-813.

Tomlinson, J. A., and Ward, C. M. (1976). Effect of carbendazim on beet western yellows disease in lettuce. *Rep. Natn. Veg. Res. Stn. 1975* p. 106.

Tomlinson, J. A., and Webb, M. J. W. (1978). Ultrastructural changes in chloroplasts of lettuce infected with beet western yellows virus. *Physiol. Plant Pathol.* **12**:13-18.

Tomlinson, J. A., Walker, V. M., Flewett, T. H., and Barclay, G. R. (1974). The inhibition of infection by cucumber mosiac virus and influenza virus by extracts from *Phytolacca americana. J. Gen. Virol.* **22**:225-232.

Tomlinson, J. A., Faithfull, E. M., and Ward, C. M. (1976). Chemical suppression of the symptoms of two virus diseases. *Ann. Appl. Biol.* **84**:31-41.

Tomlinson, J. A., Ward, C. M., Webb, M. J. W., and Faithfull, E. M. (1977). Chemotherapy of plant virus diseases. *Rep. Natn. Veg. Res. Stn. 1976* pp 105-106.

Varma, J. P. (1977). Effect of 8-azaguanine on tomato leaf curl infection and protein constitution of leaves of two varieties of tomato *Lycopersicon esculentum* Mill. *Indian J. Exp. Biol.* **15**:408-409.

Wain, R. L., and Carter, G. A. (1972). Historical Aspects. *In* "Systemic Fungicides" (R. W. Marsh, ed.). pp 1-31. Longman, London.

Way, J. M. (1968). Effects of MCPA on lettuce pre-inoculated with lettuce mosaic virus. *Plant Pathol.* **17**:75-77.

Weiss, C., and Vaadia, Y. (1965). Kinetin-like activity in root apices of sunflower plants. *Life Sci.* **4**:1323-1326.

Woodcock, D. (1968). Agricultural and horticultural fungicides-past, present and future. *Chem. Britain* **4**:294-300.

Wyatt, S D., and Shepherd, R. J. (1969). Isolation and characterization of a virus inhibitor from *Phytolacca americana. Phytopathology* **59**:1787-1794.

Yamaguchi, I. (1969). Abaomycin A, a new antibiotic. II Biological studies on aabomycin A. *J. Antibiotics* **22**:463-466.

Chapter 3

CONTROL OF WHITEFLY VECTORS OF VIRUSES BY COLOR MULCHES

Shlomo Cohen

The Volcani Institute
Bet-Dagan, Israel

3.1 INTRODUCTION

The various fields concerned with the control of whitefly-borne viruses have been reviewed by several investigators (Varma, 1963; Nene, 1972; Bird *et al.*, 1975; Costa, 1975, 1976; Bird and Maramorosch, 1978; Muniyappa, 1980). The numerous subjects discussed include the use of plant varieties resistant to virus, or vector, or both; the use of insecticides and oils aimed at killing whiteflies prior to virus tramsmission; evasion of disease by selecting planting times or sites that coincide with low whitefly populations; and increased sanitation by eradication of virus sources and by quarantine regulations to impede introduction of infected material.

Little information is available on measures for controlling whitefly-borne viruses based on vector behavior. This approach will be discussed here.

ISBN 0-12-326440-5

3.2 EFFECT OF MULCHES ON WHITEFLY POPULATIONS AND VIRUS SPREAD

Avidov (1956) reported that sawdust mulching of tomato seedbeds reduced *Bemisia tabaci* populations to very low levels. The number of eggs laid by the whiteflies per paid of cotyledons was used to measure the rate of reduction. Thus, on the 8th day after sprouting, the average number of eggs laid by *B. tabaci* during August on unmulched seed beds reaches 62.2 per pair of tomato cotyledons, and was nil on seedlings growing in sawdust-mulched plots. The author suggested that the high air temperature immediately above the mulched soil surface—which at times reached 47° to 51°C vs. 42° to 45°C in unmulched plots—was intolerable to the whiteflies and thus repelled them from the plants. The temperature recorded in the mulched seedbeds exceeded that which is lethal (46-47°C) to whiteflies. On taller plants (10 cm) the repellent effect declined, since at that height temperatures were lower.

In the same article (Avidov, 1956) the effect on whiteflies of whitewashing seed-bed soil was also mentioned. In this regard, Avidov's findings seemed to contradict the heat-controlling effect hypothesis. A maximum temperature of only 41°C was recorded in the surface layer of whitewashed soil, compared to 45°C in the unmulched control plot. At 2, 5 and 5 days after sprouting, the average numbers of eggs per seedling for whitewashed versus unmulched soils were 0.4, 0.6 and 42.4 vs. 4.1, 17.4 and 16.2, respectively. These findings suggest that temperature is not the only factor involved in the whitefly-controlling mechanism of a mulch.

In the early 1960s the whitefly-borne cucumber vein yellowing virus (CVYV) (Harpaz and Cohen, 1965) previously designated bottle ground mosaic virus (BGMV) (Cohen and Nitzany, 1960) was one of the factors limiting the cultivation of cucumbers in the warmest regions of Israel. The spread of CVYV, which is semipersistent in its vector (Harpaz and Cohen, 1965) could not be controlled with insecticides. Hence, Nitzany *et al.* (1964) tested a mulch to control the spread of the virus. They found that straw mulch markedly reduced the whitefly population and delayed CVYV spread in the first stages of cucumber growth (Fig. 1); it also increased yields and vegetative development. Figure 1 shows that the incidence of CVYV in the mulched plots rises abruptly 21 days after sowing. In hot weather, about 3 days are needed from sowing to emergence of cotyledons, and 7 days are required from inoculation to the appearance of CVYV symptoms. Therefore, the preventive effect of straw mulch lasted only for the first ten days.

Nitzany *et al.* (1964), as well as Avidov (1956), attributed the repellent effect of the straw mulch to the rise in air temperature above it. Smith and Webb (1969) and Costa (1975, 1976) suggested that the effect of mulch on whiteflies could be ascribed to repellency due to reflected light.

Cohen *et al.* (1974) used straw mulch to control the spread of tomato yellow leaf curl virus (TYLCV) in Israel. This whitefly-borne persistent virus is a common and destructive disease of tomatoes in the summer and autumn in all regions of Israel (Cohen and Harpaz, 1964; Cohen and Nitzany, 1966). Straw mulch

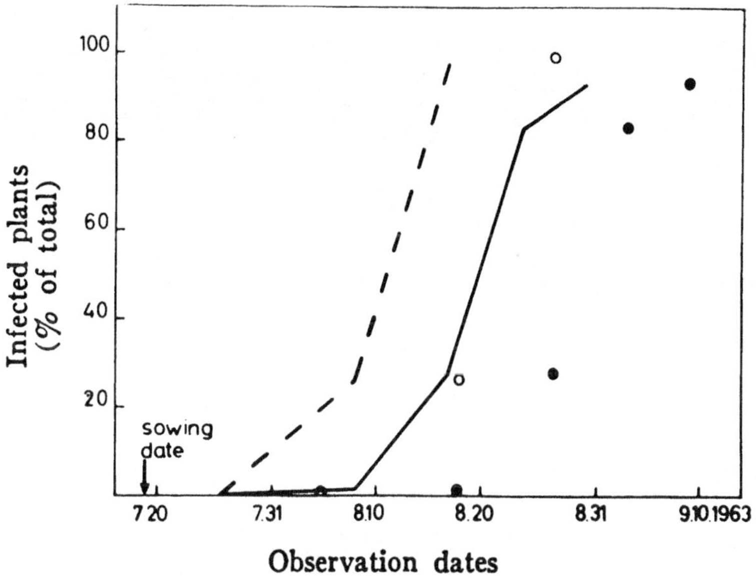

FIG. 1. Spread of bottle gourd mosaic virus (BGMV) (= cucumber vein yellowing virus, CVYV) in mulched vs. unmulched cucumber plots. ○, percentage of BGMV-infected plants in unmulched, control plots; •, percentage of BGMV-infected plants in the straw-mulched plots; ---, estimated rate of BGMV spread in control plots; –, estimated rate of BGMV spread in straw-mulched plots. (From Nitzany *et al.*, 1964.)

was very effective in preventing the spread of TYLCV (Figs. 2 and 3) during the first 18 days after germination, taking into account the time required for symptom development following inoculation; then, the protective effect sharply declined. Mulching also reduced whitefly populations.

3.3 EFFECT OF MULCHES ON WHITEFLY BEHAVIOR

3.3.1 Color Attractance

Cohen *et al.* (1974) also studied the effect of straw on the whitefly's behavior. Their experiments were conducted in a cylindrical aphid flight chamber (60cm diameter × 160cm in height) similar to that described by Kring (1966). The chamber was illuminated by two 15-W cool white fluorescent lamps. The material to be tested was glued onto a 6-cm-diameter cardboard plate and the surface was covered with a thin layer of grease before being placed on the bottom of the chamber. Whiteflies were then introduced into the apparatus from the top and two hours later the number of insects adhering to the plates was counted. It was found that more whiteflies were attracted to the straw than to tomato leaves (ratio, 2.6/1). Also, fresh (yellow) straw was more attractive to the whiteflies than old grey straw exposed to field conditions for 25 days (ratio, 2/1).

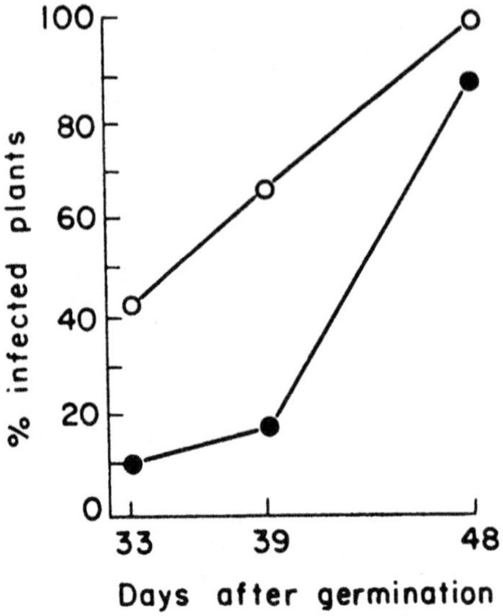

FIG. 2. Incidence of tomato yellow leaf curl virus (TYLCV) in mulched (•) and control (○) tomato plots. (From Cohen *et al.*, 1974.)

FIG. 3. Illustration of effectiveness of mulching in protecting tomato plants from whitefly-borne tomato yellow leaf curl virus (TYLCV). S, straw-mulched plot; C, unmulched plot. All plants in the unmulched plot are infected with TYLCV.

Mound (1962) found that yellow attracts whiteflies. The authors therefore suggested that the controlling effect is connected to the color attractance of whiteflies to the mulch.

Recently, Cohen and Venezia Melamed-Madjar (1978) presented direct evidence as to the connection between the controlling effect of the mulch and its attractance to the whiteflies. They tested four different colored mulches: yellow polyethylene sheets, aluminum-colored polyethylene sheets, blue-colored polyethylene sheets, and straw. All of the materials used reduced the spread of TYLCV. However, yellow polyethylene mulch was the most effective (Fig. 4) and its controlling effect extended to 20 days, in comparison with the 10 days' effectiveness of the straw mulch. The attractance of the above-mentioned materials to the whiteflies was tested in an aphid flight chamber and under field conditions. In the field experiment sticky traps consisting of petri dishes covered with polyethylene sheets of the different colors or chopped straw were used. The traps were placed on the same colored mulch treatment.

Bemisia tabaci was found to be more attracted to yellow polyethylene than to aluminum or blue-colored polyethylene or straw, in that order. The attraction of *B. tabaci* to the aluminum or blue color led the authors to conclude that the behavior of whiteflies is different from that of most aphids which are usually repelled by aluminum mulch (Smith and Webb, 1969). They suggested that ultraviolet radiation reflected from the aluminum may be the cause of the attraction of *B. tabaci* to aluminum polyethylene mulch, as was found in the case of the greenhouse whitefly *Trialeurodes vaporariorum* (Westwood) (Vaishampayan

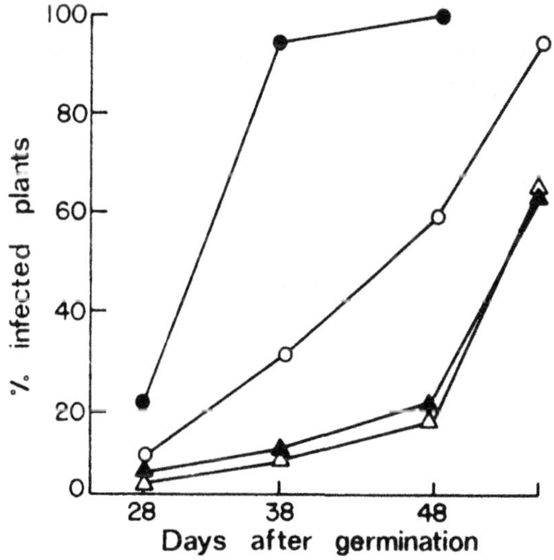

FIG. 4. The effect of different treatments on the spread of tomato yellow leaf curl virus (TYLCV). ○, Straw mulch plus sprays of azinphos-methyl starting 10 days after germination; △, mulching with yellow polyethylene sheets plus sprays of azinphos-methyl starting 10 days after germination; ▲, mulching with yellow polyethylene sheets plus sprays of azinphos-methyl starting 20 days after germination; ●, control. (From Cohen and Melamed-Madjar, 1978.)

et al., 1975). Mound (1962) determined that *B. tabaci* is also attracted by the blue/ultraviolet part of the spectrum.

Cohen *et al.* (1974) noted that the untreated plots in the experiment served as a reservoir of vast populations of virus-carrying whiteflies, thereby increasing the chances of the neighboring plants in the mulched plots to become infected. This is not the case in commercial fields. A combined treatment of yellow polyethylene mulch plus 1% sprays of azinphosmethyl starting 20 days after germination (the sprays with the insecticide were applied every 4 days) was therefore proposed to the growers. Indeed, when this formula was used in commercial tomato fields planted in September, when the whitefly population reached its peak, the incidence of TYLCV did not exceed 25% at the end of the growth period, whereas untreated fields were completely destroyed one month after planting.

Several hypotheses can be presented to explain the control effect of the whitefly-attracting mulch. Cohen *et al.* (1974) suggested that the whiteflies, when attracted to the straw mulch color, are probably killed by the high temperatures reflected by the mulch. However, no direct evidence for the effect of temperature was given. Therefore, the possibility cannot be excluded that the whiteflies, after a short stay on the straw mulch, are repelled by the high temperature, take off, and leave the field. Also, upon landing on a yellow surface, changes may occur in the whitefly's behavior, and consequently it is repelled by this color, as found with aphids (Kring, 1972). The results of experiments relating to the above-mentioned hypothesis will be discussed below.

3.3.2 Temperature Effect

In the Besor region, located in the southwest part of Israel, tomatoes grown for export to Europe in the winter, are planted in August. In order to prolong the fruit production period, it was suggested to grow the plants in screenhouses. Since in this growth season TYLCV is one of the limiting factors, we thought that by covering the plants with yellow nets, the spread of the virus could be controlled. Four screenhouses measuring 28X5X2.5 m were used, divided in the middle into two parts, each covered with a different color net. All the nets used were of 8 mesh, and thread width was 2 mm. The colors tested were yellow, white, and light grey. Thus, there were three replicates of yellow and grey and two of white net. Tomato was sown in August and sprayed regularly with azinphosmethyl 1%. Experiments were carried out in 1976 and 1977.

Figure 5, which represents the results of the experiment carried out in 1976, indicates that the spread of TYLCV was greater among the plants grown under the grey and yellow nets than among those under the white net. By the end of October the incidence of TYLCV in commercial fields next to the screenhouses did not exceed 5%. Similar results were obtained in the experiment carried out in 1977.

These findings show that the whiteflies were attracted into the screenhouses

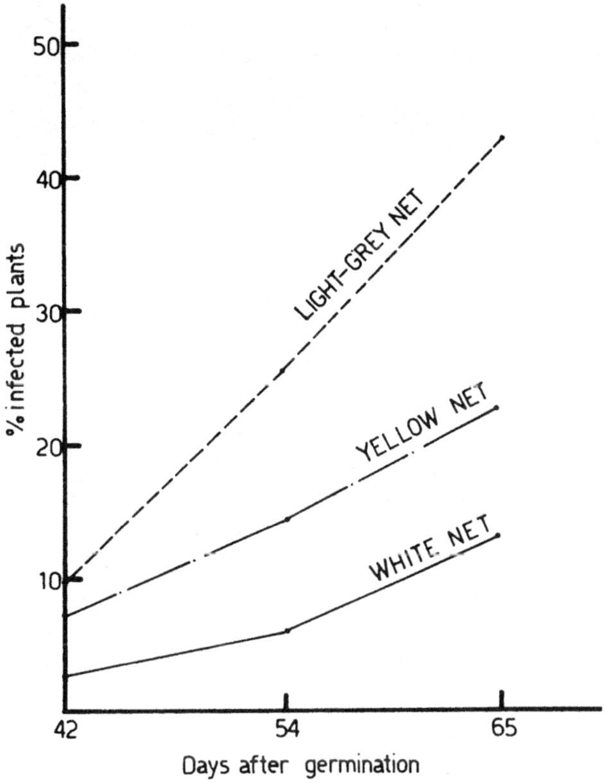

FIG. 5. The effect of colored net covers on the spread of tomato yellow leaf curl virus (TYLCV).

and that attractance is not the only factor involved in the controlling effect of the mulch. Also, in the light of these results, the previously mentioned hypothesis regarding the attraction/rejection effect of yellow can probably be rejected.

Distinct differences were found between the temperatures recorded on the surface of yellow polyethylene mulch and those on the yellow net. During the day, when the maximum air temperature at 50 cm height reached 29°C, the temperature over the yellow mulch was 4-5°C higher than that on the yellow net surface.

These results raised again the question of the involvement of temperature in the controlling effect of the yellow mulch. Experiments were therefore carried out in an aphid flight chamber, as follows: Four small (10-cm diameter) thermostatically controlled heating plates were attached to the floor of the chamber, 20 cm apart; the chamber was kept in a constant temperature room at 25° ± 2°C. Yellow-painted petri dishes covered with glue on the upper sides were placed on the heating plates. The temperature of two opposite plates was regulated to 25°C and that of the other two to 50°C. About 200 whitefly females were then introduced into the chamber from the top, and one hour later the number of insects adhering to the traps was counted (Table 1). The results were analyzed by the sign test for a 0.05 (Siegel, 1956). No significant difference in the attraction of the whiteflies was found between the yellow traps heated to 50°C or to 25°C.

This indicates that the attraction of whiteflies to a yellow surface was not reduced even when the same surface reflected a high temperature that, according to Avidov (1956), is lethal for these insects.

Hence in the subsequent experiment the combined effect of color and heat was examined. The experiment was planned as the previous one, with the exception that in this case the yellow traps were not covered with glue. One hour after the introduction of the whiteflies, the number of dead insects found in the yellow traps was recorded.

The results, summarized in Table 2, show significant differences between the number of dead whiteflies found on the yellow surfaces heated to 25°C and the number found on those heated to 35° or 45°C. The relatively high mortality recorded in the plates heated to 25°C is probably the result of extra heating of the chamber caused by the plates heated to 45°C.

These results led to the rejection of the hypothesis previously mentioned relating to the repellence of the whiteflies by the high temperatures prevailing over the mulch (Avidov, 1956; Nitzany *et al.*, 1964). They may also explain the control effect of a whitewashed soil which reflects relatively low heat of 41°C in tomato seed beds, as reported by Avidov (1956).

TABLE I. The Influence of Heat Reflected from a Yellow Surface on the Attraction of Whiteflies

	Number of whiteflies attracted to sticky yellow traps heated to:	
Test No.*	25°C	50°C
1	25	45
2	111	86
3	66	91
4	80	92
5	80	102
6	104	87
7	63	57
Total (P=0.5)	538	559

*In each experiment, 200 female whiteflies were introduced into an aphid flight chamber of the type described in section 3.3.2.

TABLE II. Influence of Heat Reflected from a Yellow Surface on the Mortality of Whiteflies

Test No.	Number of dead whiteflies in yellow traps heated to:			
	25°C	35°C	25°C	45°C
1	0	45	30	91
2	0	104	31	62
3	0	39	24	133
4	0	5	19	76
5	0	10	12	99
6	-	-	9	132
7	-	-	7	70
Total	0	203	132	662
P	0.03		0.008	

3.4 WHITEFLY VISION

When soil mulching experiments were conducted, a sharp increase was observed in TYLCV incidence in the rows of unmulched control plots bordering the mulched area. This may imply short-sightedness of the whiteflies. This hypothesis was tested in an aphid flight chamber as follows: For each test, two 6-cm-diameter yellow and two white-cardboard plates were placed alternately on the bottom of the chamber. A glass partition fitted to the inner diameter of the apparatus was then placed horizontally at different heights above the cardboards. A delicate cross mark was drawn, on the glass partition dividing the surface, into four equal parts in a way that underneath each quarter there was one yellow or white cardboard. The glass partition was then covered with transparent glue and whiteflies were introduced into the chamber from the top (above the partition). Two hours later, the number of whiteflies adhered onto the glass parts opposite the yellow and white cards was counted.

The results (Table 3) show that when the distance between the partition and the board cards was 10 cm, significantly more whiteflies were attracted to the parts opposite the yellow cards. However, the differences were not significant when the partition was located at a height of 20 cm, indicating short-sightedness of whiteflies.

3.5 DISCUSSION AND CONCLUSION

The evidence presented above suggests that a combination of two factors are involved in the whitefly-controlling mechanism of the mulches tested (Avidov, 1956; Nitzany *et al.*, 1964; Cohen and Melamed-Madjar, 1978): the color of the

TABLE III. Color Discrimination by Whiteflies from Different Distances

Test No.*	Number of whiteflies attracted to glass surface located 10 and 20cm above yellow and white cardboards 6cm in diameter			
	10cm		20cm	
	white	yellow	white	yellow
1	82	138	143	129
2	102	144	139	130
3	100	129	115	123
4	171	216	203	134
5	91	189	186	136
6	136	155	146	105
7	-	-	90	136
Total	682	971	1022	893
P	0.016		0.2	

*In each experiment, about 300 female whiteflies were introduced into the aphid flight chamber of the type described in section 3.4.

material, which attracts the whiteflies; and, its properties, which reflect heat. Thus, the whiteflies are actively attracted to the colored surfaces and probably stay there for a period of time long enough to be killed by the accumulating heat of even relatively not-so-high temperatures. The effect of heat could also be indirect, by decreasing the relative humidity in the layer immediately above the surface to a degree which is lethal for the whiteflies.

Based on this information, Cohen and Berlinger (unpublished data) were able to reduce the spread of TYLCV in glasshouses by painting the roofs, which reflect high temperatures, with a thin layer of yellow.

The control effect of the colored mulches attractive to whiteflies lasted for a relatively short period of time (20-30 days). However, during that period the mulches were much more effective than insecticides, even when applied at close intervals (Cohen and Melamed-Madjar, 1978). Moreover, the mulches protected the plants during the critical stage, since in most of the cases crop losses are greater when plants are infected in the early stages of growth (Costa, 1976; Bird and Maramorosch, 1978).

The prolongation of the mulch's control effect achieved by replacing the straw mulch with yellow polyethylene sheets (Cohen and Melamed-Madjar, 1978), should encourage further research for a better attractive dye which will remain stable under outdoor conditions (solar radiation and climatic changes). Also, the characteristic of the material surface, *viz.*, degree of smoothness, may play a role in the attraction of the whiteflies by influencing the scattering pattern of the reflected light. The way of positioning the mulch may be of importance. Thus, longer effect might be obtained by planting the crop in furrows and placing the mulch at a greater height, on heaps between the rows.

The mulch method was effective in dry and warm regions. In humid and cooler regions the temperatures may be too low to create the "lethal atmosphere" for the whiteflies. In such cases impregnation of the mulch with insecticides may replace the heat effect. Berlinger *et al.* (1979, unpublished data) studied control measures of TYLCV in glasshouses (where the humidity is high) in Israel. In their preliminary work, a distinct reduction was obtained in the spread of TYLCV in tomatoes grown in yellow polyethylene mulch, sprayed with 5% Supracide Methidathion.

As noted previously (Cohen and Melamed-Madjar, 1978), whitefly behavior differs from that of most aphids tested, when color attractance is considered. They may also differ in their landing behavior. We observed that the whitefly attack, and consequently TYLCV spread, always started from the borders of a field; when such a tomato field was surrounded with sticky yellow polyethylene sheets located vertically at 70 cm above the soil surface, neither reduction in the whitefly population nor in the spread of TYLCV was found (Cohen *et al.*, 1975). However, this method was very efficient in preventing the spread of aphid-transmitted viruses in peppers (Cohen and Marco, 1973). Better understanding of the landing behavior of whiteflies might lead to increased efficiency of the mulch, if positioned according to the whitefly landing course.

In cases where no resistant varieties are available, the mulch method may be used to prevent the spread of whitefly-borne viruses with minimal use of insecticides.

3.6 ACKNOWLEDGMENTS

The author thanks Professor I. Harpaz, The Hebrew University of Jerusalem, Faculty of Agriculture, Rehovot, Israel, for his criticism and suggestions during the preparation of the manuscript. The technical assistance of Mrs. Rachel Ben-Joseph and K. Joles is gratefully acknowledged.

3.7 REFERENCES

Avidov, Z. (1956). Bionomics of the tobacco whitefly (*Bemisia tabaci* Gennad.) in Israel. *Ktavim* 7:25-41.

Bird, J., and Maramorosch, K. (1978). Viruses and virus diseases associated with whiteflies. *Adv. Virus Res.* **22**:55-110.

Bird, J., Sanchez, J., Rodriguez, R., and Julia, F. (1975). Rugaceous (whitefly-transmitted) viruses in Puerto Rico. *In* "Tropical Diseases of Legumes" (J. Bird and K. Maramorosch, eds.), pp. 1-25. Academic Press, New York.

Cohen, S., and Harpaz, I. (1964). Periodic, rather than continual acquisition of a new tomato virus by its vector, the tobacco whitefly (*Bemisia tabaci*: Gennadius) *Entomologia Expl. Appl.* 7:155-166.

Cohen, S., and Marco, S. (1973). Reducing the spread of aphid-transmitted viruses in peppers by trapping the aphids on sticky yellow polyethylene sheets. *Phytopathology* **63**: 1207-1209.

Cohen, S., and Melamed-Madjar, V. (1978). Prevention by soil mulching of the spread of to tomato yellow leaf curl virus transmitted by *Femisia tabaci* (Gennadius) (Homoptera: Aleyrodidae) in Israel. *Bull. Entomol. Res.* **68**:465-470.

Cohen, S., and Nitzany, F. E. (1960). A whitefly transmitted virus of cucurbits in Israel. *Phytopath. Medit.* **1**:44-46.

Cohen S., and Nitzany, F. E. (1966). Transmission and host range of the tomato yellow leaf curl virus. *Phytopathology* **56**:1127-1131.

Cohen, S., Melamed-Madjar, V., and Hameiri, J. (1974). Prevention of the spread of tomato yellow leaf curl virus transmitted by *Bemisia tabaci* (Gennadius) (Homoptera: Aleyrodidae) in Israel. *Bull. Entomol. Res.* **64**:193-197.

Cohen, S., Melamed-Madjar, V., Joles, K., Joseph, R., Tomer, S., and Sender, N. (1975). Prevention of the spread of tomato yellow leaf curl virus in tomatoes by soil mulching. *Hassadeh* **55**:1466-1470 (in Hebrew).

Costa, A. S. (1975). Increase in the population density of *Bemisia tabaci*, a threat of widespread virus infection of legume crops in Brazil. *In* "Tropical Diseases of Legumes" (J. Bird and K. Maramorosch, eds.), pp. 27-49. Academic Press, New York.

Costa, A. S. (1976). Whitefly-transmitted plant diseases. *Annu. Rev. Phytopathol.* **14**:429-449.

Harpaz, I., and Cohen, S. (1965). Semipersistent relationship between cucumber vein yellowing virus (CVYV) and its vector, the tobacco whitefly (*Bemisia tabaci* Gennadius). *Phytopathol. Z.* **54**:240-248.

Kring, J. B. (1966). An aphid flight chamber: construction and operation. *J. Econ. Entomol.* **59**:1518-1520.

Kring, J. B. (1972). Flight behavior of aphids. *Annu. Rev. Entomol.* **17**:461-492.

Mound, L. A. (1962). Studies on the olfaction and colour sensitivity of *Bemisia tabaci* Genn.). *Entomol. Expl. Appl.* **5**:99-104.

Muniyappa, V. (1980). Whiteflies. *In* "Vectors of Plant Pathogens" (K. F. Harris and K. Maramorosch, eds.), pp. 39-85. Academic Press, New York.

Nene, Y. L. (1972). A survey of the viral diseases of pulse crops in Uttar Pradesh. *Res. Bull. G. B. Pant Univ. Agric. Technol., Pantnagar* **4**.

Nitzany, F. E., Geisenberg, H., and Koch, B. (1964). Tests for the protection of cucumbers from a whitefly-borne virus. *Phytopathology* **54**:1059-1061.

Siegel, S. (1956). "Nonparametric Statistics for the Behavioral Sciences," pp. 68-75. McGraw Hill, New York.

Smith, F. F., and Webb, R. E. (1969). Repelling aphids by reflective surfaces, a new approach to the control of insect-transmitted viruses. *In* "Viruses, Vectors, and Vegetation. (K. Maramorosch, ed.), pp. 631-639. Wiley (Interscience), New York.

Vaishampayan, S. M., Kogan, M., Waldbauer, G. P., and Wolley, J. T. (1975). Spectral specific responses in the visual behaviour of the greenhouse whitefly, *Trialeurodes vaporariorum* (Homoptera: Aleyrodidae). *Entomol. Expl. Appl.* **18**:344-356.

Varma, P. M. (1963). Transmission of plant viruses by whiteflies. *Bull. Natl. Inst. Sci. India* **24**:11-33.

Chapter 4

CHEMICAL CONTROL OF NEMATODE VECTORS

F. Lamberti
M. Basile

Istituto di Nematologia Agraria del C.N.R.,
Bari, Italy

4.1 INTRODUCTION

The long persistence of some plant viruses at sites of retention in their nematode vectors (Taylor and Robertson, 1975) and the possibility that nematodes can acquire the virus from weed hosts or perennial root fragments left from previous crops (Thomason and McKenry, 1975; Alphey, 1978a; Martelli, 1978) point to the use of chemicals as the most practical and economical method of controlling virus vector nematodes.

Control should aim to achieve or closely approach eradication of nematode vectors because surviving individuals may be viruliferous or may have access to

ISBN 0-12-326440-5

the various virus sources mentioned above (Thomason and McKenry, 1975). In addition several species, especially among trichodorids are very efficient as vectors (Taylor and Robertson, 1975; Roca *et al.*, 1977).

There are two types of nematicides commercially available: fumigants and nonfumigants. Their physical and technical characteristics and the interaction with environmental factors in effecting nematode control have been amply illustrated (Thomason and McKenry, 1975; Alphey, 1978a; Lamberti, 1979). Therefore, we have elected to discuss some examples of the practical application of nematicides for the control of nematode vectors in different types of soil and the evaluation of the degree of control that may be achieved under different circumstances.

4.2 FUMIGANT NEMATICIDES

4.2.1 1,3-D (1,3-dichloropropene)

The most widely used nematicides are those containing 1,3-dichloropropene (1,3-D) as the active ingredient. They have been extensively studied to determine their dynamics in different soil types in relation to environmental factors such as temperature, humidity, and content of organic matter of the treated soil (Thomason and McKenry, 1975; Lamberti, 1979). As with all the fumigants, 1,3-D is highly phytotoxic, and it would be useful to know its persistence in different soils to determine the proper time interval between application and planting of a crop. Generally and empirically an interval of one week for each 100 kg/ha of technical product applied is left before planting a crop in the treated soil. However, in certain situations such an interval could be insufficient for the elimination of phytotoxic residues from the soil; it might also be too long where soils are intensively cultivated with crops in close succession. In fact, water content and temperature of the soil can considerably modify the dispersion pattern of this chemical (Basile and Lamberti, 1978).

Studies on the determination of the amounts of 1,3-D and its metabolites in the soil are difficult when done by bioassay, but the rapid gas-chromatographic methods developed by Hannon *et al.* (1963), Williams (1968) and, more recently, Renzoni *et al.* (1976) allow accurate quantitative analyses of these compounds. Using gas chromatography, several investigations have been carried out to collect information that would permit the best use of this nematicide.

Castro and Belser (1966) found in laboratory tests that the two isomers, cis- and trans-1,3-D, are quickly hydrolyzed to their respective chloroallyl alcohols in water or wet soil. This accords with the results of Hannon *et al.* (1963) who reported the rapid disappearance of dichloropropene-dichloropropane from soil after application. However, Williams (1968) found residues of both cis and trans 1,3-D, eight months after the injection of D-D into sandy loam and muck, and assumed that this long persistence would be favored by the low pH of the soils. This assumption is confirmed by Leistra (1970, 1971), who found that 1,3-D persisted for six months after fumigation in the previous autumns, and by Basile and Lamberti (1978), who detected the two isomers up to 18 days after injection of the chemical in August, and up to 92 days after injection in January in a sandy soil of Southern Italy. Also, the fumigant persisted for up to 92 and 122 days, respectively, in sandy-clay and sandy-loamy soils in Southern Italy when applied in August, and up to 122 and 152 days, respectively, when applied in January. These data indicate that soil granular structure has very little influence on 1,3-D's persistence, which seems greatly affected by temperature and water content of the soil (McKenry and Thomason, 1974; Basile and Lamberti, 1978). However, the persistence of undecomposed 1,3-D in the soil is of little importance in the control of nematodes, since its degradation products (chloroallyl alcohols which are also highly nematicidal) are only slowly biodegraded by a *Pseudomonas* sp. (Castro and Belser, 1971). More important, in this respect, is the calculation of the percentage mortality of plant parasitic nematodes in relation to the concentration unit time of exposure to the toxicant. In laboratory tests Van Dijk (1974) demonstrated that dichloropropenes disappear at the rate of 2 to 3.5% per day in sandy soils at 15-20°C, and at about 25% per day at 20°C in clay soils. Biodegradation of chloroallyl alcohols is even faster (from 20 to over 60% per day at 15°C) but is much slower in sterilized soil (Van Dijk, 1974). According to Baines *et al.* (1977) this explains the greater persistence of these products in soil treated with high dosages of 1,3-D, which kills, along with the nematodes, the microorganisms that degrade the chemicals.

The parasite is killed when exposed to the toxicant and, therefore, knowledge of the diffusion pattern of the nematicide is helpful in evaluating the degree of control that may be expected. According to Leistra (1970), at temperatures from 2 to 20°C, 1,3-D in soil is 1% dispersed in the vapor phase, 10 to 20% dissolved in the water phase, and between 80 and 90% adsorbed to soil components. From these distribution data it should be possible to calculate, with the aid of mathematical models, the application rates of the chemical needed to achieve a satisfactory control of plant parasitic nematodes in soils whose physical and chemical characteristics are known.

Soil texture seems to have only indirect effects on the diffusion of fumigants (McKenry and Thomason, 1974), although studies in prefabricated columns simulating field conditions indicate that 1,3-D accumulates in the top 30 cm in soil with high clay content (Renzoni *et al.*, 1978). Generally, clay soils may be readily saturated with moisture at field capacity or their structure destroyed because of compaction or deflocculation. These conditions limit the movement of the vapor phase (Leistra, 1971; McKenry and Thomason, 1974). On the other hand, substantial losses of the fumigant into the atmosphere may occur when it is injected into a dry, surface-cracked clay soil, unless the soil is tilled and compacted before treatment (McKenry and Thomason, 1974). Also, organic matter may absorb a fumigant and limit its diffusion.

High soil moisture and temperature are major factors affecting the dynamics of a fumigant. The movement of the gas phase, in fact, is restricted at or near field capacity, and decreasing temperature increases the solubility of the chemical in the water phase (McKenry and Thomason, 1974). Conversely, high temperatures increase the rate of volatilization and hydrolysis (McKenry and Thomason, 1974). Field tests carried out in three soil types in southern Italy (Basile *et al.*, unpublished) have shown that of four factors causing variability, namely rate of application, soil texture, soil temperature, and soil moisture, the last two are those that mostly affect vertical distribution of 1,3-D in soil. McKenry and Thomason (1976a) report 15-20°C as the optimal temperature range for application of this fumigant, with upper and lower limits of 5° and 25°C, respectively (McKenry and Thomason, 1974). Soil preparation such as deep cultivation to break layers that may restrict the dispersal of 1,3-D and removal of residues of the previous crop, or compacting or covering the soil surface after application of the chemical may improve the efficiency of treatment (Smelt *et al.*, 1974; Thomason and McKenry, 1975; Raski *et al.*, 1976; McKenry and Thomason, 1976b; McKenry *et al.*, 1977). Furthermore, injection at two depths allows a more uniform distribution of the fumigant in deeper layers of the soil (Abdalla *et al.*, 1974a; Raski *et al.*, 1976).

Little is known about the lethal dosages for Longidoridae and Trichodoridae which represent the two families to which the nematode vectors belong. In California Raski and Lear (1962) obtained complete kill of *Meloidogyne incognita* to depths of 2 m with 1200 liters/ha of Telone or 2000 liters/ha of D-D, both nematicides containing 1,3-D. In laboratory tests, 100% kill of second stage larvae of the same nematode species required the application of 1,3-D at a rate of 2.5 μg/g of dry soil and exposure for three days (Abdalla *et al.*, 1974a). Application of 1,3-D at 2.8 μg/g of dry soil and exposure for four days gave 98% mortality of second stage larvae of *Tylenchulus semipenetrans*, the citrus nematode; complete control was achieved at concentrations of 46 μg/g of dry soil (Baines *et al.*, 1977). Longidorid and Trichodorid nematodes are at least as sensitive as free living stages of root-knot nematodes to 1,3-D. Therefore, McKenry *et al.* (1978), after having studied dosage-response of various organisms to cis- and trans-1,3-D (McKenry and Thomason, 1974), assume that the toxicant level required to control 99.9% of all stages of *Xiphinema index* is the same as that for infective stages of *Meloidogyne* spp. On the basis of their many data, gathered from several different field situations, McKenry *et al.* (1978) have constructed a table indicating the level of nematode control expected at various depths in Californian soils, following fumigation with 1,3-D at different application rates. McKenry (1978) has also suggested broadcast application rates to provide lethal dosages to soil pests at 75- or 150-cm depths in various soil types under different conditions of moisture and temperature.

To control Trichodorid nematodes, and thus prevent spread of tobacco rattle in annual crops, rates of 250/500 liters/ha of D-D (50% of 1,3 dichloropropene + 50% of 1,2 dichloropropane) may be sufficient. Cooper and Thomas (1971) almost completely suppressed the transmission of this virus in a potato field in the first year by reducing the vector population to less than 1% that of control plots. The extent of control of virus spread was much reduced in the second year of planting but was unaffected the third year. Spread of tomato black ring and raspberry ringspot viruses by *Longidorus elongatus* in a strawberry field was prevented for up to four years by autumn applications of D-D prior to planting in the following spring (Murant and Taylor, 1965). The shorter time of retention of the virus and the lower efficiency of the vector account for these long term effects of treatment (Taylor and Robertson, 1975). On the other hand, D-D fumigation of soil infested with *X. diversicaudatum* carrying the hop strain of arabis mosaic virus failed to give adequate control; and new plantings of hops became 80% infected within four years from treatment, although more than 95% of the nematode vectors had been killed initially (Pitcher and McNamara, 1973). A combination of chemical treatments and fallow was later shown to be effective (McNamara *et al.*, 1973). Numerous investigations have been made on the control of *X. index*, the natural vector of grapevine fanleaf virus (GFV). In California, Raski and Schmitt (1972) and Raski *et al.* (1976) controlled nematodes and grapevine fanleaf virus for four years or more in vineyards treated at replanting with split applications of D-D—about 2000 liters/ha injected at a depth 75 cm followed by another application of about 500 liters/ha at a depth of 20-25 cm. Boubals (1976) reported that in France there was little spread of GFV in shallow soils 16 years after application of 1000 liters/ha of D-D, whereas higher rates (1500 liters/ha) of the same chemical in deeper soils did not give satisfactory results.

4.2.2 Methyl Bromide

This chemical has been used in plant protection since the 1930's (Maw and Kempton, 1973), but Taylor and McBeth (1941) were the first to identify the dosage required to control plant parasitic nematodes in soil. With regard to the control of virus vector nematodes with this chemical, Van Gundy *et al.* (1972) reported that exposure to 600 ppm immobilized *X. index* adults after 27 hr and larvae after 43 hr. Abdalla and Lear (1975) observed that 100% kill of *X. index* was obtained by exposure to 530 ppm of methyl bromide for 24 hr. Similar results were achieved when nematodes were exposed to 350 ppm for three days, or 50 ppm for 21 days (Abdalla and Lear, 1975). This indicates the validity of the concept that the lethal dose of some nematicides is the cumulative effect of concentration and time of exposure to the toxicant. Studies on diffusion of methyl bromide in soil indicate that surface applications of 98 g/m^2 are sufficient to kill, to a depth of 2.4 m, mycelia of plant pathogenic fungi (Rackham *et al.*, 1968) known to be more tolerant to the chemical than are nematodes (Munnecke and Ferguson, 1953). The fumigant penetrated to a depth of 20 cm within one hour after its application and was detected in appreciable quantities to a depth of one meter 48 hr later (Drosihn *et al.*, 1968). In California, placement of methyl bromide in the soil at a depth of 75-80 cm resulted in gas distribution at concentrations sufficient for nematode kill to a depth of 2.44 m, even without polyethylene covering of the soil after treatment (Abdalla *et al.*, 1974b).

Penetration and persistence of methyl bromide is affected by various factors such as degree of compaction of the soil and its moisture content, temperature, and composition (Maw and Kempton, 1973; McKenry and Thomason, 1976b).

In field experiments methyl bromide gave promising control of both *X. index* and GFV (Raski and Schmitt, 1972) and general nematode control for more than four years (Raski *et al.*, 1976) when applied at a rate of 400 to 600 kg/ha.

However there are plants, such as Liliaceae (Lamberti, 1974) or carnation (Kempton and Maw, 1974), which are sensitive to the inorganic bromine persisting in the soil as a degradation product of methyl bromide (Maw and Kempton, 1973). It is possible in these cases to eliminate residues by irrigating the treated soil before transplanting (Vigodski and Klein, 1976; Basile and Lamberti, 1977).

4.2.3 EDB (Ethylenedibromide)

Ethylenedibromide is another fumigant compound widely used to control plant parasitic nematodes. Preplanting rates of 7.7 and 23 liters/ha of this chemical completely eliminated populations of *X. americanum* in strawberry plantations within two months of applications (Nemec and Malec, 1975). Studies carried out on the toxic level of EDB (McKenry and Thomason, 1974) indicate that to obtain 95% kill of free living stages of plant parasitic nematodes concentrations of 35, 81 or 290 ppm are necessary in the soil water phase at 25, 15 or 5°C, respectively, with an exposure time of one day. McKenry and Thomason (1974) also studied the movement and fate of this chemical under various pedoclimatic conditions in California and found that its rate of hydrolysis, as already demonstrated in laboratory tests by Castro and Belser (1968), is much less than that of 1,3-D at higher temperatures. Therefore EDB could well substitute for 1,3-D in field situations when the soil temperature is in excess of 25°C. The way to use EDB is illustrated in the chart of McKenry (1978) which was developed for Californian soils on the basis of experience and data obtained from various field situations.

4.2.4 DBCP (1,2-dibromo-3-chloropropane)

This nematicide is used only occasionally in preplanting treatments because better results usually are obtained with less expensive chemicals such as 1,3-D or EDB (Lamberti, 1979). Populations of *X. americanum* were eradicated with preplant applications of DBCP at the rates of 3 to 6 liters/ha (Nemec and Malek,

1975). Because of its low phytotoxicity, DBCP is more frequently used in established plantations to prevent increase of nematode population levels already reduced by more drastic preplant treatments. Dispersion of DBCP in the soil requires abundant irrigation and its penetration downward, at lethal concentration, is greatly affected by the concentration of the toxicant within the soil water and by the organic matter content of the treated soil (Johnson and Lear, 1968, 1969; Hodges and Lear, 1973a, 1973b; O'Bannon *et al.*, 1975; Basile *et al.*, 1979a). Root-knot nematodes are known to be more tolerant than trichodorids and longidorids to chemicals but were eradicated at DBCP concentrations between 20 to 25 μg/g of soil water (Johnson and Lear, 1966, 1968). However, DBCP is adsorbed rapidly by nematodes but eliminated slowly (Marks *et al.*, 1968) and, therefore, high concentrations might be unnecessary when the time of exposure is sufficient to reach a letal dosage in the nematode body. Initial concentrations of 0.25 and 1 μg of DBCP per gram of dry soil killed all nematodes within 35 and 28 days, respectively (Hodges and Lear, 1973c). This is possible because of the relatively high solubility of DBCP in the soil water phase which might function as a reservoir of the chemical. Its solubility is, in fact, reversible and inversely correlated with soil temperature (Johnson and Lear, 1969). Persistence (Basile *et al.*, 1979b) and soil sorption (Basile *et al.*, in preparation) of DBCP are also positively affected by moisture and organic matter content, factors limiting the hydrolysis of the chemical which is biologically dehalogenated in soil (Castro and Belser, 1968). In conclusion, O'Bannon and Tomerlin (1975) indicate that the optimal condition for application of DBCP is a soil temperature between 20 and 24°C followed within 2-5 days (Hodges and Lear, 1973b) by an irrigation of 10-15 cm.

4.3 NONFUMIGANT NEMATICIDES

These are usually in granular form and therefore can be spread easily with fertilizer distributors to be incorporated into the soil by disking or rotivating. Most of them are not phytotoxic and can be applied at planting or on established crops. They move through the soil in the water phase and thus their distribution in the various profiles is highly dependent on soil moisture. Their effect in preplant treatments of perennials to control nematode vectors has not shown promising results in deep soils. They are more effective when used in annual crops. Those that are systematically translocated in plants can complement the application of fumigants in perennial plantations, even if it is only to prevent the vectors from feeding on the plant roots. Two systemic nematicides, the oxime carbamates aldicarb and oxamyl, have been tested with some success against nematode vectors; but, considering the degree of control obtained against other plant parasitic nematodes (Lamberti, 1979; Greco and Thomason, 1980), good results should be obtained also with the organophosphate phenamiphos.

4.3.1 Aldicarb

The incidence of spraing in potatoes, caused by rattle tobacco virus which is transmitted by trichodorid nematodes, was very much reduced by treatment at planting time with 3.4 kg a.i./ha of aldicarb (Alphey *et al.*, 1975). This chemical is nematostatic rather than nematicidal and acts by interfering with the behavior of the nematodes. The feeding behavior of longidorids was affected (Trudgill, 1975). Moreover, 5 to 500-g/ml aqueous solutions of aldicarb inhibited hatching of *Heterodera schachtii* (Steele and Hodges, 1975); this effect was reversed by removing the chemical (Steele, 1977). Aldicarb applications also reduced the motility of males and infective larvae of *H. schachtii* and *M. javanica* (Hough and Thomason, 1975), as well egg hatch and larvae motility of the citrus nematode, *Tylenchulus semipenetrans* (Huang and Van Gundy, 1978).

Soil moisture and temperature positively affect the release of aldicarb from granular formulations (Cappedge *et al.*, 1975), whereas organic matter adsorbs it (Jamet *et al.*, 1974).

The fate of aldicarb and its breakdown products in soil and plants has been studied by Bromilow (1973) and Leistra *et al.* (1976) who indicate that its conversion to the more persistent sulphoxide and sulphone is a rapid process and is affected by soil type (Smelt *et al.*, 1978c). Both breakdown products of aldicarb are toxic to nematodes and the sulphone (Smelt *et al.*, 1978a) is much less rapidly degraded than the sulphoxide (Smelt *et al.*, 1978b).

4.3.2 Oxamyl

Various experiments (Alphey, 1978a) have demonstrated that applications of oxamyl either to the soil as a granular formulation or to the foliage as a liquid spray can limit the rate of spread of virus, even though vector nematode populations are neither eradicated nor greatly reduced. It has been shown that this chemical inhibited nematodes from acquiring and transmitting virus by preventing feeding (Trudgill and Alphey, 1976). Unfortunately, nematodes exposed to oxamyl recover when the effect of the toxicant is lost (Forer *et al.*, 1975; Bunt, 1975). Oxamyl is rapidly decomposed in the soil to nonnematoxic substances (Bromilow, 1973) and appears to be much more effective at lower soil temperatures (Bunt, 1975). These characteristics of the chemical make it a nemastatic rather than a nematicidal product (Bunt, 1975) and limit its use to annual crops only, particularly those that have a narrow period of susceptibility to a virus (Van Hoof, 1964). Various authors (Alphey, 1978b; Lamberti, 1979) indicate that foliar sprays of oxamyl provide a better nematode control than equivalent rates of granular formulation incorporated to the soil at planting time. It has been demonstrated that a nematicidal substance is exuded from roots treated with this chemical (Taylor and Alphey, 1973; Potter and Marks, 1976).

4.3.3 Phenamiphos

Phenamiphos is more persistent than oxamyl and it has been demonstrated to cause irreversible poisoning to *Ditylenchus dipsaci* (Bunt, 1975). No information is available on the direct effect of this compound on virus-vector nematodes. But Greco and Thomason (1980) have observed that at low concentrations it inhibits egg hatching, and at higher concentrations prevents root invasion of *H. schachtii* and *M. javanica*—nematodes known to be more tolerant than longidorids and trichodorids to nematicidal chemicals. In soil, phenamiphos is reduced within three weeks to sulphone and sulphoxide (Waggoner and Khasawinah, 1974) which are also nematoxic compounds. Nematode control by phenamiphos is more efficient at higher temperatures (above 25°C), but its nematicidal effect is reduced in organic soils (Bunt, 1975).

4.3.4 Quintozene

Finally it has been shown in Scotland that soil applications of the fungicide quintozene reduced populations of *Longidorus elongatus* by 95% and prevented transmission of the raspberry ringspot virus for four years (Murant and Taylor, 1965). Its nematicidal effect should therefore be considered for crops which have a short duration (2-4 years). For longer growing perennials, a combination of preplant treatments with a fumigant injected into the soil and a nonfumigant (such as dazomet applied to the soil surface) (Seinhorst, 1973) supplemented by the application of systemic nematicides on the existing crop, may sometimes give satisfactory results.

4.4 REFERENCES

Abdalla, N., and Lear, B. (1975). Lethal dosages of Methyl bromide for four plant-parasitic nematodes and the effect of soil temperature on its nematicidal activity. *Plant Dis. Rep.* **59**:224-228.

Abdalla, N., Raski, D. J., Lear, B., and Schmitt, R. V. (1974a). Movement, persistence and nematicidal activity of a pesticide containing 1,3-Dichloropropene in soils treated for nematode control in replant vineyards. *Plant Dis. Rep.* **58**:562-566.

Abdalla, N., Raski, D. J., Lear, B., and Schmitt, R. V. (1974b). Distribution of Methyl bromide in soils treated for nematode control in replant vineyards. *Pest. Sci.* **5**:259-269.

Alphey, T. J. W. (1978a). Chemical control of virus vector nematodes. *In* "Plant Disease Epidemiology" (P. R. Scott and A. Bainbridge, eds.), pp. 299-305. Blackwell Scientific Publications, Oxford.

Alphey, T. J. W. (1978b). Oxamyl sprays for the control of potato spraing disease caused by nematode-transmitted tobacco rattle virus. *Ann. Appl. Biol.* **88**:75-80.

Alphey, T. J. W., Cooper, J. I., and Harrison, B. D. (1975). Systemic nematicides for the control of trichodorid nematodes and of potato spraing disease caused by tobacco rattle virus. *Plant Pathol.* **24**:117-121.

Baines, R. C., Klotz, L. J. and De Wolfe, T. A. (1977). Some biocidal properties of 1,3-D and its degradation product. *Phytopathology* **67**:936-940.

Basile, M., and Lamberti, F. (1977). Possible impiego del bromuro di metile nella coltivazione del Garofano in provincia di Bari. *Riv. Ortoflorofrutticoltura Ital.* **61**:149-156.

Basile, M., and Lamberti, F. (1978). Distribuzione verticale e persistenza dell'1,3-dicloropropene in tre tipi di terreno dell'Italia meridionale. *Nematol. Medit.* **6**:135-145.

Basile, M., Lamberti, F., and Elia, F. (1979a). Percolazione dell'1,2-dibromo-3-cloropropano in differenti tipi di terreno. *Difesa Piante* **2**:267-272.

Basile, M., Lamberti, F., and Elia, F. (1979b). Studi sulla persistenza dell'1,2-dibromo-3-cloropropano. *Nematol. Medit.* **7**:29-36.

Belser, N. O., and Castro, C. E. (1971). Biodehalogenation. The metabolism of the nematicides cis- and trans-3-chlorallyl alcohol by a bacteria isolated from soil. *Agric. Food Chem.* **19**:23-26.

Boubals, D. (1976). Le point actuel sur les maladies a virus de la vigne. *Prog. Agric. Vitic.* **93**:625-633.

Bromilow, R. (1973). Breakdown and fate of oxime carbamate nematicides in crops and soils. *Ann. Appl. Biol.* **75**:473-479.

Bunt, J. A. (1975). Effect and mode of action of some systemic nematicides. *Mededl. Landbouw.* **75**:1-127.

Castro, C. E., and Belser, N. O. (1966). Hydrolysis of cis and trans 1,3-dichloropropene in wet soil. *Agric. Food Chem.* **14**:69-70.

Castro, C. E., and Belser, N. O. (1968). Biodehalogenation. Reductive dehalogenation of the biocides ethylene dibromide, 1,2 dibromo-3-chloropropane and 2,3-dibromobutane in soil. *Environ. Sci. Technol.* **2**:779-783.

Cooper, J. I., and Thomas, P. R. (1971). Chemical treatment of soil to prevent transmission of tobacco rattle virus to potatoes by *Trichodorus* spp. *Ann. Appl. Biol.* **69**:23-24.

Coppedge, J. R., Stokes, R. A., Kinzer, R. E., and Ridgway, R. L. (1975). Effect of soil moisture and soil temperature on the release of aldicarb from granular formulations. *J. Econ. Entomol.* **68**:209-210.

Drosihn, U. G., Stephan, B. R., and Hoffmann, G. (1968). *Z. Pflankrankh. PflPath. Pfl-Schutz.* **75**:272-287.

Forer, L. B., Trudgill, D. L., and Alphey, T. J. W. (1975). Some effects of oxamyl on the virus vector nematodes, *Longidorus elongatus* and *Xiphinema Diversicaudatum*. *Ann. Appl. Biol.* **81**:207-214.

Greco, N., and Thomason, I. J. (1980). Effect of Fenamiphos on *Heterodera schachtii* and *Meloidogyne javanica*. *J. Nematol.* **1**:91-96.

Hannon, C. I., Angelini, J., and Wolford, R. (1963). Detection of dichloropropene-dichloropropane in soil by gas chromatography. *J. Gas Chromatogr.* **7**:27-32.

Hodges, L. R., and Lear, B. (1973a). Distribution and persistence of 1,2-dibromo-3-chloropropane in soil after application by injection and irrigation water. *Nematologica* **19**:146-158.

Hodges, L. R., and Lear, B. (1973b). Effect of time irrigation on the distribution of 1,2-dibromo-3-chloropropane in soil after shallow injection. *Pest. Sci.* **4**:795-799.

Hodges, L. R., and Lear, B. (1973c). Efficacy of 1,2-dibromo-3-chloropropane for control of *Meloidogyne javanica* as influenced by concentration, exposure time and rate of degradation. *J. Nematol.* **5**:249-253.

Hough, A., and Thomason, I. J. (1975). Effects of aldicarb on the behaviour of *Heterodera schachtii* and *Meloidogyne javanica*. *J. Nematol.* **7**:221-229.

Huang, S. P., and Van Gundy, S. D. (1978). Effects of aldicarb and its sulfoxide and sulfone on the biology of *Tylenchulus semipenetrans*. *J. Nematol.* **10**:100-106.

Jamet, P., Piedallu, M. A., and Hoscoet, M. (1974). Migration et dégradation de l'aldicarbe dans différents types de sols. *Proc. Symp. Nuclear Techniques. Comparatives of Food and Environmental Contamination*, pp. 393-415. IAEA, Vienna.

Johnson, D. E., and Lear, B. (1966). The influence of nematicide exposure, time on nematode control. *Plant Dis. Rep.* **50**:915-916.

Johnson, D. E., and Lear, B. (1968). Evaluating the movement of 1,2-dibromo-3-chloropropane through soil. *Soil Sci.* **105**:31-35.

Johnson, D. E., and Lear, B. (1969). The effect of temperature on the dispersion of 1,2-dibromo-3-chloropropane in soil. *J. Nematol.* **1**:116-121.

Kempton, R. S., and Maw, G. W. (1974). Soil fumigation with methyl bromide: The phytotoxicity of inorganic bromide to carnation plants. *Ann. Appl. Biol.* **76**:217-229.

Lamberti, F. (1974). I nematodi parassiti delle liliacee da orto e loro controllo con particolare riferimento all'ambiente mediterraneo. *Proc. IV Journees Phytiatr. Phytopharm.* Circum-Medit. Montpellier, France. pp. 56-68.

Lamberti, F. (1979). Chemical and cultural methods of control. *In* "Root-knot Nematodes (*Meloidogyne* spp.) Systematics, Biology and Control" (F. Lamberti and C. E. Taylor, eds.) Academic Press, New York. pp. 405-423.

Leistra, M. (1970). Distribution of 1,3-dichloropropene over the phases in soil. *J. Agric. Food Chem.* **18**:1124-1126.

Leistra, M. (1971). Diffusion of 1,3-dichloropropene from a plane source in soil. *Pest. Sci.* **2**:75-79.

Leistra, M., Smelt, J. J., and Lexmond, T. M. (1976). Conversion and leaching of aldicarb in soil columns. *Pest. Sci.* 7:471-482.

Marks, C. F., Thomason, I. J., and Castro, C. E. (1968). Dynamics of the penetration of nematodes by water, nematicides and other substances. *Exp. Parasitol.* **22**:231-237.

Martelli, G. P. (1978). Nematode-borne viruses of grapevine, their epidemiology and control. *Nematol. Medit.* **6**:1-27.

Maw, G. A., and Kempton, R. J. (1973). Methyl bromide as a soil fumigant. *Soils Fertilizers* **36**:41-47

McKenry, M. V. (1978). Selection of preplant fumigation. *Cal. Agric.* **32**:15-16.

McKenry, M. V., and Thomason, I. J. (1974). 1,3-dichloropropene and 1,2-dibromoethane compounds: I. Movement and fate as effected by various conditions in several soils. II. Organism-dosage response studies in the laboratory with several nematode species. *Hilgardia* **42**:393-438.

McKenry, M. V., and Thomason, I. J. (1976a). Dosage values obtained following pre-plant fumigation for perennials. I. 1,3-dichloropropene nematicides in eleven field situations. *Pest. Sci.* 7:521-534.

McKenry, M. V., and Thomason, I. J. (1976b). Dosage values obtained following pre-plant fumigation for perennials. II. Using special methods of applying methyl bromide and 1,3-dichloropropene nematicides. *Pest. Sci.* 7:535-544.

McKenry, M. V., Thomason, I. V., and Naylor, P. (1977). Dosage-response of root-knot nematode-infected grape roots to cis 1,3-dichloropropene. *Phytopathology* **67**:709-711.

McKenry, M. V., Thomason, I. J., Johnson, D. E., Neja, R., and Swanson, F. (1978). Preplant fumigations with 1,3-D nematicides. *Cal. Agric.* **32**:12-13.

McNamara, D. G., Ormerod, P. J., Pitcher, R. S., and Thresh, J. M. (1973). Fallowing and fumigation experiments on the control of nettlehead and related virus disease of hop. *Proc. 7th Br. Insecti. Fungic. Conf.*, pp. 597-602.

Munnecke, D. E., and Ferguson, J. (1953). Methyl bromide for nursery soil fumigation. *Phytopathology* **43**:375-377.

Murant, A. F., and Taylor, C. E. (1965). Treatment of soil with chemicals to prevent transmission of tomato blackring and raspberry ringspot viruses by *Longidorus elongatus* de Man. *Ann. Appl. Biol.* **55**:227-237.

Nemec, S., and Malek, R. B. (1975). Effects of nematicides and strawberry growth on nematodes, especially *Xiphinema americanum*, in rootrot sites in Illinois. *J. Nematol.* **7**:328 (abstr.).

O'Bannon, J. H., and Tomerlin, A. T. (1975). Efficacy of application and dispersion of 1, 2-dibromo-3-chloropropane in a Florida soil. *Soil Crop Sci. Soc. Florida* **20**:90-93.

O'Bannon, J. H., Tomerlin, A. T., and Rasmussen, G. K. (1975). Penetration of 1,2-dibromo-

3-chloropropane in a Florida soil. *J. Nematol.* **7**:252-255.

Pitcher, R. S., and McNamara, D. G. (1973). The control of *Xiphinema diversicaudatum* the vector of arabis mosaic virus in hops. *Ann. Appl. Biol.* **75**:468-469.

Potter, J. W., and Marks, C. F. (1976). Persistence of activity of oxamyl against *Heterodera schachtii* on cabbage. *J. Nematol.* **8**:35-38.

Rackham, R. L., Wilbur, W. D., Szuszkiewicz, T. E., and Hara, J. (1968). Soil desiccation and fumigation for *Armillaria* root rot in citrus. *Cal. Agric.* **22**:16-18.

Raski, D. J., and Lear, B. (1962). Influence of rotation and fumigation on root-knot nematode populations on grape replants. *Nematologica* **8**:143-151.

Raski, D. J., and Schmitt, R. V. (1972). Progress in control of nematodes by soil fumigation in nematode fanleaf infected vineyards. *Plant Dis. Rep.* **56**:1031-1035.

Raski, D. J., Jones, N. O., Kissler, J. J., and Luvisi, D. A. (1976). Soil fumigation: one way to cleanse nematode-infested vineyard lands. *Cal. Agric.* **30**:4-7.

Renzoni, G., Elia, F., and Lamberti, F. (1976). Determinazione di picole quantitá di 1, 3-dicloropropene in terreno mediante gas-cromatografia. *Nematol. Medit.* **4**:201-209.

Renzoni, G., Elia, F., and Lamberti, F. (1978). Dinamica dell'1,3-dicloropropene in tre tipi di terreno dell'Italia meridionale in prove di laboratorio. *Nematol. Medit.* **6**:97-106.

Roca, F., Conti, M., and Lamberti, F. (1977). *Trichodorus viruliferus* (Nematoda, Trichodoridae), vettore del virus "rattle" del tabacco (TRV) nell'Astigiano. *Nematol. Medit.* **5**:185-194.

Roca, F., Lamberti, F., and Siniscalco, A. (1975). Studi sulla persistenza di alcuni nematocidi granulari nella lotta contro i nematodi galligene *Meloidogyne* spp.). *Proc. "Giornate Fitopatologiche,"* Torino, Italy, pp. 265-269.

Seinhorst, J. W. (1973). The combined effects of dichloropropane-dichloropropene injected into the soil and dazomet applied to the soil surface. *Neth. J. Plant Pathol.* **79**:194-206.

Smelt, J. H. Leistra, M., Sprong, M. C., and Nollen, H. M. (1974). Soil fumigation with dichloropropene and metham-sodium: effect of soil cultivations on dose pattern. *Pest. Sci.* **5**: 419-428.

Smelt, J. H., Leistra, M., Houx, N. W. H., and Dekker, A. (1978a). Conversion rates of aldicarb and its oxidation products in soils. I. Aldicarb sulphone. *Pest. Sci.* **9**:279-285.

Smelt, J. H., Leistra, M., Houx, N. W. H., and Dekker, A. (1978b). Conversion rates of aldicarb and its oxidation products in soils. II. Aldicarb sulphoxide. *Pest. Sci.* **9**:286-292.

Smelt, J. H., Leistra, M., Houx, N. W. H., and Dekker, A. (1978c). Conversion rates of aldicarb and its oxidation products in soils. III. Aldicarb. *Pest. Sci.* **9**:293-300.

Steele, A. E. (1977). Effects of selected carbamate and organophosphate nematicides on hatching and emergence of *Heterodera schachtii. J. Nematol.* **9**:149-154.

Steele, A. E., and Hodges, L. R. (1975). Invitro and invivo effects of aldicarb on survival and development of *Heterodera schachtii. J. Nematol.* **7**:305-312.

Taylor, A. L., and McBeth, C. W. (1941). A practical method of using methyl bromide as a nematocide in the field. *Proc. Helminthol. Soc. Wash.* **8**:26-28.

Taylor, C. E., and Alphey, T. J. W. (1973). Aspects of the systemic nematicidal potential of Dupont 1410 in the control of *Longidorus* and *Xiphinema* virus vector nematodes. *Ann. Appl. Biol.* **75**:464-467.

Taylor, C. E., and Robertson, W. M. (1975). Acquisition, retention and transmission of viruses by nematodes. *In* "Nematode Vectors of Plant Viruses" (F. Lamberti, C. E. Taylor, and J. W. Seinhorst, eds.), pp. 253-275. Plenum Press, New York.

Thomason, I. J., and McKenry, M. V. (1975). Chemical control of nematode vectors of plant viruses. *In* "Nematode Vectors of Plant Viruses" (F. Lamberti, C. E. Taylor, and J. W. Seinhorst, eds.), pp. 423-439. Plenum Press, New York.

Trudgill, D. L. (1975). The effects of five nematicides and two fungicides on six species of Longidoridae. *Proc. 8th Br. Insecti. Fungi. Conf.*, pp. 145-151.

Trudgill, D. L., and Alphey, T. J. W. (1976). Chemical control of the virus vector nematode *Longidorus elongatus* and of *Pratylenchus crenatus* in raspberry plantations. *Plant Pathol.* **25**:15-20.

Van Dijk, H. (1974). Degradation of 1,3-dichloropropenes in the soil. *Agro-Ecosystems* **1**: 193-204.

Van Gundy, S. D., Munnecke, D., Bricker, J., and Minteer, R. (1972). Response of *Meloidogyne incognita, Xiphinema index* and *Dorylaimus* sp to methyl bromide fumigation. *Phytopathology* **62**:191-192.

Van Hoof, H. A. (1964). Het Tijdstip van infective en veranderingen in de concentratie van ratelvirus (Kringeriheid) in de aardappelknol. *Meded. Landbouw. Gent.* **29**:944-955.

Vigodski, H., and Klein, L. (1976). Influence of methyl bromide soil fumigation on fungicidal efficacy and bromide residues. *Phytoparasitica* **4**:123-129.

Waggoner, T. B., and Khasawinah, A. M. (1974). New aspects of organophosphorus pesticides. VII. Metabolism, biochemical and biological aspects of Nemacur and related phosphoramidate compounds. *Pest. Rev.* **53**:79-97.

Williams, I. H. (1968). Recovery of cis-and trans-1,3-dichloropropene residues from two types of soil and their detection and determination by electron capture gas chromatography. *J. Econ. Entomol.* **61**:1432-1435.

Chapter 5

USE OF OIL SPRAYS AND REFLECTIVE SURFACES FOR CONTROL OF INSECT-TRANSMITTED PLANT VIRUSES

John N. Simons

JMS Flower Farms, Inc.
Vero Beach, Florida

5.1 INTRODUCTION

Control of aphid-borne viruses, particularly those transmitted in a nonpersistent or "stylet-borne" manner, has proven very difficult to accomplish. Such obvious approaches as vector control, cultural practices and plant breeding for resistance have largely failed to effect acceptable levels of control. In recent years researchers have started looking at procedures which are involved with interference in the transmission process and, for the first time, we are beginning to see hopeful signs that many serious virus diseases may be brought under control.

ISBN 0-12-326440-5

The two most useful techniques reported to date include the use of reflective surfaces to repel winged aphids from susceptible plants and the use of oil sprays to prevent aphids from transmitting viruses.

In this chapter an attempt will be made to explain what is required for reflective surfaces and oils in order for commercial acceptance of such practices. An extensive review of the use of oils is not included as Vanderveken's (1977) recent paper includes an excellent bibliography.

5.2 REFLECTIVE SURFACES

5.2.1 Historical

Aphids, like most insects, respond preferentially to certain wavelengths of light. Shortly after development of wings, aphids become strongly attracted to short wave light and fly towards the sky in a migratory or dispersal flight. After flying for several hours their response to short wave light is reversed and they are repelled by it. At this time they become attracted to yellow and green wavelengths and start searching for a suitable host plant (Kennedy, 1960; Johnson, 1969; Moericke, 1950). This knowledge led Smith *et al.* (1964) to investigate the possibility of using a reflective surface such as aluminum foil to repel winged aphids from virus susceptible crops and thus prevent spread of viruses.

5.2.2 Commercial Use of Reflective Surfaces

The use of reflective surfaces does provide significant protection to young plants when at least 50% coverage of the ground is used. However, the technique has not achieved widespread commercial acceptance.

There are three principal problems which have limited the use of reflective surfaces. First, the repellant effect is lost as the plants increase in size and cover the reflective surface. Second, the cost of reflective materials is rather high. Third, in the case of aluminum foil, disposal after the growing season is difficult.

Growers in south Florida use plastic mulches to grow many of their crops. In the early fall when temperatures are high they use white plastic (reflective) to help keep the soil temperature down and during the winter months they use black plastic (nonreflective) to increase the soil temperature. Unfortunately, aphids are not numerous during the early fall and virus spread generally does not become a problem until cool weather has set in. Thus the use of reflective surfaces is limited to a time span when aphid populations (and virus diseases) are not a problem.

Irrespective of these drawbacks, the demonstration that manipulation of aphid behavior could significantly affect virus spread makes the work of Smith and coworkers (1964) noteworthy.

5.3 OIL SPRAYS

5.3.1 Historical

Since Bradley *et al.* (1962) first reported that oil interfered with aphid transmission of potato virus Y, many researchers have investigated the possibility of using oils for control of certain aphid-borne viruses. It was soon established that the antitransmission phenomenon extended to every nonpersistent virus tested and Vanderveken *et al.* (1966) showed that transmission of the semipersistent beet yellows was suppressed by oil. Published reports indicate that oils do not affect aphid transmission of persistent viruses including pea enation mosaic virus (Vanderveken, 1968) and potato leafroll virus (Hein, 1971). Recently, however, Zitter and Everett (1979) have found that spread of the persistently transmitted tomato yellow virus was reduced under field conditions in Florida. The tomato yellows virus appears to be closely related to potato leafroll virus (Zitter, personal communication).

With this amount of evidence as to the potential for using mineral oil sprays for virus control one would think that their use would be commonplace some 17 years after the discovery by Bradley *et al.* (1962). With the exception of limited use in Europe on seed potatoes and lilies there has been little commercial utilization of mineral oil for virus control. In the United States the first commercial use of oil did not occur until 1977 when JMS Stylet-Oil^R became available in Florida. There are several reasons why oils have been slow to gain commercial acceptance including 1) lack of adequate efficacy, 2) phytotoxicity, 3) lack of research effort to develop oils for virus control, and 4) lack of interest on the part of oil companies and agricultural chemical manufacturers. There has also been a general attitude of disbelief that a treatment which only interfered with transmission of virus by aphids; and which had no direct toxic effect on either aphids or viruses, could be sufficiently efficaceous to cause suppression of virus spread on a practical level.

In this chapter an attempt will be made to explain what is required for an oil to be useful for commercial control of aphid transmitted viruses as well as what some of the limitations are.

5.3.2 Epidemiological Aspects

5.3.2.1 Effect of Oil on Primary and Secondary Virus Spread. Before beginning a discussion of oils and their use for virus control it seems worthwhile to discuss certain aspects of the epidemiology of aphid-borne virus diseases as these parameters are critical in establishing limitations on the potential usefulness of oils. It is well established that oils interfere both with the acquisition of nonpersistent viruses as well as with the inoculation of them by aphid vectors (Bradley,

1963). It is also well established that in most field situations the majority of transmissions occur between crop plants (secondary spread) and a relatively small part of the diseased population is comprised of plants infected by virus brought in from the outside (primary spread) (Broadbent, 1969). One would expect that a treatment which interfered significantly with introduction of virus into a field (primary spread) would prove highly efficaceous under field conditions. This indeed has been the case with JMS Stylet-Oil^{R} where small plot field trials have consistently shown 3 to 8 fold reductions in virus spread (Zitter and Ozaki, 1978), whereas under large-scale field situations it has been common to observe almost total suppression of virus spread.

One major difference between small experimental plots and large fields has been that the small plots were established with primary inoculum present in them whereas the grower's fields were essentially free of disease at the time oil treatment was begun. Where growers have left unsprayed blocks to monitor the effectiveness of the oil treatment we have observed the following without exception: 1) virus infected plants did not appear in the unsprayed areas until several weeks later than would normally have been expected, and 2) once disease was established in the unsprayed area, the spread was very rapid and ultimately 100% of these unsprayed plants became infected. The delay in appearance of disease was undoubtedly associated with the location of the unsprayed blocks in the middle of large areas of oil-treated blocks of plants which resulted in protection being provided for several weeks beyond normal. Interestingly, the plants in the oil-sprayed blocks on either side of the unsprayed blocks remained almost completely free of virus even many weeks after the unsprayed blocks had become totally infected. Such an occurrence could only be attributable to a strong suppression of primary spread by the oil. And this of course makes a strong case for growers to start oil treatments as soon as winged aphids make their appearance rather than waiting for virus diseases to appear in the crop.

There obviously is a significant effect of oil reducing secondary spread of virus as well; otherwise, we would not have observed the level of control seen in small plot trials. And, since most spread under field conditions is secondary in nature, it is essential that oil be effective in limiting plant to plant spread.

5.3.2.2 Inoculum Potential. The observations reported above indicate that oil sprays are going to lose their effectiveness as the inoculum potential increases. At this time we do not know just what level of infection must be present in a field before oil sprays become ineffective, but under Florida conditions the level is probably somewhere between 10 and 20% infection. Such factors as numbers of winged aphids present, the titer of transmissible virus present in infected plants and plant density (stand) are all important. One thing that appears highly probable, however, is that disease incidence increases at a rather slow rate until 1 to 2% infection is reached, after which spread can be very rapid, often reaching two to three-fold each week. I have observed one instance with pepper and tobacco etch virus where oil sprays were initiated when disease incidence had reached 3.5% and the incidence increased to only 8% after 10 weeks. During this time unsprayed areas left as controls in the same field became 100% infected in less than 6 weeks.

Some of the most striking evidence for the importance of suppressing inoculum potential involves double cropping of peppers in Florida. The use of oil on peppers has suppressed virus diseases to where it has become feasible to produce two crops of fruit from the same plants. The first pickings are made during November and December; the plants are cut back to a height of about 15 cm in late December to early January; and a second crop of fruit is produced on the regrowth for harvest in March and April. This results in a roughly 50% increase in total yield, with the second crop costing about only 15% as much to grow as did the first crop. This technique succeeds largely because of the use of plastic mulch culture in which adequate fertilizer for a second crop is available and season long weed and soil pest problems are eliminated by fumigation of the plant beds.

Inoculum potential is particularly important where seed-borne virus diseases are involved. Two agronomic crops which suffer significant losses from seed-borne virus diseases but whose cash value is too low to warrant applications of oil on the production plants are peanut (peanut mottle) and cowpea (cucumber mosaic). Demski (personal communication) in Georgia is testing oil on the seed crops and his results are very encouraging. The cost to protect a seed crop with oil (0.75% concentration applied at weekly intervals) is about \$35 per acre (\$80 per hectare) for material. Kuhn and Demski (1975) have estimated that losses from peanut mottle in Georgia amount to \$11,000,000 annually. The cost for oil to produce the 30,000 acres needed for seed production to plant the 512,000 acres of commercial peanuts would be about \$1,000,000, in effect providing the Georgia peanut growers an 11:1 return on investment in virus control. In the case of peanut mottle, seed transmission in commercially used seed lots occurs at a rate of from 0.1 to 1.0% (Kuhn and Demski, 1975). This low level of incidence should be controllable through the use of oil sprays.

Where higher levels of inoculum potential exist, such as with seed potatoes, oil sprays may prove less effective. Work in Maine with JMS Stylet-Oil^{R} has been encouraging, however. Hammond (personal communication) has tested the oil on several varieties of potatoes and found once-a-week applications to be quite effective in reducing spread of potato virus Y (Table I). Interestingly, when oil was used in combination with temik (a systemic insecticide applied at planting time) the oil was not as effective. The possible reasons for this are discussed below.

5.3.2.3 Effects of Insecticides on Efficacy of Oil. The use of insecticides for control of nonpersistent aphid-borne viruses is recognized as being ineffective, primarily because insecticides do not act quickly enough to prevent transmission of virus. In spite of the scientific evidence available, there exists a common practice among growers to use aphicides, even though there are insufficient populations to consitute an entomological threat. Psychologically, growers feel that the use of aphicides cannot do any harm with respect to virus control.

TABLE I. Results of 1978 Spray Trials for Control of Potato Virus Y in Maine[a]

Potato	% Infection: Oil at 1 Week	Oil at 2 Weeks	Temik + Oil at 1 Week	Temik + Oil at 2 Weeks	Control
Russet Burbank	2.97	5.57	10.63	9.99	7.26
Pungo	1.59	5.25	7.28	4.86	4.31
Ontario	0.96	6.05	2.17	1.89	4.75
Atlantic	0.91	2.04	4.77	3.33	3.71
Norchip	1.11	3.90	2.09	4.67	2.15
Mean	1.51	4.56	5.39	4.95	4.87

[a]JMS Stylet-Oil was used at 0.75% concentration and sprayed at 400 psi pressure.

In addition, they find it difficult to accept the experimental evidence which shows insecticides to be ineffective in limiting virus spread.

Where quick acting, highly toxic, and short residual aphicides are used, there probably is no danger of aggravating virus spread. But, where potential for sublethal doses of insecticide being administered to aphids exists, a scenario may be created in which there is real enhancement of virus transmission. Sublethal doses of insecticides can cause increased flight activity in winged aphids and thus enhancement of virus transmission in the crop field. One of the first reports of this phenomenon involved the use of DDT and increased spread of potato virus Y in potatoes (Fernow and Kerr, 1953). During the past year additional reports (personal communications) of this phenomenon have come to my attention, including a report on spread of maize dwarf mosaic in sweet corn following use of soil applications of systemic insecticides and of cucumber mosaic in cantaloupe which was associated with the use of endosulfan as an aphicide. Evidence of this phenomenon has not been observed in Florida, probably because growers in this area generally use insecticides which are highly effective, quick acting and usually have short residual activity.

The use of oil along with an insecticide has been reported to provide better virus control than with either treatment along (Zschiegner *et al.*, 1971). Oil was used to control potato virus Y and a systemic insecticide provided control of potato leafroll virus in these trials. There may indeed be merit in using such a combination approach but the potential for certain insecticides to cause hyperactivity in alate aphids should not be overlooked. Certainly the use of aphicides should be associated with the presence of winged aphids and the value of placing a systemic insecticide in the soil at planting time when no significant aphid flight activity is expected for 6 to 8 weeks seems dubious at best.

All of the reports in the literature indicate that transmission of persistent viruses is reduced by use of insecticides but not by oil (Broadbent, 1969; Vanderveken, 1977). Recent work done in Florida, however, has shown that weekly applications of oil (JMS Stylet-Oil^{R}) to tomatoes effectively prevented spread of the tomato yellows virus (Zitter and Everett, 1979). This virus is transmitted in a persistent manner and appears to be very closely related to the potato leafroll virus. In light of this finding, tests for effects of oil on transmission of persistent viruses should be reevaluated. Most of the testing to date has been done using techniques which optimized the chances for transmission to occur (several aphids per plant and long feeding periods) thus it was possible to have missed detection of antitransmission activity. It is also possible that transmission of virus by alate aphids is affected by oil in a manner different from that found with apterous aphids. Laboratory testing almost invariably utilizes apterous aphids. It should be noted that the activity of oil with tomato yellows virus was first observed by the author in commercial tomato fields, and this observation was confirmed by experimentation using small plot field trials. Further experimentation is needed to establish whether the effect of the oil is on acquisition of virus, inoculation of virus, or both.

5.3.2.4 Effect of Plant Density on Usefulness of Oils. In Florida, oil is used to control virus spread in pepper, tomato and squash. These crops are all highly susceptible to nonpersistent aphid-borne viruses and outbreaks approaching 100% infection are frequently observed. Virus spread in squash (zucchini or crookneck types) can be particularly rapid with total involvement (watermelon mosaic virus 1) often taking place within 4 weeks after disease is first observed. Under laboratory conditions watermelon mosaic virus 1 is readily transmitted by aphids, but not nearly as efficiently as potato virus Y or tobacco etch virus. Yet these latter two viruses generally require at least 6 to 8 weeks to totally infect a pepper planting. And, in the case of watermelon as a crop, spread of watermelon mosaic is even more rapid than in squash. The green peach aphid is the principal vector species in all of these cases. Oil, applied under laboratory conditions, is highly effective in preventing inoculation of virus to all three crops, yet under test plot conditions, the oil is most effective on pepper, less effective on squash, and almost ineffective on watermelon. It is possible that differences in aphid behavior account for these differences in effectiveness of oil, but it seems more likely that differences in plant density (stand) are involved. Pepper is planted at about 20,000 plants per acre (44,000 per hectare); squash at from 6,000 to 8,000 plants per acre (13,000 to 18,000 plants per hectare), and watermelon at from 500 to 700 plants per acre (1,200 to 1,800 plants per hectare). When one is dealing with a plus and minus phenomenon such as is the case with systemic plant virus infections, there is going to be a significant interaction between the number of available susceptibles and the probability of obtaining infection in a given susceptible. When the procedure used to limit the spread of virus is marginally effective, which is the case with oil, the parameter of plant stand can be expected to be of real importance in the epidemiological picture.

5.3.3 Physical Properties of Oils and Emulsifiers

5.3.3.1 Viscosity of Oils. Mineral oils are not the only oil materials which have been found to interfere with virus transmission by aphids. Such diverse substances as plant lipids (Bradley *et al.*, 1962), milk fat (Hein, 1964), and silicones (Vanderveken, 1972) have all been shown to provide antitransmission activity. Mineral oils appear to be a bit more inhibitory than other materials but this conclusion is conjectural as differences in formulations and application procedures have varied so much as to make direct comparisons impossible.

One of the more obvious possibilities that might account for differences among oily substances has to do with differences in coverage of plant leaves, and viscosity of the oily material certainly does affect the distribution of the material on the leaf. Dewijs *et al.* (1979) have reported that paraffinic oils of low viscosity (below 40 SUS) and of rather high viscosity (above 200 SUS) are not as effective in preventing aphid inoculation of viruses as oils with intermediate viscosities (50 to 180 SUS). These results indicate that the commercially available mineral oils with viscosities ranging from 60 SUS to 120 SUS are optimum for use in virus control.

Our own tests have been done with oils of 60, 70, and 110 SUS viscosities. These are commercially available as Sunoco 6E, 7E, and 11E, respectively. All are formulated similarly with 1.2% emulsifier being used. Figure 1 shows results of trials using these oils at a concentration of 0.75% and spray pressures ranging from 100 to 600 psi. Pepper plants were sprayed the day before inoculation with potato virus Y using two green peach aphids per plant. Volume of spray applied was kept consistent at each pressure by changing the speed of the spray boom. Except at a pressure of 100 psi Sunoco 7E oil was more effective than either the 6E or 11E oils. Efficacy was affected by spray pressure with pressures of 400 psi or greater being better than lower pressures. These results indicate that viscosity of the oil is an important parameter in choosing oils for virus control. They also show that spray pressure has considerable significance in determining efficacy.

5.3.3.2 Emulsifiers. Emulsifiers are added to mineral oils in order to facilitate a more even distribution of oil on the surface of plant leaves. There are many emulsifiers used for this purpose and the choice by oil formulators usually involves the cost of the emulsifier as much as the effectiveness of the material in making a good emulsion. The emulsions that are produced generally "break" rather rapidly and within a few minutes after application the oil has separated from the water phase of the emulsion. Very little has been done to investigate the importance of the emulsifier component for oils used for virus control.

Emulsifiers used for formulating agricultural spray oils are generally nonionic although both anionic and cationic emulsifiers are available. Anionic and cationic emulsifiers are generally more expensive than nonionic emulsifiers, and there seems to be a tendency for them to be phytotoxic to plants. Our own work has

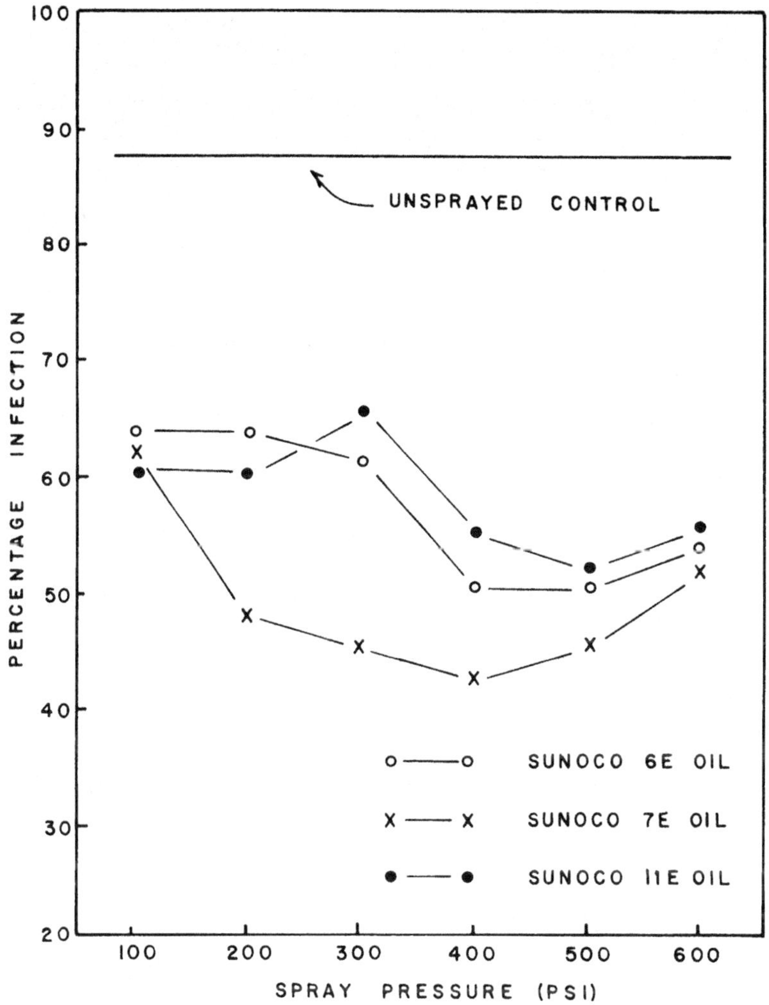

FIG. 1. Effect of Sunoco 6E, 7E, and 11E oils on aphid transmission of potato virus Y in pepper. Oils were applied at 0.75% concentration the day before inoculation using two green peach aphids per plant. Amount of spray applied was kept constant at each spray pressure. Results are based on six replications of 10-15 plants per treatment.

been done largely with nonionic materials. We have investigated several, using them at differing concentrations, and mixing them with oils of 60, 70, and 110 SUS viscosities.

Results of a few of these trials are shown in Table II. The emulsifiers were used at recommended (manufacturer's) rates in combination with several different oils. Oils were applied at an 0.75% concentration using the same pressure (400 psi) and nozzles (Teejet TX-4) for each formulation. It is quite apparent that significant differences in efficacy existed among the several formulations.

TABLE II. Efficacy of Different Oil Formulations in Preventing Inoculation of Potato Virus Y to Pepper[a]

Spray Material Oil (Viscosity)	Emulsifier	% Infection
Sunspray (70 sec)	Ninate 401 & Poe 5	36
Sunspray (70 sec)	Triton X-363M	36
Sunspray (70 sec)	T-Mulz A02	46
Brit. Pet. (100 sec)	Span-80	62
Brit. Pet. (120 sec)	Span-80	42
Unsprayed		88

[a]Oils were applied 24 hr prior to inoculation using two green peach aphids per plant. Oils were used at a concentration of 0.75% and were sprayed at 400 psi pressure.

5.3.4 Application Technology

5.3.4.1 Spray Pressure. There has been little work done on the possible influence of spray pressure on efficacy of oil formulations. The results shown in Fig. 1 indicate very clearly that spray pressure is critical in optimizing efficacy. We have taken results from our laboratory studies and carried out small plot field trials to see whether such results could be confirmed in the field (Table III).

In this study Sunoco 7E oil was compared with JMS Stylet-Oil[R] for control of watermelon mosaic virus in zucchini squash. Oils were applied weekly at an 0.75% concentration. Each oil was applied at pressures of 200 psi and 400 psi. The speed of the spray rig was varied so that equal amounts of spray were applied at both pressures. Results were in good agreement with those obtained from laboratory testing (Table III). A pressure of 400 psi was more effective than 200 psi with both oils and the JMS Stylet-Oil[R] formulation was superior to the Sunoco 7E oil.

TABLE III. Effect of Oils on Transmission of Watermelon Mosaic Virus in Zucchini Squash under Laboratory and Field Conditions[a]

		% Infection in			
			Small plot field test		
Treatment	Spray Pressure (psi)	Lab	Apr 28	May 4	May 11
Sunoco 7E oil	200	55.0	11.3	17.0	33.8
Sunoco 7E oil	400	28.0	3.8	11.3	20.4
JMS Stylet-Oil[R]	200	23.0	6.0	9.7	17.8
JMS Stylet-Oil[R]	400	16.0	4.8	8.2	11.3
Unsprayed		74.0	9.8	16.5	35.1

[a]Oils were applied as 0.75% emulsions. Equal volumes of emulsions were applied at each spray pressure.

5.3.4.2 Nozzle Orifice. There has been little work done on the possible influence of nozzles on efficacy of oil formulations as they are used for virus control. Both spray pattern and droplet size are determined by the type of nozzle used in a spray machine. It seems likely that the choice of nozzles could be very important in optimizing spray efficacy.

Our work has been done exclusively with hollow cone spray patterns as produced by the Teejet TX (Spraying Systems Co.) series of nozzle tips. We have found that nozzle orifice is a critical component in designing an optimum application configuration. This can be illustrated by results shown in Figures 2, 3

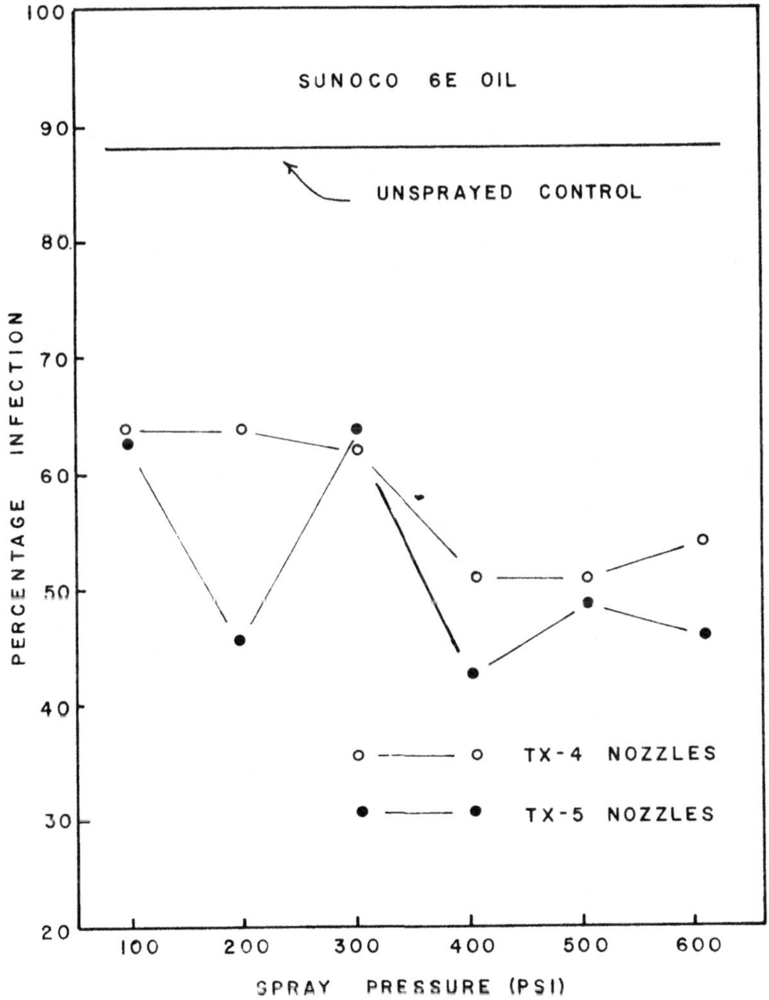

FIG. 2. Effect of nozzle orifice and spray pressure on efficacy of Sunoco (a) 6E, (b) 7E, and (C) 11E oils in preventing transmission of potato virus Y to pepper by green peach aphids. Oil was applied at 0.75% concentration the day before inoculation. The amount of spray applied was kept constant at each spray pressure. Results are based on 6 replications of 10-15 plants per treatment.

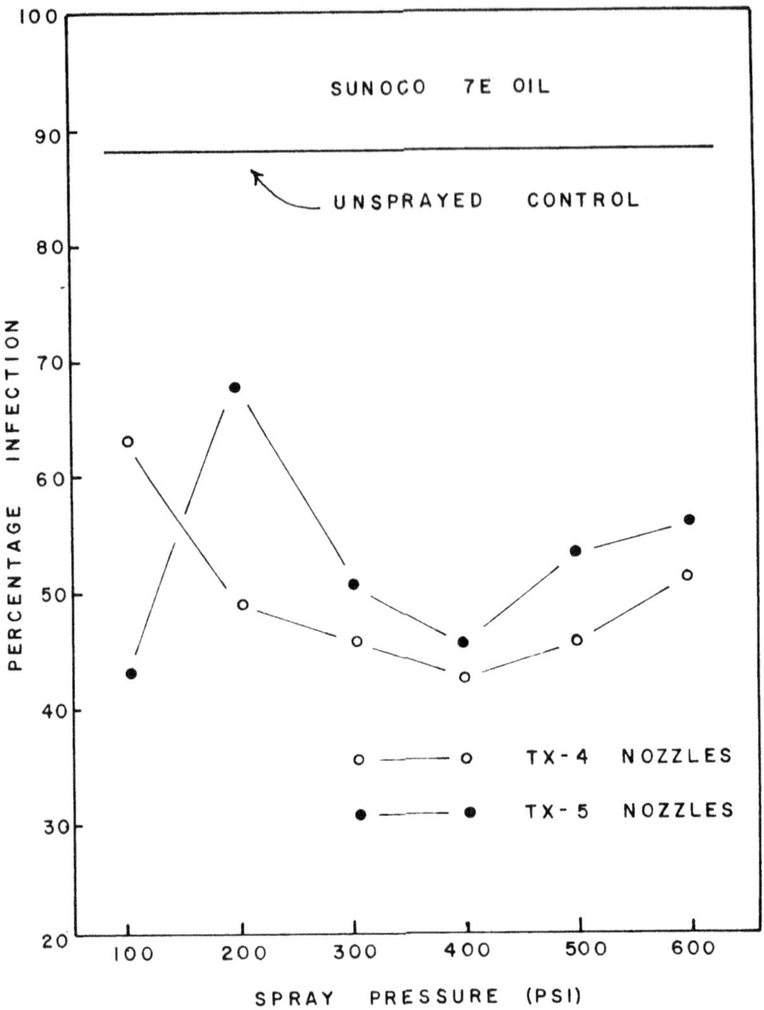

FIG. 3. Effect of nozzle orifice and spray pressure on efficacy of Sunoco 7E oil in preventing transmission of potato virus Y to pepper by green peach aphids. Oil was applied at 0.75% concentration the day before inoculation. Amount of spray applied was kept constant at each spray pressure. Results are based on six replications of 10-15 plants per treatment.

and 4 where Sunoco 6E, 7E and 11E oils were sprayed at pressures of 100 to 600 psi on peppers and two nozzle sizes, TX-4 and TX-5, were used. As in all trials, the amount of spray emulsion applied was kept constant by varying the rate of travel of the spray boom. Relatively little difference was found with the Sunoco 6E oil (Fig. 2) although the TX-5 nozzles were slightly superior at pressures of 400 psi and higher. The large difference at 200 psi pressure is likely anomalous. With Sunoco 7E oil (Fig. 3) there was a consistent improvement

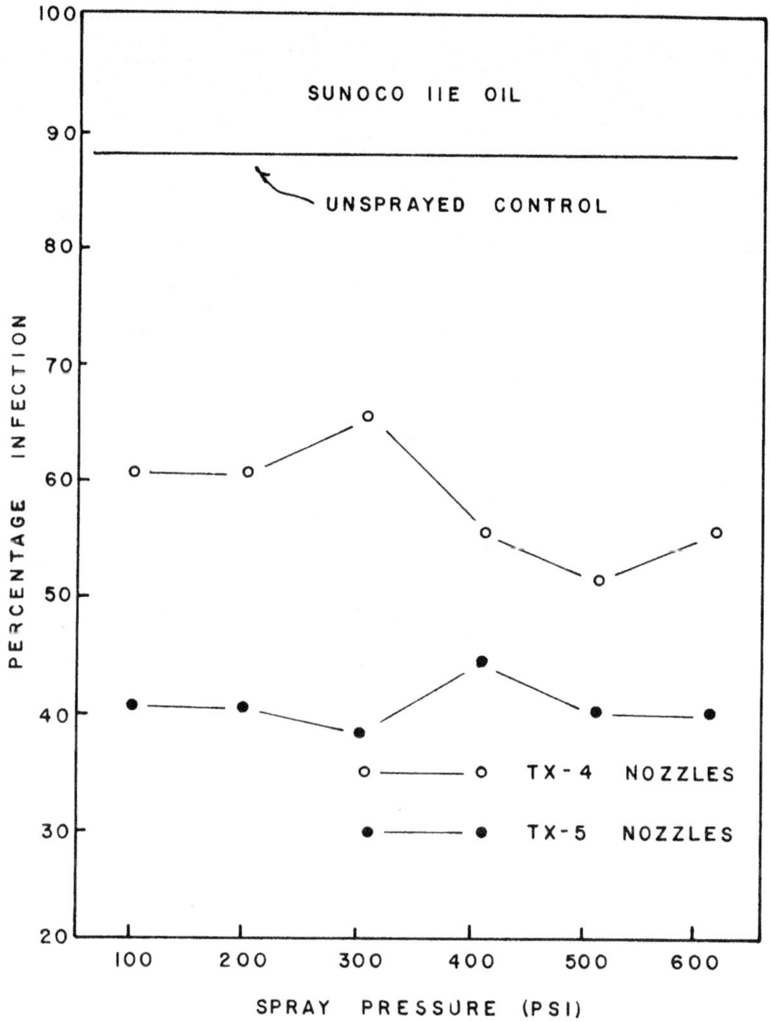

FIG. 4. Effect of nozzle orifice and spray pressure on efficacy of Sunoco 11E oil in preventing transmission of potato virus Y to pepper by green peach aphids. Oil was applied at 0.75% concentration the day before inoculation. Amount of spray applied was kept constant at each spray pressure. Results are based on six replications of 10-15 plants per treatment.

in performance using the TX-4 nozzles which held through pressures of 200-600 psi. On the other hand, with the Sunoco 11E oil the TX-5 nozzles consistently provided the best results (Fig. 4). These results show clearly that subtle changes in application procedure can have dramatic effects on efficacy of oil formulations. The differences cannot be accounted for in terms of overall effects on coverage, but it is entirely possible that they are related to effects on droplet size as influenced by the viscosity of the oils interacting with nozzle orifice and spray pressure. It is doubtful that the droplet size of the emulsions is changed substantially as the emulsions are comprised of 99.25% water, but it does seem possible that the droplet size of the oil within the emulsion droplet may be significantly affected.

5.3.4.3 Aerial Application of Oils. There has been little done with respect to applying oils by air for control of virus diseases. There is an inherent problem in obtaining good coverage with protectant-type chemicals where aerial applications are made, and this difficulty is going to be no less serious in the case of oils.

We have done a considerable amount of testing using oils applied by means of a sintered sleeve type applicator (Beecomist). This type of applicator produces a very high percentage of uniform sized particles, and sleeves that produce particles of several sizes are available.

We have compared sleeves that produce particles of 20, 35, and 60 microns, respectively, using various experimental formulations of 70 sec oil. Under laboratory conditions the best efficiency we have obtained is about half that obtainable with the TX series nozzles. Emulsions of oil cannot be expressed through the sleeves as serious foaming problems are encountered. Thus one must use pure oils, with or without emulsifier added.

Even though laboratory results indicated that poorer activity would result from aerially applied oil, there still are interesting possibilities with this technique. Whereas ground spray rigs direct almost all of the oil on the target crop, aerially applied material should collect on non-target plants as well. Since interference with virus transmission occurs no matter where an aphid contacts the oil, there is a possibility that spraying oil on the entire crop area (ditchbanks, weedy borders, etc.) would cause reduction in virus transmission which would overcome the inherent lowered efficacy when oil is used on just the crop plant.

During the fall of 1977 two pepper fields were treated with aerially applied oil utilizing an aircraft which had been fitted with four Beecomist applicators (35 μm size particles). Half of each field (about 25 acres) was sprayed by air and the other half was treated using conventional ground equipment. Within three weeks it had become apparent that control of virus spread was not being achieved where the aerial applications were being made, whereas with the ground equipment almost no spread of virus was occurring. The experiment was discontinued at this point and both fields were sprayed only with ground equipment.

The most obvious reason for the lack of control when oil was applied by air was the relatively small amount of oil (3 quarts per acre) which could be applied with one spray pass of the airplane. This was about the same amount of oil which was being applied by the ground equipment and of course it was dispersed over the entire cropping area instead of being directed principally at the crop plants. An increase in gallonage could only be achieved by resorting to two or more passes over the field, but this was not considered to be an economically feasible procedure.

5.3.4.4 Air Blast Sprayers. Loebenstein *et al.* (1966) have reported that low volume air blast spraying of cucumber with oil emulsions was highly effective in reducing spread of cucumber mosaic in Israel. The type of air blast sprayer they used would not lend itself to application where a high spray pressure was used. However, air blast sprayers equipped with high pressure pumps do exist, and we have made some attempts to utilize such equipment in spraying with TX-type nozzles.

We have found that in order to obtain adequate coverage it is necessary to double the number of nozzles used in conventional air blast spraying. This can be done easily by installing tees and street els in the sprayer manifold. Approximately 24 nozzles can be used on one side of an air blast sprayer, which, under zero wind conditions, will deliver adequate material to give adequate coverage to a distance of about 20 ft (6 m). Inadequate coverage was obtained beyond this distance, thus it proved necessary to spray from both directions to provide adequate coverage where 40 to 50-ft wide plantings were used.

The major problem that we could not overcome was interference by wind. Even a very low wind velocity (<3 mph) significantly reduced the area that was covered. In Florida this severely limits the potential for use of such a machine. Even under ideal weather conditions the air blast sprayer will not provide coverage equal to that obtained from a properly outfitted ground spray rig.

5.3.4.5 General Comments on Coverage. The effectiveness of any protectant-type spray meterial is closely related to the thoroughness of coverage of the foliage. In the case of oils, one is dealing with areas of plant canopy which are subject to probing by alate aphids. Since alate aphids generally are found only in the outer canopy of the plant it is not necessary to obtain coverage deep within the plant as would be the case with a protectant-type fungicide. Coverage of the underside of leaves is not considered to be essential, since alate aphids do not normally probe except on the upper leaf surface unless they have settled permanently on the plant.

Oils persist on sprayed leaves for at least 10 to 14 days (Simons *et al.*, 1977). The necessity for weekly sprays, or in the case of rapidly growing plants, biweekly sprays, is because new foliage needs protection. Thus spray booms should be designed to insure thorough coverage of the uppermost parts of the plants.

Nozzles should be placed about 10 in (25 cm) apart and we have found that doubling of overhead nozzles (those spraying straight down) by using tees and two short nipples placed at a 45° angle to the spray boom is useful. Nozzles should be kept from 12 to 15 in (30 to 40 cm) from the plant canopy. Adequate coverage can be obtained using sufficient nozzles to spray 40 gal per acre on young plants and 100 gal per acre on mature plants. Spray machines should be driven no faster than 3 mph.

It is important not to spray oil on wet foliage as poor coverage will result. In addition, application of oil to wet leaves can cause protectant-type chemicals to be redistributed resulting in loss of efficacy of such chemicals. Oils that have been specially formulated for virus control should be applied separately from other chemicals. The surfactant systems used in other chemicals will interfere with distribution of oil from a specially formulated virus control product. We recommend applying oil prior to spraying other chemicals and to wait at least 24 hr after applying oil before spraying other chemicals.

5.3.5 Testing with Oils for Antiviral Activity

The author has been doing small plot field testing with oils for 11 years and laboratory testing for 2 years. This section is included to try to give the reader some appreciation of certain problems inherent in doing experimental work with oils for control of virus diseases.

5.3.5.1 Laboratory Testing. In the case of nonpersistent viruses there are no special problems in demonstrating antitransmission activity of oils under laboratory conditions. Obtaining data which are of comparative value in terms of field trials is another matter, however. We have found it necessary to use plants which have been grown under outdoor conditions for at least two weeks prior to use in tests and to use application technology similar to that used under field conditions. In our own testing program this has meant construction of a special spray machine in which nozzles, spray pressure and rate of travel of the spray boom could all be controlled.

Virus inoculum sources should be maintained using only aphid inoculated viruses. Virus cultures which are maintained for any great length of time by mechanical inoculation generally lose their aphid transmissibility. Source plants should be of a uniform age and have been infected for the same length of time. For many viruses with which we have worked, two weeks after inoculation has proven to be an optimum time.

5.3.5.2 Field Testing. When laboratory work has been carefully conducted we have found that small plot field trials generally provide results which confirm the laboratory studies.

During the past 11 years considerable evolution has occurred in the design and execution of our small plot field trial work. The following procedures have been found to give consistently good results.

Plot Design. We have tried a good many designs and have found the randomized block to be the most satisfactory with one major change. The unsprayed control plots should be located a short distance (50 to 75 m) from the treated plots. The presence of unsprayed blocks next to treated ones will cause considerable bias in test results. We have also found it helpful to establish our plots at right angles to the prevailing wind direction. If sufficient land is available we recommend laying out the plots as pairs in one direction and extending as far as necessary in the long direction. Each plot (treatment) should be approximately square in configuration within this elongated layout. Four replications of each treatment are adequate.

Another scheme which has worked well is one used by Zitter (personal communication) in which each treatment is isolated by at least 100 m from any other treatment. This of course will only be practical when small numbers of treatments are being used. It is useful when several crops (viruses) are being used simultaneously and the only chemical treatment being used is oil. Zitter (personal communication) has found that 3 replications are adequate with this procedure . It is necessary to introduce virus into the plots to make this approach successful.

Plot Size. Because systemic infections can be recorded only once, it is necessary to use somewhat larger plant samples than in the case of diseases which show varying amounts of incidence. We have found that minima of 150 to 250 plants per replication are needed.

Virus Inoculum. We have not been able to utilize naturally occurring inoculum without encountering large within replication variability as a problem. Placing a border row of infector plants outside the plot area works well but is no more satisfactory than introducing infector plants into the plots. We have tried a number of ways of doing this and have found that placement of a single point source of inoculum in the center of each treatment results in rapid virus spread and provides the uniform initial incidence of disease so necessary to minimize within replication variation. We generally provide two infector plants located side by side to protect against accidental loss of a single plant.

The use of virus sources which are readily aphid-transmissible is extremely important. Many laboratory cultures of viruses are maintained by sap inoculation and frequently these cultures are poorly transmitted by aphids. The inocu lum plants in the field plots can be inoculated by sap, providing the virus has been maintained by aphid transmission. If mechanical inoculations is done in the field, it is important to keep the inoculum cold and away from light.

Spraying. The application technique used for applying oil is very important. Use a spray machine which provides good coverage of the entire outer canopy of leaves. Penetration of spray into the center of the canopy is not necessary as migrant aphids tend to restrict their activity to the exposed portion of the plant.

In order to obtain adequate coverage with oil emulsions it is necessary to use at least 35 gal per acre on small plants and up to 150 gal per acre on large plants such as trellised tomatoes. Increasing the amount of oil applied should always be done by increasing the number of nozzles used and not by increasing the concentration of oil in the emulsion. A tractor speed of about 3 mph is recommended.

5.3.6 Phytotoxicity of Oils

One of the more serious drawbacks to the use of oils for virus control has been the inherent toxicity of oils to plants. Vegetable species are particularly sensitive. We have found that under commercial conditions there are no serious problems with phytotoxicity on any crop tested to date provided that certain precautions are taken. The following comments are made to give the reader some general information on the phytotoxicity problem.

5.3.6.1 Physical and Chemcial Properties of Oils. There are two physical parameters of particular importance in selecting oils that are safe for use on vegetable crops. These are chemical inertness, usually expressed as the per cent unsulfonated residue, and the viscosity. Oils used as agricultural sprays generally have an unsulfonated residue of no less than 90%. For use on vegetables this value should be no less than 94%. We have done some work in which a 94% UR oil was compared with a 99% UR oil for toxicity to crookneck squash, a particularly sensitive species of plant. Both oils were similar in viscosity (70 sec) and both were formulated similarly with respect to their emulsifier constituents. They were applied to squash using concentrations of 0.75% and 1.5% on a weekly schedule. We found slight injury on seedling squash with the 94% oil when used at 1.5%. However, after two weeks this injury disappeared and no further evidence of damage was apparent as the plants grew. Yields of squash were not affected by any of the treatments. And, virus control was no better using 1.5% concentrations of oil than the recommended label concentration of 0.75%.

We have compared oils (all 94% UR and formulated similarly) with viscosities of 60, 70 and 110 sec (SUS) for toxicity to peppers, squash and tomatoes. Concentrations of oils used were 0.75, 1.0 and 1.5%. Only the 110 sec oil showed symptoms of injury and only at the 1.0 and 1.5% concentrations. Injury showed in all three plant species tried. Under both field and laboratory conditions we have found that 60 sec oil is not as effective in limiting virus spread as a 70 sec oil, thus only the 70 sec oil is being used commercially.

5.3.6.2 Effects of Concentration of Oil. After having selected a safe and effective oil, the next factor affecting possible phytotoxicity is the concentration of oil to be used. As indicated above, we have found that optimum effectiveness can be achieved with a concentration of 0.75% providing that proper application technology is used. Oils have been used experimentally by a number of workers at concentrations of from 2 to 5%. Often the oils used were in the 110 to 120 sec viscosity range thus it is not surprising that some degree of injury has been observed on sensitive crops. Potato is quite sensitive to oil and the usefulness on this crop has been somewhat reduced. Yields in European seed potatoes have been reported to be reduced by as much as 20% (Van der Want, personal communication) when 2 to 4% concentrations of oil have been used. In our own tests, weekly applications of oil (0.75% concentration) have caused only a 3% yield reduction among some eight varieties tested. Similarly, Conover (personal communication) has found that 1.5% oil caused very serious injury to tomatoes in Florida; yet we have never found any sign of injury or depression in yields of tomato from the use of oil.

5.3.6.3 Compatibility between Oil and Other Chemicals. Oils which have been specifically formulated for virus control should not be used in combination with other agricultural chemicals in the spray tank. The addition of other chemicals means that additional surfactant systems will be incorporated in the mixture, and this should be avoided.

Most insecticides and several fungicides are compatible with oils when they are applied separately from the oil. However, there are certain materials which should not be used in combination with oils as severe phytotoxicity will result. Generally, chemicals which tend to be injurious to plants when used alone will be more injurious if used in combination with oil. Examples include sulfur (wettable forms more so than flowable preparations), dinitros, karathane, dichlone and daconil. It would appear that oil enhances the penetration of many chemicals into leaves and thus aggravates inherent toxicity problems.

The use of aerially applied chemicals can result in problems not found when the same chemical is applied by ground equipment. This is because aerially applied chemicals are sprayed with very little water resulting in deposition of rather large droplets of concentrated chemical. When oil is sprayed over these droplets of concentrated pesticide, injury may appear which would not be observed if the pesticide had been dispersed more uniformly over the leaf by use of a dilute spray technique. I have observed this in the case of cantaloupe where aerially applied maneb proved quite toxic when used with oil, whereas the same amount of maneb applied by ground equipment as a dilute spray was completely compatible with oil.

Incompatibility problems with oil and other pesticides are usually worse during periods of hot weather. Oils are not injurious when used alone at high temperatures, but they definitely aggravate injury problems if used with chemicals which are more injurious by themselves at high temperatures. Again, this is probably related to enhanced foliar penetration as the result of the oil being on the leaves.

5.3.6.4 Effect of Weather on Oils. Oil, once applied to a dry leaf, will persist for at least 2 weeks under dry conditions. Light rain (<1 cm) does not remove oil from foliage but heavy rains will.

Under laboratory conditions, Simons *et al.* (1977) found that continuous exposure to a temperature of 32°C caused a significant loss in activity of oil after 6 days. Plants held at 24°C showed much less loss of oil, and plants held at 16°C were about as protected as on the day they were treated. Growth of the leaves was not significantly involved in the loss of activity. Whether the oil disappeared because of evaporation or absorption into the leaf (or both) is not known.

Oils should not be applied to plants when temperatures are below 10°C. The viscosity of oil changes drastically as the temperature drops, thus considerable changes in emulsion quality can be encountered. Phytotoxicity from oils is often found at low temperatures. This is probably because oils tend to collect in larger droplets on leaves thus resulting in injury.

5.4 CONCLUSIONS AND DISCUSSION

Mineral oil can be used as an effective, economical and safe means of controlling a number of plant virus diseases. To accomplish this, however, requires the use of oil specifically formulated for virus control along with the use of appropriate application techniques.

The success we have had in using low concentrations of oil and achieving good virus control is probably attributable to several factors. Whereas most oil formulations make good emulsions when combined with water, they tend to collect on leaves as intermittent films unless rather large quantities of oil are used. In this case the large deposits of oil result in phytotoxicity to the plant.

In order to use small quantities of oil and still achieve control of virus spread, it is necessary to cause the oil to be rather selectively deposited on the leaf surface. It is particularly desirable to place the oil in the deep grooves which border each epidermal cell and in which aphids usually initiate their feeding probes. This requires that oil sprayed on leaves should have some degree of mobility on the leaf surface.

It appears that this can be accomplished by using the proper oil formulation coupled with the proper application technique. We have recently been able to measure the size of oil droplets which are formed in JMS Stylet-Oil^{R} when sprayed through Teejet TX series nozzles and have found them to be in the 0.1 to 0.3 μm size range. Droplets of this size are so small they do not coalesce on contact and thus should have some mobility on the leaf surface. It should be noted that the oil droplets are contained within the emulsion droplets which average about 40 to 50 μm in diameter. In other words the 99.25% water component in the emulsion droplet serves as a carrier for the several thousand small oil droplets within each emulsion droplet.

Our work has shown an interaction between nozzle orifice size and efficacy on various plant species, with TX-4 nozzles being better on peppers and TX-5 nozzles superior on squash and tomatoes. There are substantial differences in leaf hirsuteness among these species, and perhaps this is involved. Simons *et al.* (1977) showed that oil on the surface of the leaf is directly involved with anti-transmission activity. There may be subtle differences in oil droplet size formation associated with small changes in orifice size which are reflected in distributional effects on leaf surfaces.

Because oil and reflective mulches have different mechanisms of action their use in combination should be more effective than with either one alone. Nawrocka *et al.* (1975) compared oil and reflective mulches for control of cucumber mosaic in lettuce but did not combine the two treatments. In their work, reflective mulches were more effective than oil sprays but these results may have been biased by the use of rather narrow plots (4 rows × 9m long). Toscano *et al.* (1979) have compared oil sprays and reflective mulches for the control of watermelon mosaic in zucchini squash and found that reflective mulches (aluminum foil or white plastic) were more effective than the oil. They reported difficulty in obtaining adequate coverage with oil and phytotoxicity when 4% oil was used. Neither of these studies utilized oil formulations specifically developed for virus control, nor were application procedures optimized for virus control.

Although oils and reflective surfaces are mostly known for their effects on reducing transmission of nonpersistent aphid-borne viruses, there are strong indications that transmission of viruses by insects other than aphids can also be affected. Nawrocka *et al.* (1975) reported a lowered incidence of aster yellows in mulched lettuce, apparently because leafhoppers were repelled by the mulch. There was also some indication that the oil reduced the spread of aster yellows. In this regard, Nault (personal communication) has found that oil sprays, under laboratory conditions, reduced transmission by leafhoppers of the maize chlorotic dwarf virus in corn. This virus is unusual in having a semipersistent relationship with its leafhopper vectors.

Singh *et al.* (1973) have reported that oil sprays reduced the spread of the whitefly transmitted tomato leaf-curl virus under field conditions. The oil caused significant reductions in whitefly populations through its effect on the sessile immature form, but it is doubtful that this was the reason for reduced virus spread.

The use of control procedures which exert their effects through such subtle mechanisms as causing changes in insect behavior and interference with virus transmission mechanisms has attributes beyond just control of diseases. These techniques can be readily incorporated into pest/management schemes which depend on preservation of beneficial insect species. And, because of the rather indirect nature of the modes of action of reflective surfaces and oils, it seems likely that resistance to the effects of these procedures should be slow to develop.

5.5 REFERENCES

Bradley, R. H. E. (1963). Some ways in which a paraffin oil impedes aphid transmission of potato virus Y. *Canad. J. Microbiol.* 9:369-380.

Bradley, R. H. E., Wade, C. V., and Wood, F. A. (1962). Aphid transmission of potato virus Y inhibited by oils. *Virology* **18**:327-328.

Broadbent, L. (1969). Disease control through vector control. *In* "Viruses, Vectors, and Vegetation" (K. Maramorosch, ed.), pp. 593-630. Wiley (Interscience), New York.

Dewijs, J. J., Strum, E., and Schwinn, F. J. (1979). The viscosity of mineral oils in relation to their ability to inhibit the transmission of stylet-borne viruses. *Neth. J. Plant Pathol.* **85**:19-22.

Fernow, K. H., and Kerr, S. H. (1953). Leafroll controlled by use of insecticides. *Am. Potato J.* **30**:187-196.

Hein, A. (1964). Die Wirkung eines Milchfilms auf die Übertragung eines nicht-persistenten Virus durch Blattläuse. *Z. Pflanzenkrankh.* **71**:267-270.

Hein, A. (1971). Zur Wirkung von öl auf die Virusübertragung durch Blattläuse. *Phytopathol. Z.* **71**:42-48.

Johnson, C. G. (1969). "Migration and Dispersal of Insects by Flight." Methuen, London.

Kennedy, J. S. (1960). The behavioral fitness of aphids as field vectors of viruses. *Rep. 7th Commonw. Entomol. Conf., 1960,* pp. 165-168.

Kuhn, C. W., and Demski, J. W. (1975). The relationship of peanut mottle virus to peanut production. *Res. Rep. Georgia Agric. Exp. Sta., No 213.*

Loebenstein, G., Deutsch, M., Frankel, H., and Sabar, Z. (1966). Field tests with oil sprays for prevention of cucumber mosaic virus in cucumbers. *Phytopathology* **56**:512-516.

Moericke, V. (1950). Über das Farbsehen der Pfirsichblattlaus *(Myzodes persicae). Z. Tierpsychol.* **7**:265-274.

Nawrocka, B. Z., Eckenrode, C. J., Uyemoto, J. K., and Young, D. H. (1975). Reflective mulches and foliar sprays for suppression of aphid-borne viruses in lettuce. *J. Econ. Entomol.* **68**:694-698.

Peters, D., and Lebbink, G. (1973). The effect of oil on the transmission of pea enation mosaic virus during short inoculation probes. *Ent. Exp. Appl.* **16**:185-190.

Simons, J. N., and Beasley, C. W. (1977). Visualization of oil on leaf surfaces—a technique. *J. Econ. Entomol.* **70**:307-308.

Simons, J. N., McLean, D. L., and Kinsey, M. G. (1977). Effects of mineral oil on probing behavior and transmission of stylet-borne viruses by *Myzus persicae. J. Econ. Entomol.* **70**:309-315.

Singh, S. J., Sastry, K. S. M., and Sastry, K. S. (1973). Effect of oil spray on the control of tomato leaf-curl virus in field. *Indian J. Agric. Sci.* **43**:669-672.

Smith, F. F., Johnson, G. V., Kahn, R. P., and Bing, A. (1964). Repellancy of reflective aluminum to transient aphid virus vectors. *Phytopathology* **54**:748 (Abstr.).

Smith, F. F., and Webb, R. E. (1969). Repelling aphids by reflective surfaces, a new approach to the control of insect-transmitted viruses. *In* "Viruses, Vectors, and Vegetation" (K. Maramorosch, ed.), pp. 631-639. Wiley (Interscience), New York.

Toscano, N. C., Wyman, J., Kido, K., Johnson, H., Jr., and Mayberry, K. (1979). Reflective mulches foil insects. *Calif. Agric.*, July-Aug., pp. 17-19.

Vanderveken, J. (1968). Importance des relations vecteur, virus dans l'inhibition de la transmission aphidienne des phytovirus par des pulverisations d'emulsions huileuses. *Ann. Epiphytol.* **19**:141-146.

Vanderveken, J. (1972). Contribution a l'étude de l'inhibition de la transmission aphidienne de phytovirus par des substances huileuses. Dissertation, Faculté des Sciences agronomiques de l'État à Gembloux.

Vanderveken, J. (1977). Oils and other inhibitors of nonpersistent virus transmission. *In* "Aphids as Virus Vectors" (K. F. Harris and K. Maramorosch, eds.), pp. 435-454. Academic Press, New York.

Vanderveken, J., Bourge, J. J., and Semal, J. (1966). Effets d'huiles minerales sur la transmission de phytovirus par pucerons. *Meded. Rijksfakulteit Landbouwwetenschappen Gent.* **31**:974-980.

Zitter, T. A., and Everett, P. H. (1979). Use of mineral oil sprays to reduce the spread of tomato yellows virus disease in Florida. *Univ. Fla., Immokalee ARC Res. Rep.* **SF79-1**.

Zitter, T. A., and Ozaki, H. V. (1978). Aphid-borne vegetable viruses controlled with oil sprays. *Proc. Fla. State Hort. Soc.* **91**:287-289.

Zschiegner, H. J., Kramer, W., Sass, O., Fritsche, R., and Dubnik, H. (1971). Successful restriction of the spread of nonpersistent viruses by methods of virus vector control. *Proc. 6th Br. Insect. Fung. Conf.*, pp. 319-323.

Chapter 6

CONTROLLING SEED AND INSECT-BORNE VIRUSES

J. B. Quiot
G. Labonne

I.N.R.A.
Pathologie Vegetale
Petit-Bourg, Guadeloupe FWI

J. Marrou

I.N.R.A.
Paris, France

ISBN 0-12-326440-5

6.1 INTRODUCTION

Until now, virus diseases have been considered practically incurable. And the first aim in controlling this kind of disease is to reduce or eliminate virus spread within the field. This approach will reduce disease incidence and crop production loss. The second aim is to avoid the appearance of virulent strains that are able to produce unexpected epidemics that disrupt harvesting forecasts. Because the transmission cycles of viruses are complex and generally involve vectors such as insects, mites, nematodes, fungi, or man, many opportunities can be found by virologists to break the "virus-vector-plant" cycle and stabilize virus populations.

During the last ten years, more than two thousand publications have been referenced concerning control of virus diseases. Since the subject is large and the possible answers varied, it is not possible here to give a complete case by case study. On the other hand, a certain number of general reviews about a part or the whole of this subject have been or soon will be published. In particular, ones by Stevens (1960), Broadbent (1964), Swenson (1968), Leclant (1968), Agrios (1969), Matthews (1970), Duffus (1971), Carter (1973), Thresh (1974, 1976), Gibbs and Harrison (1976), Wheeler (1976), Berger (1977), Harris and Maramorosch (1977, 1980, 1981), Zitter (1977), and Maramorosch and Harris (1979, 1981) can be mentioned.

In this review, we have tried to regroup the different ways of controlling virus diseases according to the level at which they can be applied: field or farm, district and, finally, the international level.

6.2 CONTROL OF VIRUS DISEASES AT THE FIELD LEVEL

The aim of a control method is to maintain the damages produced by the pathogen below an economically acceptable level. In practice, this aim can be reached only when resistant varieties, the resistance of which is stable in time and space, can be used. Unfortunately, such resistances do not exist against most viral diseases. One must then be satisfied with searching for control methods which delay the introduction of the pathogen into the field or slow the rate of disease progression within the field. Indeed, it has frequently been observed that late infection does not result in serious reductions of yield. Thus Rose (1978) observes that maize yield is not altered when the crop is infected by maize streak after the eighth week of culture. In other crops, susceptibility to viral infection decreases drastically when the plants become older. For example, Messiaen *et al.* (1963) observed that tomatoes are very rarely infected by cucumber mosaic virus once they have reached the bloom stage.

However, there are exceptional instances where early infection can be less severe than late infection. Broadbent (1976), for example, reported that, before the mild strain of tobacco mosaic virus selected by Rast (1972) had been used in agricultural practice, it had seemed better for the growers of the Isle of Wight to inoculate their young tomatoes with a common strain in order to avoid the period of crisis which follows infection from taking place during the period of fructification.

The other way of controlling virus disease, that is to say to reduce the speed of disease spreading within the field, can be realized by altering host sensitivity or the number and nature of virus sources and vectors. Practically, controlling the incidence of virus disease in a field can be approached on four different levels: the virus source, the virus, the virus vector, and the susceptible crop.

6.2.1 Virus Sources

Virus sources can be located in the crop itself, when virus is seedborne, or in surrounding areas, if weeds, volunteers or other fields are infected.

6.2.1.1 Planting Virus-Free Seed or Propagative Organs.

Seed-borne Viruses. During recent years, more and more viruses have been found to be seed-transmitted. Bennett (1969) reported 53 seed-transmitted viruses in 1969. In 1974, Phatak pointed out 85 seed-transmitted viruses, 69 of which can be found in marketable seed.

Two kinds of seed transmission can be distinguished (Bennett, 1969). Some viruses are retained at the surface of the seeds or in parts of them besides the embryo, in which cases a mechanism allowing for infection of the new seedling at germination is necessary. Other viruses infect the embryo inside the seed directly, and a young plant issued from an infected seed is always diseased. In the latter case, every virus detected in the seed will correspond to a field infected plant.

Control of seed-transmitted virus is sometimes obtained by using healthy seed, if ecological studies have shown that the main sources of viruses are the infected seeds. This presupposes an ability to produce healthy seeds, to detect virus infection in seeds sold to growers, and to convince growers to use these preferred healthy seeds.

Production of virus-free seeds can be obtained when mother plants are bred in an uncontaminated area and if care is taken to avoid virus contamination during seed-processing. For some viruses, such as tobacco mosaic virus in tomatoes, which are carried at the surface of seeds, disinfection with chemicals (e.g., sodium phosphate, sodium hypochlorite, hydrochloric acid, etc.) can significantly reduce the rate of infected seeds (Bennett, 1969). Heat therapy can sometimes eliminate virus without affecting germination. Fletcher *et al.* (1969) observed that cucumber green mottle virus can be eliminated from seeds by heat treatment at 70°C for more than one day. However, these well known methods have been too much neglected by seed growers. Furthermore, hybrid production and seed multiplication are more and more achieved abroad, far from the control of government and scientific staffs of seed companies.

The test of commercialized seeds for virus-transmission presents some technical problems: what rate of infected seed may be an acceptable risk for the grower, how to separate with accuracy the good and bad seed lots, which test must be used, what size samples must be analyzed, and so on.

The maximal tolerable virus rate in seed has been particularly studied in the case of lettuce mosaic virus where it has been estimated to be less than one contaminated seed per thousand tested (Zink *et al.*, 1956). Nevertheless, in some areas even this rate can be too high (Zitter, 1977), and a level of zero contamination per 30,000 seeds tested must then be obtained (Kimble *et al.*, 1975).

Types of tests used to check seed-lots have been extensively studied by Phatak (1974). He noted that visual inspection of seeds is not a reliable test, even if it can bring some useful information in some cases (e.g., soybean infected by soybean mosaic virus).

Biological tests are more useful. These include the growing of test seedlings under controlled conditions, which generally gives good results. However, sometimes symptoms are not clearly visible because of mild virus strains or tolerant plant varieties. And Lister and Murant (1967) pointed out that several seed-transmitted nepoviruses do not give clear symptoms on seedlings. Indexing of seed lots on differential hosts is very effective in some cases, such as the check of lettuce seed for lettuce mosaic virus on *Chenopodium quinoa* (Pelet, 1965; Marrou and Messiaen, 1967).

The use of highly specific serological tests permits the detection of seed-borne viruses. The Ouchterlony agar double diffusion test and the latex agglutination test have been successfully used for different viruses (see Phatak, 1974). Improvements in serology techniques, e.g., enzyme linked immunosorbent assay (ELISA) and immune or serologically specific electron microscopy (SSEM), should further improve test sensitivity. The sensitivity of this kind of testing may be increased by sampling methods which concentrate infected seeds. Ryder and Johnson (1974) suggest that testing be applied to the lightest fraction of lettuce seed that has been separated by an air stream.

Common use of healthy seeds by growers of a district is not always an easily attained goal. In developed countries, the diffusion of new techniques is generally satisfactory. Producer associations are sometimes able to obtain complementary testing of proposed seed lots, as in the U.S.A. for lettuce seed (Kimble *et al.*, 1975). In other less developed countries, however, traditional practices have contributed to the continued use of untested seeds. Thus, growers are accustomed to produce their own seed for periods of several years without any virus testing. Also, there can be, besides seed growers that sell selected seeds, a traditional trade on local markets where untested seeds are for sale. Thus, Lockhart and Fischer (1974) pointed out that, in Morocco, bean common mosaic virus is endemic in traditionally planted bean seeds. Education of growers, then, is the only solution. The problem is the same for tomato seeds that are infected by tobacco mosaic virus, but disinfection techniques are simple enough to be used by growers who want to produce their own seeds (Broadbent, 1976).

Vegetatively Propagated Species. When a vegetatively propagated species is susceptible to a given virus, progressive infection of the whole species is generally observed, unless sanitary selection measures are used. The use of virus-free plants is possible at a commercial level only if some plants can be guaranteed free of virus and if these plants can be propagated in an environment free of contamination. Such virus-free mother plants sometimes can be found in old cultivars or carefully tested natural populations. However, virus-free clones are more and more often being obtained either by heat therapy, meristem tip culture, or a combination of both techniques.

Heat therapy can be realized either by high temperature treatment, from 45°C to 54°C for dormant plant material (sugarcane setts, dormant trees, or tubers), during limited periods from a few minutes to some hours, or by long-term treatments at lower temperatures for growing plants. These techniques have been used to eliminate viruses from a large number of infected species (see Hollings, 1965; Nyland and Goheen, 1969; Carter, 1973).

Meristem tip cultures also have been put into practice (Morel and Martin, 1952). They are based on the observation that, in systemically infected plants, meristematic tissues are not invaded by the virus. It is possible to cut off these organs under aseptic conditions and to propagate them in glass tubes on nutrient agar or wicks of filter paper dipped into a nutrient solution. Finally, some complete young seedlings are obtained, a certain number of which are found to be without any disease symptoms. By additional biological or physical tests, it can be established that some of the seedlings are no longer virus-infected. The combination of both techniques, thermotherapy and meristem tip culture has been found effective in removing virus from several species.

Multiplication of virus-free plants involves two Problems: first, to multiply plants as quickly as possible and, second, to keep healthy material free from further contamination. The multiplication rate of a given species defines the time necessary to produce, from one healthy plant, an amount of healthy material large enough to be marketable. When this rate is low (as is the case with garlic, *Allium sativum* L.), 8 to 10 years may be necessary to produce a marketable number of plants. Some techniques have been proposed to increase the multiplication rate of healthy clones, for instance, potato propagation by cuttings (Slack, 1980) or division of regenerated plants by culturing in liquid agitated medium (Walkey and Woolfitt, 1968; Roca *et al.*, 1978).

In order to avoid new contamination during multiplication of regenerated clones, a program of certification must be established. Then, drastic rules must be established to define the sanitary conditions that are acceptable for the different levels of multiplication from the first generation of mother plants obtained through meristem culture to certified seeds which will be sold to the growers.

Sanitary selection is often applied to a species for different pathogens (viruses, fungi, bacteria, and nematodes) and, if possible, tests must be combined to reduce the number of controls. In addition, some species show phenotypic variations and need to be regularly recloned; this can also be included in the sanitary selection schedule (for instance, in sanitary selection of garlic in France [Messiaen *et al.*, 1966]. To check for the presence of contaminated plants, there are different kinds of control tests. Mother plants are controlled as accurately as possible: thanks to the small number of plants to be controlled, the cost of testing is of little importance. Above all, tests must be precise: repeated inoculations on several differential hosts, repeated serological tests, and observations by electron microscopy are often used. In some cases, such as detecting potato spindle tuber viroid in potato mother plants, polyacrylamide gel electrophoresis is used (Slack, 1980).

Mass production of virus-free seeds requires other control measures: fields for increasing clonal lines are visited by trained inspectors who are able to detect infected plants by visual examination and eliminate them. However, the effectiveness of such practices as roguing is limited because neither late infections nor latent virus strains can be so detected (Hill, 1978). Furthermore, oil-sprays, insect attacks, or even senescence of leaves can mask virus disease symptoms. It is necessary to perform complementary tests on representative samples using reliable and rapid tests. Searching for symptoms on plantlets grown under defined conditions, inoculation of differential hosts, or serological tests are frequently used.

It should be noted that the establishment of a sanitary selection program on a country-wide scale is an expensive and difficult operation. And it will only be profitable if truly virus-free certified seeds can be regularly produced in quantities large enough to satisfy growers' needs. Before starting a sanitary selection program, an epidemiological study is advisable to check that restored material will not be recontaminated too quickly because of too numerous virus sources in the area. The benefits expected from such programs are often sufficient to justify their establishment. Nevertheless, there is a necessity for a constant survey of the whole multiplication scheme by one or several qualified pathologists; control tests can be improved in pace with improvements in testing technology.

Unexpected contaminations of clonal lines during multiplication can also occur. Pathologists must identify the causes and sources of contamination to improve the sanitary selection scheme. For instance, in several countries of Western Europe, those involved in potato selection programs experienced difficulties during the years 1975 and 1976. Several explanations for this were proposed by pathologists: climatic variations altered the dynamics of vector populations; adaptation of vectors to organo-phosphorus insecticides; increased seed prices leading growers to keep the same seed during two successive years and thus multiply the number of virus-sources; and introduction of mechanical harvesting which leaves numerous little tubers in the field which can become new sources of viruses (Hill, 1978). Spraying was not effective against potato leaf roll virus infection because it often came either too early or too late (Turl, 1978); massive increases in the numbers of maize crops near seed potato fields may have modified the levels of some vector populations such as *Rhopalasiphum padi* L.

Perennial Species. Production of virus-free ornamentals and fruit trees is similar to that of species with a vegetative multiplication. It mainly supposes that nurseries are carefully surveyed. Because several harmful viruses are often present in the same tree, it is necessary to use very specific tests.

6.2.1.2 Protection of Seedbeds. The setting up of seedbeds far from any virus sources and their protection against vectors are easy measures for reducing the number of virus sources at the moment of planting. For instance, Wellman, cited by Duffus (1971), noticed that the elimination of weeds for at least 75 feet around seedbeds decreases the losses caused by cucumber mosaic virus in celery. In the same way, Robert and Maury (1970) recommend that potato germinators be kept clear to avoid the development of aphid vector populations. These simple measures are, nevertheless, far from common practice.

6.2.1.3 Elimination of Virus Sources. Control of virus diseases is also abetted by decreasing the number of available virus sources. Actions to reduce the latter will vary depending on whether the cultures concerned are perennial or seasonal.

For perennial cultures, the way most frequently used is to plant healthy material as far as possible from local sources of infection. This can help to delay the onset of an epidemic.

For seasonal cultures, systematic removal of the virus sources when the crop is absent can help to break the virus cycle and delay disease epidemics in the crop to be planted the following year. These sources can be inside the field or in surrounding areas.

Virus Sources in the Field. When ploughing is not done immediately after harvesting a crop, virus-infected volunteer plants can contribute to maintain virus late in the season and thus allow the virus to bypass its annual cycle (Louie, 1968; Bruckhart and Lorbeer, 1977; Howell and Mink, 1977).

Susceptible weeds which grow between rows can also be important virus sources: during a recent survey of the infection of such weeds by cucumber mosaic virus in southeastern France (Quiot *et al.*, 1979a), *Senecio vulgaris* and *Portulaca oleracea* were found to be 67% and 85% infected, respectively.

Virus can be maintained in a field by poor cultural practice. Thus, Heie and Petersen (1956), cited by Jadot (1976), and Thresh (1976) pointed out the role played by poorly covered beet-root silos as sources of beet yellow mosaic virus. Hill (1978) suggests that mechanical potato harvesters may leave behind little tubers that contribute to the maintenance of virus in the field.

Virus Sources in the Vicinity of the Crop. The importance of weeds in the ecology of virus diseases has been largely reviewed by Duffus (1971) who emphasized their role as virus or vector reservoirs. In the case of persistent viruses such as beet western yellows virus, he pointed out the role of weed species in providing shelter for both virus and vectors.

During recent years, ever increasing numbers of species have been recognized as potential hosts for plant viruses (Thornberry, 1966; Edwardson, 1974). Some viruses appear to have an extremely large range of potential hosts. For instance, Price (1940) pointed out that 190 species belonging to 41 families were susceptible to infection by cucumber mosaic virus. These numbers were raised to 307 species and 42 families by Thornberry (1966). A recent bibliographical review (Douine *et al.*, 1979) records 775 species belonging to 86 families.

In order to define ways of control, it is necessary to establish which susceptible species are the most important virus reservoirs under natural conditions. Studying cucumber mosaic virus ecology in American lettuce fields, Bruckhart and Lorbeer (1977) pointed out that weeds were important as virus reservoirs and that it would be of some interest to conduct a quantitative study on the importance of such virus sources around a field.

Such a study is difficult because of the frequently observed asbence of symptoms in infected weeds. Thus, natural infection cannot be quantified by visual observation. Nevertheless, Tomlinson and Walkey (1973) compared three isolates of cucumber mosaic virus from *Stellaria media* (L.) Vill. collected from three different countries. They noticed that no symptoms occurred on *S. media* inoculated with their homologous strains (from the same country) but that symptoms did occur when a given CMV strain was inoculated to a heterologous *S. media*.

Contamination by CMV of weeds surrounding a vegetable field was studied from 1974 to 1977 in southeastern France by monthly sampling and inoculation to several differential hosts. Of 137 species of weeds that had been inventoried in the area, 39 appeared to be naturally infected (Quiot *et al.*, 1979a, 1979b); twelve of the 39 were perrenials. It was also found that CMV was seed-transmitted by weeds, as previously shown by Tomlinson *et al.*, (1970, 1973) and Hani (1971). In such a case, the breaking of the virus cycle by reservoir destruction appears unlikely.

A study on the redistribution of infected weeds showed that the infection rate varied according to the particular weed species, the growing season, and the weeds location. The highest infection rates were observed among isolated plants growing on access tracks and ploughed soils. This phenomenon is probably linked with the greater attractiveness of such plants to aphids (Quiot *et al.*, 1979a).

Virus Sources Far from the Susceptible Field. Besides weeds growing in ditches and along roads bordering fields, there are other sources of contamination such as other cultivated species susceptible to the same virus disease, aromatic or ornamental plants in home gardens, etc. Control of such virus sources is feasible only by applying control measures against both vector(s) and virus on a district scale.

Roguing. When a crop must be kept free of virus (e.g., seed crop protection from seed-borne virus), removal of early infected plants can sometimes be effective. Thus, the number of virus sources within the field is decreased; this slows the spread of virus and forestalls epidemics of virus diseases which can spread like wildfire.

When the rate of spread of a disease is very slow, and the disease endemic, a careful program of roguing may lead to local eradication (van der Plank, 1970).

6.2.2 Viruses

Prevention of plant infection by a virus or blockage of virus multiplication in the plant are methods of control that have been sought after for a long time. Till now, however, few results have been obtained which can be of practical use. A review of the subject was recently presented by Gupta (1977). It appears that the best use of antiviral agents is in combination with other methods in effecting the cure of infected material by such techniques as thermotherapy or meristem tip culture.

However, two techniques representing more or less direct action against the virus itself have been successfully tested in the field: these are cross-protection and the use of carbendazym. Cross-protection is a mechanism which consists in infecting a plant with a mild virus strain to protect it against a more pathogenic one. Although used initially for comparing virus isolates, this technique is now used as a method of control at the field level, specially for protecting tomatoes against tobacco mosaic virus—thanks to the isolation by Rast (1972) of a TMV mutant that does not provoke any symptom on tomatoes. This strain of TMV and some subcultures are now widely used to protect tomatoes that are cultivated in greenhouses; for instance, during the winter of 1978, (Anonymous, 1979) ca. 12,600,000 tomato plants were protected this way in France. The technique is satisfactory; nevertheless, it must be noted that protection is effective only against TMV strains that are closely related to the protecting strain (Migliori, 1973). On the other hand, there is a risk involved in using such a technique to protect field crops: it seems hazardous to introduce large quantities of a foreign virus into a given country, and the protective effect against an homologous virus can be outweighed by the possibility of a mixed, highly pathogenic, infection with other viruses that might infect the same plant. Premunition must be considered as a stopgap measure that buys time for the production of resistant varieties rather than as a method of control that must be developed (Migliori, 1973).

An efficient treatment, not against the virus itself but against the symptoms that it induces, has been proposed by Tomlinson *et al.* (1976). They showed that carbendazym reduces the severity of symptoms produced by tobacco mosaic virus in tobacco. In a field test, this substance, a byproduct from benomyl, also significantly reduced the importance of beet western yellows virus symptoms on lettuce. However, this technique does not eliminate virus from the field, and the disappearance of symptoms has only been observed for some plant viruses.

Accurate knowledge of the biology and the ecology of viral populations under natural conditions can contribute to the research of new ways of control. For instance, the characterization of virus strains within a given area provides data needed to define a strategy for the use of resistant varieties (Leroux *et al.*, 1979). In the same way, knowledge of the distribution of the different strains of a given virus in the world and the study of their ecology (specially their pathogenicity and their adaptability to climatic conditions) enable one to forecast the risks involved in bringing an introduced strain into a new area, and the dangers thereby encountered.

6.2.3 Vectors

In the virus-vector-plant chain, it is the vector which acts as the disseminating element allowing for the spread of the virus; so it can be tempting to try to break the chain at this link.

However, if one is to take action against the vector, it is first necessary to identify with certainty the vector or vectors. When a list of viruses is consulted, it appears that the vectors are unknown for many of them. Moreover, when a virus-vector combination is discovered, it is necessary to find the answers to two questions: first, is the combination the only one that exists and, second, if others exist, what is the relative epidemiological role of each of them under field conditions. The difficulties involved in answering questions such as these can best be illustrated by the following two examples.

It is still possible to find new vectors, even when several others are already known. For instance, Zitter and Tsai (1977) found that a fly (*Diptera*: Agromizidae) was able to transmit three potyviruses for which many species of aphid have been previously identified as vectors. In this case, the perspective of disease control may have to be modified in countries where vector Diptera are prevalent.

When many species have been identified as vectors of a given virus, as is generally the case for nonpersistent aphid-borne viruses, it can be expected that they will differ widely in their roles as spreaders of the virus under field conditions. And it is often difficult to appreciate these differences. Numerous researchers have attempted to find correlations between the catches of potential vectors and disease epidemics. But it appears that even if the catches give an evaluation of the relative activity of the different species, generally these results are greatly influenced by the trapping technique that has been used. Indeed, because of trap selectivity, the results are biased and, in many cases, the bias cannot be estimated (Eastop, 1955; Robertson and Klostermeyer, 1958; Costa and Lewis, 1967; Moericke, 1967; Zettler *et al.*, 1967; A'Brook, 1973a; Robert *et al.*, 1974; Rabasse *et al.*, 1977). Moreover, the catching of a potential vector does not in itself inform one as to whether or not the insect was a virus carrier at the moment of the catch. This can only be verified by directly testing the trapped insects. This kind of testing has been performed by some researchers, such as Knoke *et al.* (1976) working with corn viruses or Plumb (1976) with barley yellow dwarf virus. These experiments have shown that substantial differences may occur between the actual situation observed in the field and an extrapolation based on vector population activity and transmission assays performed under laboratory conditions.

In the same way, Tables I and II represent aphids that were caught in the south of France by a suction trap and immediately placed on a susceptible host to detect the presence of cucumber mosaic virus. It appears that several aphid species, some of which have not been previously recorded, effectively carry the virus. This suggests that more data are needed about the period of pullulation and the population structure of aphid species in order to define rational methods of controlling virus spread.

Another unanswered question involves primary and secondary infections. If these two kinds of infection can be easily distinguished in some cases, it is more generally difficult to distinguish the relative influence of each vector or, for a given vector, of different morphs on each type. Such knowledge would be very useful in determining which control method must be used and when it must be put into practice.

Nevertheless, a lot of measures have been tested for controlling virus vectors. Chronologically these measures range from those taken against viruliferous insects arriving in the vicinity of a cultivated field to ones specifically aimed at preventing vectors from inoculating the cultivated plants.

6.2.3.1 Avoiding Vector Arrival. When there is a temporal coincidence between the cultivated plants and the vector, increasing the distance between virus source and susceptible crop may be a crucial factor in reducing the number of viruliferous insects entering the field. This solution which appears efficient on a regional level (Gibbs and Harrison, 1976; Thresh, 1976) seems to be adaptable to the field level too, particularly in the case of the nonpersistent viruses where short retention periods by most individual aphids further dilutes the inoculative

TABLE I. Records of aphid species which were found bearing cucumber mosaic virus (CMV) during field trials performed between 1975 and 1979 in southeastern France. Aphids were caught in suction traps operating at 0.25 m to 1.50 m above the ground, near tomato and muskmelon plots. The results represent about 23,000 aphids that were checked for CMV inoculativity on *Cucumis melo* plantlets.

Acyrthosiphon pisum Harris	*Cavariella aegopodii* Scopoli
Acyrthosiphon sp.	*Hayhurstia atriplicis* Linnaeus
Aphis citricola Van der Goot	*Hyadaphis passerinii* Del Guercio
Aphis craccivora Koch group	*Hyperomyzus lactucae* Linnaeus
A. fabae Scopoli group	*Macrosiphum euphorbiae* Thomas
A. gossypii Glover group	*Myzus persicae* Sulzer
A. nerii Boyer de Fonscolombe	*Rhopalosiphum maidis* Fitch
A. pomi De Geer	*R. padi* Linnaeus
Aphis sp.	*Schizaphis graminum* Rondani
Brachycaudus amygdalinus Schouteden	*Staegeriella necopinata* Borner
B. helichrysi Kaltenbach	*Therioaphis trifolii* Monell
Capitophorus eleagni Del Guercio	*Uroleucon sonchi* Linnaeus
C. hippophaes Walker	

TABLE II. Percentage of CMV-inoculative Aphids among the Main Aphid Species Caught from May 27 to July 29 of 1978 by a Suction Trap Operating at 0.25 m above the Ground.[a]

Aphid species	Number of aphids tested[a]	% inoculative aphids
Aphis citricola	79	2.6
A. craccivora group	954	6.0
A. fabae group	1027	2.4
A. gossypii group	2279	10.0
Brachycaudus amygdalinus	74	7.8
Brevicoryne brassicae	155	0.0
Capitophorus eleagri	258	1.6
C. hippophaes	244	1.7
Hayhurstia atriplicis	28	20.0
Hyalopterus pruni	155	0.0
Macrosiphum euphorbiae	108	2.0
Myzus persicae	124	5.0
Staegeriella necopinata	62	3.6
Therioaphis trifolii	192	0.5

[a] See Table I heading.

potential of the vector (Simons, 1957; Karl *et al.*, 1972; Proeseler and Weidling, 1975). For instance, Simons (1957) and Havranek and Laska (1972) have shown that a distance of about 100 m between virus source and susceptible plants can effectively protect the latter. For some high grade crops such as Elite strawberry runners in England, distances between source and crop as great as five miles are recommended (Thresh, 1976). However, the efficiency of separating a source from a susceptible crop can be influenced by several factors such as the level of vector populations, the prevailing wind, the aspect of the interlying landscape, etc. And it seems better, before defining secure distances in a given area, to conduct a careful study of the disease gradients as proposed by Thresh (1976).

The contamination of a field by a virus transmitted by flying insects generally begins along windward, border rows. This is the well known edge effect; however, it is also well known that the presence of a windbreak modifies the course of epidemics in areas where there is a prevailing wind direction (Johnson, 1969; Thresh, 1976). Heterogeneity and sometimes a delay in contamination are observed on the leeward edge. It appears that the windbreak mainly affects the spatial distribution of flying insects by acting on the wind velocity (Lewis, 1966; Robert, 1976). Windbreaks can constitute a way of reducing the spread of viruliferous aphids (Bonnemaison, 1961; Marrou *et al.*, 1979), and their efficiency can be strongly influenced by their physical features, such as length, height, thickness, and porosity (Lewis and Dibley, 1970; Guyot, 1970), or by the frequency of the prevailing wind. Robert (1976) has also shown that different insects, even different species of aphids, are not equally influenced by a given windbreak.

In addition to the classical windbreaks made of trees or bushes, efficient windbreaks have also been made with lines of annuals such as sunflowers, corn, etc., or even with dried canes (*Arundo donax*) or P.V.C. nets (Simons, 1957; Toba *et al.*, 1977; Rose, 1978). However, the plants used as windbreaks must not become reservoirs for either viruses or vectors.

6.2.3.2 Avoiding Vector Alighting. Knowledge about visual stimuli, such as yellow traps which were first developed and used for catching insects by Moericke, has now been applied to field protection in the form of reflective mulches that partially or completely cover the soil. The mulches used are either attractive or repellent. In the former case, insects must be killed or immobilized immediately when they alight.

Both methods have been proved efficient in reducing virus spread and increasing crop yield when applied against aphid vectors. The first experiments, reviewed by Smith and Webb (1969), were conducted with aluminum foils or black or transparent polyethylene sheets. All these mulches have a repellent effect—aluminum more so than others. Maximal repellency was obtained when at least 50% of the area was covered. Since then, this technique has been investigated by many researchers and varied kinds of repellent mulches have been tested: e.g., sawdust and huskrice (Costa, 1972; Kemp, 1978). Cohen and Marco (1973), on the other hand, have obtained good protection by using attractive, sticky, yellow polyethylene sheets.

Good protection has also been obtained by several researchers in Israel against the bottle gourd mosaic virus and the tomato yellow leaf curl virus transmitted by the whitefly *Bemisia tabaci* Genn. (Avidov, 1956, cited by Nitzany *et al.*, 1964; Cohen and Melamed-Madgar, 1974, 1978). Several mulches were used (sawdust, straw, yellow or blue polyethylene sheets) and all of them were proven to be efficient (Cohen and Melamed-Madjar, 1978). In addition to the attractive effect of the mulch that allows for the catching of the whiteflies, it seems that there is also a temperature effect. According to Avidov and Nitzany *et al.* (1964) temperatures above the mulch are too high for the development of whitefly populations on young plants; thus, this contributes to a decrease in virus spread through a decrease in vector numbers.

Though the efficiency of these methods has been proven, there are limitations in some cases. For example, Loebenstien *et al.* (1975) noticed a negative effect on a pepper yield in a plot protected by aluminum foil. They suggested that the mulch induced a lower soil temperature compared to other treatments by reflecting the sun, thus retarding plant maturation. (But this phenomenon has a beneficial effect in hot regions [Zitter 1977].) Esparza-Duque and Verhoyen (1975) have shown that increases in vector numbers decrease the effectiveness of the mulch as a protectant.

Surely, improvements in the use of such control methods can be obtained if there is coordination between investigations on visual behavior of the vectors and ones on precise identification of the actual vectors acting in the field. Reports such as those by Smith and Webb (1969) relating preferential development of *Aphis gossypii* Glover on plants protected by aluminum foils underscore the need for this type of coordination.

Several researchers have reported lowered incidence of virus in high density crops which completely cover the soil surface, and this phenomenon was related to a decrease in vector numbers effected by a unified surface (A'Brook, 1968; Farell, 1976). The decrease can be partially explained by a lower attractiveness of such covered plots to aphids, compared with other plots where less numerous cultivated plants contrast with the naked soil. Such an effect has been observed by A'Brook (1973b) in peanut fields and by Smith (1969, 1976) in Brussel sprouts fields. A homogeneous plant cover can also be obtained by cultivating a mixture of two species, when they are susceptible to different viruses.

Aphid alarm pheromones, especially *trans*-B-farnesene, have been heralded as new potential weapons against plant infestation and virus spread by aphids (Nault and Montgomery, 1977). The chemical *trans*-B-farnesene has the unique characteristic of being specific to aphids yet relatively nonspecific within the Aphidinae; furthermore, it is nontoxic. However, since its discovery in the early 1970's (Bowers *et al*., 1972; Edwards *et al*., 1973; Wientjens *et al*., 1973), initial optimism regarding its applicability in reducing virus spread has not been justified. Till now, *trans*-B-farnesene has proven ineffective in laboratory and field testing, even when it has been formulated in time-release, polyvinylchloride bars to overcome problems stemming from its instability and volatility (Yang and Zettler, 1975; Hille Ris Lambers and Schepers, 1978).

6.2.3.3 Avoiding Plant Inoculation. Finally, after it has alighted on and probed a plant, a viruliferous insect may feed on that plant and produce progeny. At this point, there are a number of control approaches that might be applied, depending on whether the virus is transmitted by the vector in a nonpersistent or persistent manner.

Spraying with insecticides is a technique which has been tried for a long time for controlling the spread of virus diseases (Bawden, 1964; Broadbent, 1969; Carter, 1973). But results have shown that insecticides are generally not efficient for controlling the spread of nonpersistent viruses. Sometimes some chemicals worsen the situation by promoting greater vector activity before death (Shanks and Chapman, 1965; Lehmann **et al.**, 1976). On the other hand, some good results have been obtained against the spread of persistent viruses, specially when the timing of spray is integrated with a well adapted system of vector monitoring. However, the use of insecticides presents some disadvantages which contribute to their limited use: the price of field spraying is rather high; there is a real danger of enhancing the occurrence of resistant vector biotypes, and most vector species possess a huge potential for multiplication; frequent sprayings of large areas contribute to the elimination of the vectors' natural enemies, which can increase the difficulties of limiting vector populations, even if the action of the natural enemies on virus spread is not so obvious (Frazer, 1977).

Other chemicals such as analogues of juvenile hormone appear less noxious; experiments have shown that the use of these chemicals against aphids results in interesting decreased levels of aphid populations and quasi-elimination of winged aphids able to fly (see the review of Stockel, 1975). Unfortunately such chemicals have no action on insects coming from outside the sprayed area, and the role of such insects in spreading virus is generally not negligible.

Oil sprays are also effective in reducing the spread of aphid-transmitted, nonpersistent viruses. Many researchers have confirmed the efficiency of this technique (Vanderveken, 1977; Simons, 1981). However, it also presents some disadvantages: phytotoxicities are sometimes observed and a yield reducing effect has appeared in some experiments (Bosschaert *et al.*, 1974; Bos *et al.*, 1978); others have observed that the oil and resulting discolored leaves can mask virus symptoms and hamper visual observations by field inspectors in sanitary fields (Bokx and Cuperus, 1978). However, complete coverage of the plants, including the young leaflets of the tip, which is needed for oil treatment efficiency, seems to be the main problem. Actively growing young crops need frequent and costly spraying which can be possible only for high value crops.

Another method of controlling the spread of nonpersistent viruses is based on the alighting behavior of aphids and the fact that these viruses are lost by their vectors in a few probes. Studies have shown that aphids alight on a plant chosen at random (Kring, 1972). Generally, aphids recognize the suitability of plants only after having probed long enough to transmit nonpersistent viruses; if a plant is found to be unsuitable, the aphid undertakes a new flight to visit other plants. Thus, probing on unsuitable host plants which were also immune or resistant to virus infection would reduce the inoculative potential of the vector. One method of reducing the number of infections produced by viruliferous, alate aphids is to plant a given percentage of immune plants along with the susceptible species in the field that must be protected; the two species can be arranged in alternate rows. Furthermore, it is possible to choose immune species which, thanks to their shape (Broadbent, 1969; Dunn, 1969) or color (Muller, 1964), may provide additional protection to the susceptible plants by acting as windbreaks or by producing a masking effect, respectively. Such a phenomenon might explain, at least in part, the multiline effect.

6.2.4 Cultivated Plants

Virus diseases being incurable at the field level, the best way to control them is to avoide infection either as previously described, by preventing the introduction of the virus into a field, or by using selected varieties that are resistant to virus infection. The latter method is very efficient and widely used when available.

Several kinds of resistance have been described. Recently, Russell (1978), using a classification formerly proposed by Bjorling in 1966, distinguished six different kinds of resistance:

Immunity—It is generally the case when no infection is possible.

Resistance to virus infection—Plants are susceptible but show a propensity to escape from virus infection.

Resistance to virus spread–Plants shown an infection which is generally limited to a few cells surrounding the entry point of the virus, *i.e.*, hypersensitivity.

Resistance to virus multiplication–Only a low concentration of virus particles occurs in the infected plant.

Tolerance–There are several kinds of tolerance. Most commonly, there is either no apparent symptom on the plant or no yield decrease.

Resistance to the vector–Three kinds can be distinguished: no preference of the vector for the host plant; antibiosis or decreased vector growth and multiplication on the host, and host plant tolerance to mild infestations by the vector (Painter, 1951; Kennedy, 1976).

The aim of the selection is to introduce one or several of these resistances into a selected variety to protect it against the main virus or viruses which might infect it; and, as far as possible, these introduced resistances must remain effective for several years. Thus, the rational elaboration of such combined resistance requires close collaboration between plant breeders, plant virologists and vector entomologists: first, to detect the appropriate sources of resistance; second, to incorporate them into a suitable variety, third, to define the best ways of using the newly obtained variety.

The first stage, the identification of the genes for resistance, can be achieved by different methods (Nelson, 1973; Russel, 1978). One commonly used method is to observe infections within experimental plots under field conditions, the diseased plants being detected by indexing them on a range of differential host plants. Then, the accuracy of this search for genes for resistance can be enhanced by repeating the essays under varied geographic and climatic conditions, and by manipulating the sources of virus or the vector populations to standardize the inoculum pressure.

Another method, which is also often practiced, consists in artificially inoculating seedlings grown under controlled conditions. In this way, a collection of strains or isolates covering the variability of the pathogen can be used as inoculum, alone or mixed, and the inoculum pressure can be defined very accurately.

When the search is oriented towards the detection of partial resistance, it may be necessary to design elaborate tests to clearly pull out the sources of such resistance. Such tests have been used by Pochard (1977a, 1979b) in breeding pepper for resistance to viruses: they include inoculations with atypical strains or the use of a stress such as plant decapitation before virus inoculation. In some cases, several distinct partial resistances may appear in mix; then a study on the mechanism of resistance may be the only way of differentiating them. For instance, *Cucumis melo* PI 161375 was shown to contain some resistance to cucumber mosaic virus (Risser *et al.*, 1977), but a more thorough study has shown that it contains at least two kinds of resistance: a vertical resistance against some strains of the virus and a resistance against the vector, *Aphis gossypii* Glover (Lecoq *et al.*, 1979).

The second stage, introduction of virus resistance into a given variety, also involves a number of difficulties that must be overcome. One is to design a reliable and easy method of testing for and recognizing resistant plants in the progeny of a hybridization. Such tests may be easily planned and carried out when the selection concerns major genes; generally, an inoculation of the seedlings with a given virus strain is convenient. They are less easy when several partial resistances must be introduced; sometimes, recognition and understanding of the different resistance mechanisms that are involved is necessary.

Another sometimes encountered difficulty is a more or less narrow linkage between the resistance to be introduced into the commercial variety and undesirable genes affecting the growth of the plant, the quality of the crop, or resistance to other pathogens. For example, resistance of tomatoes to tobacco mosaic is sometimes associated with poor fertility (Laterrot, 1973).

Nevertheless, the use of resistant varieties to control virus diseases has improved the production of many cultivated species (Nelson, 1973; Russell, 1978). In some cases, tremendous gains have been realized. For example, since the beginning of the twentieth century, increases in sugar-beet production in the U.S.A. have mainly resulted from the uses of varieties resistant to curly top virus and beet yellows virus (Duffus, 1977).

However, the use of resistant varieties presents some limitations. The main risk is the occurrence of a new pathotype able to infect the resistant variety in a new area. This was sometimes observed when tomato hybrids protected against TMV by the single dominant gene TM-1 were cultivated in England (Pelham *et al.*, 1970). Pelham (1968) was able to isolate the TMV pathotypes that can infect tomato varieties having the different genes for resistance. The same situation is now being encountered in Europe, with respect to pepper varieties resistant to TMV (Rast, 1977).

What are the factors which define the efficiency of a resistant variety at the field level? This efficiency seems to depend on the interaction between intrinsic characteristics of the type of resistance used, other specific characteristics of the virus which is to be controlled, such as its capacity for variation and adaptability, and the type of agricultural management, e.g. crop rotations or monocultures. The first point is linked to the efficiency limits shown by the resistance genes introduced into the commercial varieties. The efficiency of a given resistance, at the field level, may be altered through the action of several factors. The best known factors are high temperatures which allow a virus to overcome resistance of the hypersensitivity type (for example, peppers or tobaccos that are resistant to tobacco mosaic virus), large numbers of viruliferous vectors which may break down the type of resistance described as a "tendency to escape virus infection," and growing conditions (e.g., the availability of water, nutrients, and nitrogen) which might alter some types of resistance. The boundaries of a new resistant variety's efficiency can be defined by preliminary experiments performed under controlled conditions for all the aforementioned factors. This is not the case for viral properties and agricultural management practices that might alter resistance efficiency.

The appearance of new pathotypes adapted to a new resistant variety seems to be less common for viruses than for fungi, as pointed out by Russel (1978). But some examples can be found which show either an accumulation of adapted strains following the cultivation of a new resistant variety (Pelham *et al*., 1970, have pointed out such a case in a field planted with a tomato variety resistant to tobacco mosaic virus) or even the pre-existence of such strains before the cultivation of the resistant variety (Leroux *et al*., 1979, demonstrated this for cucumber mosaic virus in melons).

The capacity for variation seems to be an important characteristic in the process of adaptation to a new resistant variety, and, beyond mutations, other mechanisms may be effective in the appearance of new biological properties: pseudorecombination in the case of a divided genome (Marchoux, 1976) or the addition of a supernumerary RNA (Kaper *et al*., 1976; Kaper and Waterworth, 1977). Thus the rational use of a resistant variety is dependent on knowledge of the capacity of the virus for variability, a phenomenon which can be studied either by extensive observation and experimentation on viral population ecology or on artificial mutagenesis, but not without risk in the latter case.

Finally, the system of cultivation might decrease or increase the length of time during which a resistant variety is effective. Cultivation of the same variety, or varieties that possess the same resistance genes, without breaking either in space or in time, presents well known dangers (Russel, 1978). Hence, control methods have to research the use of distinct genes of resistance among different plant varieties and the practice of culture rotations.

6.3 CONTROL OF VIRUS DISEASES AT THE REGIONAL LEVEL

The word region is used here to mean an area which is relatively homogeneous with respect to agricultural and climatic conditions. At this level, it is possible to propose some methods of control which may be inefficient or too expensive on the field or farm level.

6.3.1 Actions to Avoid Virus Spread

6.3.1.1 Eradication. Theoretically, a virus which spreads slowly, has a limited host range, and is not propagative in its vectors, can be eradicated from an area by careful elimination of infected plants, if no new inoculum comes from outside the system. Practically, such an operation is costly and hazardous; there are few reports of successful eradication.

For instance, in recent years, a strain of plum pox virus has spread across the orchards of Western Europe (Sutic, 1971). The involved strain appeared to be mainly disseminated by the planting of diseased material, and efforts were made to eradicate the virus from nurseries and orchards. These efforts were finally halted and now the main control method consists of carefully surveying the nurseries to reduce the risk of planting new orchards with infected material (Adams, 1978).

In the case of annual crops, eradication can be tried, if a possibility of breaking the annual cycle of the virus appears. Such a case was reported by Chamberlain and Bayliss (1948) who were able to eliminate onion yellow dwarf virus by discarding the biannual crops of onion seed plants from areas where market onions are produced.

6.3.1.2 Control Based on Knowledge of Vector Dynamics. If no way has been found to control a given virus disease at the field level, it is sometimes possible to research other new control methods based on knowledge of vector dynamics during the period of crop susceptibility. Such a study needs to be effected on a regional scale because the range of most insect vectors is greater than field-size, and specially primary infections can sometimes be initiated by incoming, long-flying vectors.

For many years, trapping of vector populations using yellow pans or suction traps has been undertaken in England (Heathcote *et al*., 1969; Taylor, 1974) and is now tried in other countries. The traps are placed in an array across the country. Suction traps are designed to catch flying insects at the height of 12.5 meters. The main trapped species which belong to Noctuidae and Aphididae are identified daily and data from the various collection stations are collated.

Such data can furnish valuable information about the dynamics of potential vector populations. Furthermore, it can be hoped that the increasing sensitivities of serological techniques such as ELISA will permit detection of virus even in a single aphid (Gera *et al*., 1978; Clarke *et al*., 1980). By combining the two techniques (trapping and serology), periods of viruliferous vectors flight activity and frequency can be measured and predicted. Such data can be used to enhance several control methods. Combined with field observations, they can enhance the accuracy of the warning programs used to protect high value crops such as seed potatoes. For example, the times for insecticide sprays and burning of the leaves could be defined with far greater accuracy.

Data on vector dynamics can also contribute to the establishment of more accurate prediction models than those previously based on the relationship between climatic conditions and the recording of field outbreaks (Hurst, 1965; Watson, 1966).

6.3.1.3 Control Based on Optimal Management of Time and Space. Even before the production of improved models, it has been found that some modifications of cultural conditions were effective in preventing severe virus outbreaks at the regional level.

Sowing earlier, sometimes with the help of polyethylene covers, can result in increased yields by helping to make crops less susceptible to virus infection during later periods of vector flight activity (Johnstone and Rapley, 1979). A'Brook (1974), on the other hand, found that, under local conditions, delaying wheat or barley sowing towards the end of autumn resulted in crops less frequently infected by barley yellow dwarf virus because the seeds begin to grow after the last flights of vectors have occurred.

It can be thought also that the extension of an epidemic to the whole of a given region can be avoided if susceptible fields are arranged in such a way that the distance between any two fields is greater than the flight range of viruliferous vectors. Such isolation techniques are satisfactorily used in the production of many healthy stocks of planting material. The safe distance between fields that are susceptible to a particular virus can be deducted from the study of virus infection gradients. As clearly reviewed by Thresh (1976), this emphasizes the need for ecological studies of virus diseases to define regional methods of control.

6.3.2 Survey of Changes in the Surrounding Conditions

Changes in agricultural practices or in the surrounding environment may indirectly produce changes in the evolution of virus epidemics. Such an effect is generally unexpected. For instance, Rose (1978) has pointed out that the new irrigation of large areas in Africa has resulted in conspicuous increases in the population levels of some species of *Cicadulina* (leafhoppers) vectors of maize streak virus.

In other cases, the deleterious effect is almost predictable. In southeastern France, for example, the development of winter vegetable growing in greenhouses has allowed an aleyrodid, *Trialeurodes vaporarium* Westw., to survive in the greenhouses during the cold season and spread to vegetable fields in the summer; it can be feared that another aleyrodid, *Bemisia tabaci* which is an efficient vector of viruses will behave similarly. At that time the risk of introduction of whitefly-borne viruses from the eastern Mediterranean cannot be ignored.

6.4 CONTROL OF THE VIRUS DISEASES AT THE INTERNATIONAL LEVEL

The growing development of international exchanges involves an important risk of worldwide spread of a great number of viruses, vectors, and host weeds which are now only present in limited parts of the world. Even in the case of already vastly spread viruses, these exchanges can contribute to the accumulation of several strains of these viruses in every country where ecological conditions are suitable. Furthermore, there is also the possibility of new strains occurring by the mechanism of pseudo-recombination, in the case of viruses having multipartite genomes. The reality of such risks is difficult to appreciate because, generally, it is not possible to determine if a newly identified virus in a country was native and had remained unidentified until then or if its presence is due to a recent introduction.

Seed-borne viruses are of particular interest because, if careful border controls are not enforced, large quantities of viruses can be introduced in contaminated seed lots to as yet uninfected countries. When the seeds are planted, the virus carried with them can find optimal conditions to provoke epidemics if a suitable vector is available. The same thing can be observed in the case of the fruit tree viruses: for instance, the plum pox which has invaded Western Europe during recent years was probably carried in contaminated planting material (Sutic, 1971). Even if the suitable vector is not available in the country where the virus has been introduced, there is the possibility that virus will be maintained in self-propagated or perennial material till a strain adapted to a locally available vector occurs. Bar-Joseph (1978) and Raccah *et al.* (1978) have described the case of tristeza virus which remained in Israel for 30 years after its introduction without dissemination before a strain appeared which is efficiently transmitted by *Aphis gossypii* which is very common.

The prevention of the disease introduction sometimes seems a hopeless fight, but its importance and its need cannot be denied. The techniques used for frontier health control of plants and seeds are well defined and have been extensively described by Neergaard (1979). The improvement of such measures appears rather difficult. Of course, frontier inspection could encompass more viruses, and the detection techniques used can be improved (Phatak, 1974). But, it would also be necessary to reduce the importance of exchanges which escape any control, such as the traditional trades between islands of an archipelago or between developing countries. And what can be done regarding the fad, enhanced by the development of overseas tourism, of growing in the home garden more and more exotic plants collected as "souvenirs" in foreign countries?

6.5 CONCLUSION

The control of virus diseases is preeminantly the aim of applied virology. Production of healthy seed is effective and generally does not present any insurmountable technical problem. But programs for healthy seed production require constant attention from pathologists. Control of virus diseases by the use of resistant varieties also appears very efficient. To date, the occurrence of adapted virus strains appears less frequent than observed in the case of fungus diseases. Unfortunately, resistant varieties are not yet available for every cultivated plant, and breeding for resistance is a long term job even if sources of resistance have been identified. There are also numerous other approaches to control which have been recognized as rather efficient in some situations and which give satisfying results in limited cases.

It can be expected that new problems will occur in the future due to the development of international trade which will scatter viruses and virus strains into every area where conditions suitable to their establishment are encountered. Furthermore, some new cultural practices may enhance the probabilities of a new virus, a new host plant, or a new vector becoming established in a given area.

As pointed out by Bos (1976) control of virus diseases in a country can be enhanced by maintaining suitable virus laboratories. The staffs of such facilities must be able to make careful identifications of the viruses involved in disease and also to conduct thorough investigations of the ecology of the identified viruses (identification of virus populations, weed hosts, and vectors, as well as their interactions) in order to rationally choose the best ways of control to be applied.

6.6 ACKNOWLEDGMENTS

The authors are specially grateful to Dr. K. F. Harris and M. C. Massali for correcting the English text.

6.7 REFERENCES

A'Brook, J. (1968). The effect of plant spacing on the number of aphids trapped over the groundnut crop. *Ann. Appl. Biol.* **61**:289-294.

A'Brook, J. (1973a). Observations on different methods of aphid trapping. *Ann. Appl. Biol.* **74**:263-277.

A'Brook, J. (1973b). The effect of plant spacing on the number of aphids trapped over cocksfoot and kale crops. *Ann. Appl. Biol.* **74**:279-285.

A'Brook, J. (1974). Barley yellow dwarf virus: what sort of a problem. *Ann. Appl. Biol.* **77**:92-96.

Adams, A. N. (1978). The incidence of plum pox virus in England and its control in orchards. *In* "Plant Disease Epidemiology" (P. R. Scott and A. Baimbridge, eds.), pp. 213-220. Blackwell Scientific, Oxford.

Agrios, G. N. (1969). "Plant Pathology." Academic Press, New York.

Anonymous, (1979). CTIFL–Documents. *C. R. Activite CTIFL* **61**:35.

Bar-Joseph, M. (1978). Cross protection incompleteness: A possible cause for natural spread of citrus tristeza virus after a prolonged lag period in Israel. *Phytopathology* **68**:1110-1111.

Bawden, F. C. (1964). "Plant Viruses and Virus Diseases," 4th edition. Ronald Press, New York.

Bennett, C. W. (1969). Seed transmission of plant viruses. *Adv. Virus Res.* **14**:221-261.

Berger, R. D. (1977). Application of epidemiological principles to achieve plant disease control. *Annu. Rev. Phytopathol.* **15**:165-183.

Bokx, J.A. de, and Cuperus, C. (1978). Effect of spraying with mineral oil on seed potatoes. *Med. Fac. Landbouww. Rijks. Univ. Gent.* **43**:1027-1035.

Bonnemaison, L. (1961). Protection des betteraves-racines et des portegraines contre la jaunisse basée sur la lutte contre les vecteurs. *Ann. Epiphyties* **12**:155-218.

Bos, L. (1976). Applied plant virus research: an analysis and a scheme of organization. *Neth J. Plant Pathol.* **82**:24-38.

Bos, L., Huijberts, N., Huttinga, H., and Maat, D. Z. (1978). Leek yellow stripe virus and its relationships to onion yellow dwarf virus; characterization, ecology and possible control. *Neth. J. Plant Pathol.* **84**:185-204.

Bosschaert, L., Esparza-Duque, J., and Verhoyen, M. (1974). Etude systematique des aphides captures en pieges jaunes de Moericke en relation avec l'epidemiologie et l'elimination du Virus de la mosaique du celeri. *Med. Fac. Landbouw-Wetenschappen Gent* **39**: 1229-1238.

Bowers, W. S., Nault, L. R., Webb, R. E., and Dutky, S. R. (1972). Aphid alarm pheromone: isolation, identification, synthesis. *Science* **177**:1121-1122.

Broadbent, L. (1964). Control of plant virus diseases. *In* "Plant Virology" (M. K. Corbett and H. D. Sisler, eds.), pp. 330-364. Univ. of Florida Press, Gainesville.

Broadbent, L. (1969). Disease control through vector control. *In* "Viruses, Vectors and Vegetation" (K. Maramorosch, ed.), pp. 593-630. Wiley (Interscience), New York.

Broadbent, L. (1976). Epidemiology and control of tomato mosaic virus. *Annu. Rev. Phytopathol.* **14**:75-96.

Bruckart, W. L., and Lorbeer, J. W. (1976). Cucumber mosaic virus in weed hosts near commercial fields of lettuce and celery. *Phytopathology* **66**:253-259.

Carter, W. (1973). "Insects in Relation to Plant Disease."Wiley (Interscience), New York.

Clarke, R. G., Converse, R. H., and Kojima, M. (1980). Enzyme-linked immunosorbent assay to detect potato leafroll virus in potato tubers and viruliferous aphids. *Plant Disease* **64**:43-45.

Chamberlain, E. E. and Bayliss, G. T. S. (1948). Onion yellow dwarf. Successful eradication. *N. Zeal. J. Sci. Technol.* **A29**:300-301.

Cohen, S., and Marco, S. (1973). Reducing the spread of aphid-transmitted viruses in peppers by trapping the aphids on sticky yellow polyethylene sheets. *Phytopathology* **63**: 1207-1209.

Cohen, S., and Melamed-Madjar, V. (1974). Prevention of the spread of tomato yellow leaf curl virus transmitted by *Bemisia tabaci* (Gennadius) (Homoptera: Aleyrodidae) in Israel. *Bull. Ent. Res.* **64**:193-197.

Cohen, S., and Melamed-Madjar, V. (1978). Prevention by soil mulching of the spread of tomato yellow leaf curl virus transmitted by *Bemisia tabaci* Genn. (Hemiptera: Aleyrodidae) in Israel. *Bull. Ent. Res.* **68**:465-470.

Costa, C. L. (1972). Controle de molestias de virus de plantas con superficies reflectivas repelentes ao vector. *Rev. Peruana Entomol.* **15**:135-139.

Costa, C. L. and Lewis, T. (1967). The relationship between the size of yellow water traps and catches of aphids. *Ent. Exp. Appl.* **10**:485-487.

Douine, L., Quiot, J. B., Marchoux, G., and Archange, P. (1979). Recensement des espèces végétales sensibles au virus de la mosaique du concombre. Étude bibliographique. *Ann. Phytopathol.* **11**:439-475.

Duffus, J. E. (1971). Role of weeds in the incidence of virus diseases. *Annu. Rev. Phytopathol.* **9**:319-340.

Duffus, J. E. (1977). Aphids, viruses and the yellow plague. *In* "Aphids as Virus Vectors" (K. F. Harris and K. Maramorosch, eds.), pp. 361-383. Academic Press, New York.

Dunn, J. A. (1969). The colonization by *Cavariella aegopodii* Scop. carrot plants of different sizes. *Ann. Appl. Biol.* **63**:318-324.

Eastop, V. F. (1955). Selection of aphid species by different kinds of insect-traps. *Nature* 76-936.

Edwards, L. J., Siddall, J. B., Dunham, L. L., Uden, P., and Kislow, C. J. (1973). Trans-B-farnesene, alarm pheromone of the green peach aphid, *Myzus persicae* (Sulzer). *Nature* **242**:126-127.

Edwardson, J. R. (1974). "Host Ranges of Viruses in the PVY-group." *Florida Agricultural Experiment Station Monogr. Ser.*

Esparza-Duque, J., and Verhoyen, M. (1975). Réduction des infections de virus de la mosaique du céleri par des pulvérisations d'huile et des surfaces réfléchissantes répulsives pour les vecteurs. *Med. Fac. Landbouww. Rijks. Univ. Gent* **41**:779-789.

Farrell, J. A. K. (1976). Effects of groundnut crop density on the population dynamics of *Aphis craccivora* Koch in Malawi. *Bull. Entomol. Res.* **66**:317-329.

Fletcher, J. T., George, A. J., and Green, D. E. (1969). Cucumber green mottle mosaic virus its effect on yield and its control in the Lea Valley England. *Plant Pathol.* **18**:16-22.

Frazer, B. D. (1977). Plant virus epidemiology and computer simulation of aphid populations. *In* "Aphids as Virus Vectors." (K. F. Harris and K. Maramorosch, eds.), pp. 413-434. Academic Press, New York.

Gera, A., Lobenstein, G., and Raccah, B. (1978). Detection of cucumber mosaic virus in viruliferous aphids by enzyme linked immunosorbent assay (ELISA). *Plant Dis. Reptr.* **62**:189-192.

Gibbs, A., and Harrison, B. (1976). "Plant Virology. The Principles." Edward Arnold, London.

Gupta, B. M. (1977). Inhibition of plant virus infections by antiviral agents. *In* "Aphids as Virus Vectors" (K. E. Harris and K. Maramorosch, eds.), pp. 455-472. Academic Press, New York.

Guyot, G. (1970). Les effets des brise-vents en matiére plastique. *Proc. 4th Coll. In. Plastiques Agric. Paris, Juin 1970.* pp. 293-303.

Hani, A. (1971). Zur Epidemiologie des Gurkenmosaikvirus im Tessin. *Phytopathol. Z.* **72**: 115-144.

Harris, K. F., and Maramorosch, K., eds. (1977). "Aphids as Virus Vectors." Academic Press, New York.

Harris, K. F., and Maramorosch, K, eds. (1980). "Vectors of Plant Pathogens." Academic Press, New York.

Harris, K. F., and Maramorosch, K., eds. (1981). "Pathogens, Vectors, and Plant Diseases: Approaches to Control." Academic Press, New York.

Havranek, P. H., and Laska, P. (1972). The course of the migration of cucumber mosaic virus in a vegetable locality. *Sci. Agric. Bohemoslov* **3**:213-225.

Heathcote, G. D., Palmer, J. M. P., and Taylor, L. R. (1969). Sampling for aphids by traps and by crop inspection. *Ann. Appl. Biol.* **63**:133-166.

Hill, S. A. (1978). Current virus infection of seed potatoes in Britain. *In* "Plant Diseases Epidemiology" (P. R. Scott and A. Baimbridge, eds.), pp. 229-234. Blackwell, Oxford.

Hille Ris Lambers, D., and Schepers, A. (1978). The effect of trans-B-farnesene, used as a repellent against landing aphid alatae in seed potato growing. *Potato Res.* **21**:23-26.

Hollings, M. (1965). Disease control through virus-free stock. *Annu. Rev. Phytopathol.* **3**: 367-396.

Howell, W. E., and Mink, G. I. (1977). The role of weed hosts, volunteer carrots, and overlapping growing seasons in the epidemiology of carrot thin leaf and carrot motley dwarf viruses in central Washington. *Plant Dis. Reptr.* **61**:217-222.

Hurst, G. W. (1965). Forecasting the severity of sugar beet yellows. *Plant Pathol.* **14**:47-53.

Jadot, R. (1976). Aspects desépidémies de jaunisse et de mosaïque de la betterave. II source d'infection, allure des épidémies, lutte. *Rev. Agri.* **4**:29-62.

Johnson, C. G. (1969). "Migration and Dispersal of Insects by Flight." Methuen, London.

Johnstone, G. R., and Rapley, P. E. L. (1979). The effect of time of sowing on the incidence of subterranean clover red leaf virus infection in broad-bean *(Vicia faba). Ann. Appl. Biol.* **91**:345-351.

Kaper, J. M., Tousignant, M. E., and Lot, H. (1976). A low molecular weight replicating RNA associated with a divided genome plant virus: Defective or satellite RNA? *Biochem. Biophys. Res. Commun.* **72**:1237.

Kaper, J. M., and Waterworth, H. E. (1977). Cucumber mosaic virus associated RNA 5: causal agent for tomato necrosis. *Science* **196**:429-431.

Karl E., Lehmann, W., and Proeseler, G. (1972). Untersuchungen zur Dauer der Infektiosität hungern der Blattläuse verschiedener Arten nach Akquisition nichtpersistenter Viren. *Tag. Berlin Akad. Landwizksch. Wiss.* **121**:49-56.

Kemp, W. G. (1978). Mulches protect peppers from viruses. *Can. Agri.* **23**:22-24.

Kennedy, G. G. (1976). Host plant resistance and the spread of plant viruses. *Environ. Entomol.* **5**:827-832.

Kimble, K. A., Grogan, R. G., Greathead, A. S., Paulus, A. O., and House, J. K. (1975). Development, application and comparisons of methods for indexing lettuce seed for mosaic virus in California. *Plant Dis. Reptr.* **59**:461-464.

Knoke, J. K., Anderson, R. J., and Louis, R. (1976). Virus disease epiphytology: developing field test for disease resistance in maize. *Proc. Intern. Maize Virus Dis. Coll. Workshop.* Ohio Agri. Res. Dev. Center, Wooster.

Kring, J. B. (1972). Flight behavior of aphids. *Ann. Rev. Entomol.* **17**:461-492.

Laterrot, H. (1973). Résistance de la tomate au virus de la mosaique du tabac. Difficultés rencontrées pour la sélection de variétés résistantes. *Ann. Amelior. Plantes* **23**:287-313.

Leclant, F. (1968). Connaissances actuelles sur les pucerons dans leurs relations avec les maladies á virus des plantes. *Ann Epiphyties* **19**:455-482.

Lecoq, H., Cohen, S., Pitrat, M., and Labonne, G. (1979). Resistance to cucumber mosaic virus transmission by aphids in *Cucumis melo. Phytopathology* **69**:1223-1225.

Lehmann, W., Claus, S., and Karl, E. (1976). Der Einfluss verschiedener Pflantzenarten und ihrer chemischen Behandlung auf die Anzahl der Probesaugstiche und die Dauer des ersten Einstichs von Aphiden. *Arch. Phytopathol. Pflanzen.* **12**:345-354.

Lehmann, W., Claus, S., and Karl, E. (1977). Der Einfluss verschiedener Pflanzenarten und ihrer chemischen Behandlung auf die Laufaktivität von Aphiden. *Arch. Phytopathol. Pflanz.* **13**:137-144.

Leroux, J. P., Quiot, J. B., Lecoq, H., and Pitrat, M. (1979). Mise en évidence et répartition dans le Sud-Est de la France d'un pathotype particulier du virus de la mosaique du concombre. *Ann. Phytopathol.* **11**:431-438.

Lewis, T. (1966). Artificial windbreaks and the distribution of turnip mild yellows virus and *Scaptomyza apicalis* (Diptera) in a turnip crop. *Ann. Appl. Biol.* **58**:371-376.

Lewis, T., and Dibley, G. C. (1970). Air movement near windbreaks and a hypothesis of the mechanism of the accumulation of airborne insects. *Ann. Appl. Biol.* **66**:477-484.

Lister, R. M., and Murant, A. F. (1967). Seed transmission of nematode borne viruses. *Ann. Appl. Biol.* **59**:49-62.

Lockhart, B. E. L., and Fischer, H. U. (1974). Chronic infection by seed-borne Bean Common Mosaic Virus in Morocco. *Plant Dis. Reptr.* **58**:307-308.

Loebenstein, G., Alper, M., Levy, S., Palevitch, D., and Menagen, E. (1975). Protecting peppers from aphid-borne viruses with aluminium foil or plastic mulch. *Phytoparasitica* **3**: 43-53.

Louis, R. (1968). Epiphytology of onion yellow dwarf. PHD thesis. Cornell University, Ithaca, New York.

Marchoux, G. (1975). Propriétés biologiques et génétiques des A.R.N. du virus de la mosaique du concombre. Thése Dr. Sc. Fac. Marseille.

Maramorosch, K., and Harris, K., eds. (1979). "Leafhopper Vectors and Plant Disease Agents." Academic Press, New York.

Maramorosch, K. and Harris, K., eds. (1981). "Plant Diseases and Vectors: Ecology and Epidemiology." Academic Press, New York.

Marrou, J., and Messiaen, C. M. (1967). The *Chenopodium quinoa* test: a critical method for detecting seed transmission of lettuce mosaic virus. *Proc. Int. Seed Test. Assoc.* **32**:49-57.

Marrou, J., Quiot, J. B., Duteil, M., Labonne, G., Leclant, F., and Renoust, M. (1979). Ecologie et épidémiologie du virus de la mosaique du concombre dans le Sud-Est de la France. VIII Influence des brise-vents et des cultures environnantes sur la dissémination du virus de la mosaïque du concombre. *Ann. Phytopathol.* **11**:375-392.

Matthews, R. E. F. (1970). "Plant Virology." Academic Press, New York.

Messiaen, C. M., Maison, P., and Migliori, A. (1963). Le virus I du Concombre dans le Sud-Est de la France. *Phytopathol. Mediter.* **II**:251-260.

Messiaen, C. M., Marrou, J., and Leroux, J. P. (1966). Sélection sanitaire chez les allium cultivés et reproduits par voie végétative. *Actes Congr. U.P.M. Bari,* pp. 434-441.

Morel, G., and Martin, C. (1952). Guérison de dahlias atteints d'une maladie ä virus. *C. R. Acad. Sci. Paris* **235**.1324-25.

Migliori, A. (1973). Etude d'une méthode biologique: la prémunition des tomates contre le virus de la mosaique du tabac. Thése d'Université Montpellier USTL.

Moericke, V. (1967). Host plant specific colour behaviour by *Hyalopterus pruni. Entomol. Exp. Appl.* **12**:524-34.

Muller, H. J. (1964). Über die Anflugdichte von Aphiden auf Fardige Salatpflanzen. *Entomol. Exp. Appl.* 7:85-104.

Nault, L. R., and Montgomery, M. E., (1977). Aphid pheromones. *In* "Aphids as Virus Vectors" (K. F. Harris and K. Maramorosch, eds.), pp. 527-545. Academic Press, New York.

Neergaard, P. (1979). "Seed Pathology." Macmillan, New York.

Nelson, R. R. (1973). "Breeding Plants for Disease Resistance." Pennsylvania State Univ. Press, Philadelphia.

Nitzany, F. E., Geisenberg, H., Koch, B. (1964). Tests for the protection of cucumbers from a white-fly borne virus. *Phytopathology* **54**:1059-1061.

Nyland, G., and Goheen, A. C. (1969). Heat therapy of virus disease of perennial plants. *Ann. Rev. Phytopathol.* 7:331-354.

Painter, R. H. (1951). "Insect Resistance in Crop Plants." Univ. Press of Kansas, Kansas City.

Pelet, F. (1965). Dosage du virus de la mosaique de la laitue par indexage de la graine sur *Chenopodium quinoa. Rev. Hort. Suisse* **38**:7-10.

Pelham, J. (1968). Tomato and tomato mosaic; a dynamic relationship. *Abst. 1st Congr. Intern. Plant. Pathol.* London, 1968.

Pelham, J., Fletcher, J. T., and Hawkins, J. H. (1970). The establishment of a new strain of tobacco mosaic virus resulting from the use of resistant varieties of tomato. *Ann. Appl. Biol.* **65**:293-297.

Phatak, H. C. (1974). Seed-borne plant viruses–identification and diagnosis in seed health testing. *Seed Sci. Technol.* **2**:3-155.

Plumb, R. T. (1976). Barley yellow dwarf virus in aphids caught in suction traps. 1969-1973. *Ann. Appl. Biol.* **83**:53-59.

Pochard, E. (1977a). Méthodes pour l'étude de la résistance partielle au virus de la mosaique du concombre chez le piment. *3rd Meeting Eucarpia "Capsicum 1977," Avignon-Montfavet, 5-8 Juillet, 1977,* pp. 93-104.

Pochard, E. (1977b). Etude de la résistance aux souches Européennes du virus de la pomme de terre (PVY) chez le piment. *3rd meeting Eucarpia "Capsicum 1977, Avignon-Montfavet, 5-8 Juillet, 1977,* pp. 109-118.

Price, W. C. (1940). Comparative host range of six plant viruses. *Am. J. Bot.* **27**:530-541.

Proeseler, G., and Weidling, H. (1975). The retention period of strains of PVY in several aphid species, and the effect of temperature. *Arch. Phytopathol. Pflanzen.* **11**:335-345.

Quiot, J. B., Marchoux, G., Douine, L., and Vigoroux, A. (1979a). Ecologie et Epidémiologie du Virus de al Mosaique du Concombre dans le Sud-Est de la France. V-Rôle des espéces spontanées dans la conservation du virus. *Ann. Phytopathol.* **11**:325-348.

Quiot, J. B., Devergne, J. C., Marchoux, G., Cardin, L., and Douine, L. (1979b). Ecologie et epidémiologie du virus de la mosaique du concombre dans le Sud-Est de la France. VI–Conservation de deux types de populations virales dans les plantes sauvages. *Ann. Phytopathol.* **11**:349-358.

Rabasse, J. M., Brunel, E., Delecolle, R., and Rouze-Jouan, J. (1976). Influence de la dimension des piéges á eau colorés en jaune sur les captures d'aphides dans une culture de carottes. *Ann. Zool. Ecol. Anim.* **8**:39-52.

Raccah, B., Bar-Joseph, M., and Loebenstein, G. (1978). The role of aphid vectors and variation in virus isolates in the epidemiology of tristeza disease. *In* "Plant Disease Epidemiology" (P. R. Scott and A. Baimbridge, eds.), pp. 221-228. Blackwell, Oxford.

Rast, A. T. B. (1972). M II–16, an artificial symptomless mutant of tobacco mosaic virus for seedling inoculation of tomato crops. *Neth. J. Plant Pathol.* **78**:110-112.

Rast, A. T. B. (1977). Introductory remarks on strains of TMV infecting peppers in the Netherlands. In Capsicum 77 C. R. *3rd Meeting Eucarpia "Capsicum 1977," Avignon-Montfavet, 5-8 Juillet, 1977*, pp. 83-84.

Risser, G., Pitrat, M., and Rode, J. C. (1977). Etude de la résistance du melon *(Cucumis melo L.).* au virus de la mosaique du concombre. *Ann. Amelior. Plantes.* **27**:509-522.

Robert, Y. (1976). Distribution spatio-temporelle des pucerons de la pomme de terre en zone bocagére et en zone arasée. *C. R. Table Ronde C.N.R.S. "Ecosystemes Bocagers," Rennes*, pp. 419-425.

Robert, Y., and Maury, Y. (1970). Capacités vectrices comparées de plusieurs souches de *Myzus persicae* Sulz., *Aulacorthum solani* Kltb. et *Macrosiphum euphorbiae* Thomas dans l'étude de la transmission de l'enroulement de la Pomme de Terre. *Potato Res.* **13**:199-209.

Robert, Y., Rabasse, J. M., and Rouze-Jouan, J. (1974). Sur l'utilisation des piéges jaunes pour la capture de pucerons en culture de pomme de terre. I. Influence de la hauteur du piégeage. *Ann. Zool. Ecol. Anim.* **6**:349-372.

Robert, Y., and Rouze-Jouan, J. (1976). Neuf ans de piégeages de pucerons des céréales: *Acyrthosiphon (Metopolophium) dirhodum* WLK, *A. (M.). festucae* WLK, *Macrosiphum (Sitobion) avenae* F., *M. (S.) fragariae* et *Rhopalosiphum padi* L. en Bretagne. *Rev. Zool. Agr. Pathol. Veg.* **75**:67-80.

Robertson, R. S., and Klostermeyer, E. C. (1958). Aphid populations on field beans in Washington. *J. Econ. Entomol.* **51**: 178-181.

Roca, W. M., Espinoza, N. O., Roca, M. R., and Bryan, J. E. (1978). A tissue culture method for the rapid propagation of potatoes. *Ann. Potato J.* **55**:691-701.

Rose, D. J. W. (1978). Epidemiology of Maize streak disease. *Ann. Rev. Entomol.* **23**:259-282.

Russel, G. E. (1978). "Plant Breeding for Pest and Disease Resistance." Butterworths, London.

Ryder, E. J., and Johnson, A. S. (1974). A method for indexing lettuce seeds for seedborne lettuce mosaic virus by airstream separation of light from heavy seeds. *Plant Dis. Reptr.* **58**:1037-1039.

Shanks, C. H., and Chapman, R. K. (1965). The effects of insecticides on the behavior of *Myzus persicae* Sulzer and its transmission of potato virus Y. *J. Econ. Entomol.* **58**:79-83.

Simons, J. N. (1957). Effects of insecticides and physical barriers on field spread of pepper vein banding mosaic virus. *Phytopathology* **47**:139-145.

Simons, J. N. (1981). Use of oil sprays and reflective surfaces for control of insect-transmitted plant viruses. *In* "Pathogens, Vectors, and Plant Diseases: Approaches to Control" (K. F. Harris and K. Maromorosch, eds.), Chap. 5. Academic Press, New York.

Slack, S. A. (1980). Pathogen-free plants by meristem-tips culture. *Plant Dis.* **64**:14-18.

Smith, F. F., and Webb, R. E. (1969). Repelling aphids by reflective surfaces, a new approach to the control of insect-transmitted viruses. *In* "Viruses Vectors and Vegetation" (K. Maramorosch, ed.), pp. 631-640. Wiley (Interscience), New York.

Smith, J. G. (1969). Some effects of crops back ground on populations of aphids and their natural enemies on Brussels sprouts. *Ann. Appl. Biol.* **63**:326-329.

Smith, J. G. (1976). Influence of crop back ground on aphids and phytophagous insects on Brussels sprouts. *Ann. Appl. Biol.* **83**:1-13.

Stevens, R. B. (1960). Cultural practices in disease control. *In* "Plant Pathology. An Advanced Treatise." (J. G. Horsfall and A. E. Dimond, eds.), Vol. 3, pp. 357-430. Academic Press, New York.

Stockel, J. (1975). Les analogues de l'hormone juvénile ou régulateurs de croissance chez l'insecte. Possibilités d'applications dans la lutte contre les espéces nuisibles. *Ann. Zool. Ecol. Anim.* **7**:91-119.

Sutic, D. (1971). État des recherches sur le virus de la sharka. (8éme Symposium Européen sur les Maladies á Virus des Arbres Fruitiers, Bordeaux 1970). *Ann. Phytopathol.* 161-170.

Swenson, K. G. (1968). Role of Aphids in the Ecology of Plant Viruses. *Annu. Rev. Phytopathol.* **6**:351-374.

Taylor, L. R. (1974). Monitoring change in the distribution and abundance of insects. Rep. Rothamsted Exp. Stn., 1973, part 2.

Thornberry, H. H. (1966). "Index of Plant Virus Diseases." USDA Agriculture Handbook No. 307.
Thresh, J. M. (1974). Temporal patterns of virus spread. *Annu. Rev. Phytopathol.* **12**:111-128.
Thresh, J. M. (1976). Gradients of plant virus diseases. *Ann. Appl. Biol.* **82**:381-406.
Toba, H. H., Kishaba, A. N., Bohn, G. W., and Hield, H. (1977). Protecting muskmelons against aphid-borne viruses. *Phytopathology* **67**:1418-1423.
Tomlinson, J. A., and Carter, A. L. (1970). Studies on the seed transmission of cucumber mosaic virus in chick weed *(Stellaria media)* in relation to the ecology of the virus. *Ann. Appl. Biol.* **66**:381-386.
Tomlinson, J. A., and Walker, V. M. (1973). Further studies on seed transmission in the ecology of some aphid-transmitted viruses. *Ann. Appl. Biol.* **73**:293-298.
Tomlinson, J. A., Faithfull, E. M., and Ward, C. M. (1976). Chemical suppression of the symptoms of two virus diseases. *Ann. Appl. Biol.* **84**:31-41.
Turl, L. A. D. (1978). Epidemiology of potato aphids in 1975-1977 with regard to the incidence of potato leaf roll virus in Scotland. *In* "Plant Disease Epidemiology" (P. R. Scott and A. Baimbridge, eds.), pp. 235-242. Blackwell, Oxford.
van der Plank, J. E. (1975). "Principles of Plant Infection." Academic Press, New York.
Wientjens, W. H. J. M., Lakwijk, A. C., and van der Marel, T. (1973). Alarm pheromones of grain aphids. *Experientia* **29**:658-660.
Yang, S. L., and Zettler, F. W. (1975). Effects of alarm pheromone on aphid probing behavior and virus transmission efficiency. *Plant Dis. Reptr.* **59**:902-905.
Zink, F. W., Grogan, R. G., and Welch, J. E. (1956). The effect of the percentage of seed transmission upon subsequent spread of lettuce mosaic virus. *Phytopathology* **46**:662-664.
Zitter, T. A. (1977). Epidemiology of aphid-borne viruses. *In* "Aphids as Virus Vectors" (K. F. Harris and K. Maramorosch, eds.), pp. 385-412. Academic Press, New York.
Zitter, T. A., and Tsai, Jo (1977). Transmission of three potyviruses by the leafminer, *Liriomyza sativae* (Diptera: Agromyzidae). *Plant Dis. Reptr.* **61**:1025-1029.
Zitter, T. A., and Simmons, J. N. (1980). Management of Viruses by Alteration of Vector Efficiency and by Cultural Practices. *Ann. Rev. Phytopathol.* **18**:289-310.

Chapter 7

THE HOST AS A VECTOR: EXCLUSION AS A CONTROL

Robert P. Kahn

Plant Protection and Quarantine Programs
Animal and Plant Health Inspection Service
U.S. Department of Agriculture
Hyattsville, Maryland

7.1 INTRODUCTION

Legal Basis of Quarantines. Quarantines are regulations promulgated by governments to reduce the chances of man inadvertently introducing harmful pests and pathogens on articles imported from foreign areas. The legal basis of such quarantines is: 1) the legislation passed by national and sometimes state or provincial governments as acts, orders or regulations, or 2) enabling legislation that authorizes, usually the Minister of Agriculture, to issue rules, orders or directives.

ISBN 0-12-326440-5

Standards or guidelines for establishing regulations are often suggested by international plant protection organizations (European and Mediterranean Plant Protection Organization, 1977; Inter-African Phytosanitary Council, 1962, 1980), but these have no legal status. In addition, 75 countries have ratified or adhere to the International Plant Protection Convention (IPPC) of 1951 known as the "Rome Convention." The IPPD provides for the international model phytosanitary certificate which accompanies certain articles in international movement.

The degree to which articles moved by man are regulated should be determined by known or estimated pest risk associated with the movement of these articles from one country to another. If the regulations are matched to risk, then it appears that the regulations are based on biology. However, if political, social or economic influences prevail, then such regulations are not based on biology alone (Kahn, 1979). Frequently the actions taken under regulations are influenced by administrative factors or legal problems. The regulations and procedures under which countries import articles have been summarized for most countries (U.S. Department of Agriculture, 1933-1974; Ling, 1952, 1954).

The articles regulated by some or all countries include, but are not restricted to, the following: plants and plant parts or materials; agricultural and non-agricultural cargoes, soil, containers, dunnage, packing materials, plant growing media accompanying plants, baggage, mail, carriers or means of conveyance, and garbage produced on carriers or means of conveyance.

Plant quarantine regulations have the following features in common: 1) specify prohibitions; 2) grant exceptions to prohibitions for scientific or other specified purposes; 3) require import permits; 4) specify authorized ports or points of entry; 5) frequently require phytosanitary certificates (based on the IPPC model certificate) or certificates of origin, or both; 6) stipulate inspection on arrival; 7) prescribe or authorize treatment upon arrival to eliminate a risk, or deny entry to infested or infected articles, and 8) prescribe safeguards after entry such as quarantine, isolation, utilization, or consumption.

An example of a model national regulation has been published (Gram, 1960).

Plant Quarantine as a Control Measure. Plant quarantine is traditionally listed among major categories of control methods. For example, Agrios (1978) lists: 1) regulatory (quarantine), 2) cultural, 3) biological, 4) physical, and 5) chemical. Walker (1969) lists: 1) exclusion (quarantine), 2) eradication, 3) protection, and 4) resistance. A National Academy of Science subcommittee report (1968) on plant pathogens lists: 1) avoidance, 2) exclusion (includes quarantine), 3) eradication, 4) protection, 5) resistance, and 6) therapy.

Yet, in spite of the recognition of quarantines as a control measure, their effectiveness is controversial (McCubbin, 1946, 1956; Stakman and Harrar, 1957; Mathys and Baber, 1980; Walker, 1969; MacGregor, 1978). The argument stems from the fact that we cannot measure how many pests and pathogens would have entered a new region had quarantine measures not been in effect. Pests and pathogens do enter new regions, in spite of quarantines, as evidenced by reports

TABLE I. List of 125 Countries whose Plant Quarantine Regulations were Surveyed[1]

North America	Europe	Southwest Pacific	South America	Africa	Asia
Antigua	Austria	Australia	Argentina	Algeria	Bangladesh
Barbados	Belgium	Br. Solomon Is.	Bolivia	Benin	Cyprus
Belize	Bulgaria	Cook Is.	Brazil	Cameroon	Hong Kong
Bermuda	Czechoslovakia	Fiji	Chile	Egypt	India
Canada	Denmark	New Caledonia	Colombia	Gambia	Indonesia
Costa Rica	Finland	New Hebrides	Ecuador	Ghana	Iran
Cuba	France	New Zealand	Guyana	Guinea	Iraq
Dominica	Germany (GDR)	Papau and New Guinea	Paraguay	Ivory Coast	Israel
Dominican Republic	Germany (FRG)	Tahiti	Peru	Kenya	Japan
El Salvador	Greece	Tonga	Surinam	Liberia	Jordan
Guadeloupe	Guernsey		Uruguay	Libya	Korea
Guatemala	Hungary		Venezuela	Madagascar	Lebanon
Haiti	Iceland			Malawi	Malaysia
Honduras	Ireland			Mauritania	Pakistan
Jamaica	Italy			Morocco	Philippines
Martinique	Jersey			Mozambique	Saudi Arabia
Mexico	Luxembourg			Niger	Sri Lanka
Nicaragua	Malta			Nigeria	Syria
Panama	Netherlands			Senegal	Taiwan
St. Vincent	Norway			Seychelles	Thailand
Trinidad & Tobago	Poland			Sierra Leone	Turkey
United States	Portugal			South Africa	Viet Nam
	Romania			Southern Rhodesia	
	Soviet Union			Sudan	
	Spain			Tanzania	
	Sweden			Tunisia	
	Switzerland			Uganda	
	United Kingdom			Upper Volta	
	Yugoslavia			Zaire	
				Zambia	
TOTAL					
22	29	10	12	30	22

[1] Countries arranged by geographic region.

in the scientific literature *each year*. Critics tend to charge these occurrences to the activities of man and the failure of quarantine. In doing so, they discount the role of nature.* Whereas, in some cases, there is biological evidence to implicate man, it is often not possible to charge the crime to either man or nature. Did man or nature move the coffee rust fungus to the Western Hempsphere? Sugarcane smut, known to be present in the Caribbean area, was reported in Florida in 1978 (U.S. Department of Agriculture, 1978b). The United States' plant quarantine regulations (U.S. Department of Agriculture, 1978a) prohibit the importation of sugarcane plant materials, except that vegetative propagation materials may enter through the U.S. Department of Agriculture's (USDA) Sugarcane Quarantine Facility in Beltsville, Maryland, located more thatn 1,000 miles (1600 km) from the nearest commercial sugarcane field (Gillaspie and McKnew, 1975). Did man or nature introduce the sugarcane smut fungus into Florida?

Whereas we cannot measure what would have happened in terms of pest traffic if we did not have prohibitions, we can measure the frequency and variety of pests and pathogens that have been detected at ports of entry (USDA, 1979; Canada Department of Agriculture, 1978) or at quarantine stations (Kahn, *et al.*, 1967, 1980). However, we cannot measure the number of times these organisms passed through undetected at ports of entry.

Charges are also made by critics that quarantines are often based on, or influenced by, nonbiological rather than biological factors (MacGregor, 1978). The relative roles of biological, economic and political factors in policy, regulations, or entry status determinations have been reviewed (McCubbin, 1946, 1956; Gram, 1960; Kahn, 1979; University of California, 1969). I have proposed a concept of pest risk analysis by which it is possible to infer, but not prove, when quarantine actions are biological rather than economic, political, or social (Kahn, 1979).

Exclusion as a Plant Quarantine Control Strategy. Exclusion procedures are implemented against articles moved by man to prevent pests and pathogens from entering the area in which the host is grown. Exclusion procedures are usually regulatory actions to reduce the chances of spreading these harmful organisms either within or between countries.

Objectives. This chapter has three objectives concerning exclusion as a control. The first is to review the different types of control methods that are considered as exclusion. The second is to measure the extent to which two of these methods, i.e., 1) exclusion of the host as a vector, and 2) exclusion of pests and pathogens, are practiced as a control measure on a worldwide basis. The third is

*"Nature," as used herein, is a term of convenience that refers to the summation of all the natural factors that influence or relate to the movement of pests or pathogens to new areas and their subsequent establishment, including climate, inoculum potential or population density, presence of susceptible hosts, life cycles, etc.

to determine whether biology appears to be the *modus operandi* in the implementation of these two exclusion methods.

7.2 A REVIEW OF EXCLUSION METHODS AND PROCEDURES

Exclusion of pests and Pathogens by Certification at Origin. Plants are grown in the country of origin under appropriate phytosanitary regimes. Prior to shipment, they, or their mother plants, are subjected to indexing and other pest and pathogen detection tests. The exporting country certifies that the plant material meets the health standards for pests and pathogens specified by the importing country which may also stipulate that specific pathogen detection tests (e.g., virus indexing) be conducted. The certification required is more extensive than the model phytosanitary certificate (see Section 7.1, *"Legal Basis of Quarantines"*) statement that the consignment "is apparently free of . . . etc." The understanding that authorizes this is between the official plant protection and quarantine services of each country involved.

Countries which lack well-developed quarantine services are more inclined to accept "certification elsewhere" than countries with well-developed quarantine services. The latter usually prefer to rely upon their own inspections and the use of their own quarantine stations and testing than on "certification elsewhere."

Exclusion of Pests and Pathogens at Ports of Entry. Plant materials and other imported articles are usually inspected, or at least their samples are inspected, at ports of entry. Countries may list organisms which are to be excluded if encountered on imported articles at ports of entry. Sometimes an eradicant treatment is available which will eliminate the pest; more often the consignment is denied entry or destroyed.

The organisms listed for exclusion are usually those that are not established in the importing country, but often there are extenuating circumstances under which countries list already-established pests and pathogens for exclusion. These circumstances include, but are not limited to, the following: 1) the government desires or requires that the growers plant certified or healthy planting stock, 2) eradication, suppression or containment programs are underway against named pests or pathogens, and 3) there are exotic races of domestic pest species in other countries.

Many countries do not publish a list of organisms they intend to exclude. Often when pests or pathogens are found on articles at ports of entry, quarantine officials may make a determination of the entry status of such articles on a case-by-case basis. In doing so, they may or may not be guided by unpublished or in-house lists or policies. Many countries consider that the threat to agriculture of invading pests and pathogens is too unpredictable to justify all-inclusive lists of specific pests.

Exclusion by Prohibition of the Host as a Vector. Regulations are promulgated which list, usually by genus but sometimes by species or common name, plants which are prohibited when they originate from one or more named countries. These plants are prohibited because they are hosts of harmful organisms which may or may not be specified by name in the same regulation which prohibits the host. Exceptions to the prohibition are often made for scientific purposes under appropriate safeguards such as passage through a quarantine station. However, a special permit to authorize the entry of prohibited plants for scientific purposes is usually required.

Exclusion of the Host During Inspection. If a prohibited host is found in mail or baggage or is presented for entry in cargo, the host is usually denied entry. If a prohibited host is found mixed with non-prohibited plants, the entire consignment is usually denied entry. If a plant is unidentified at the time of entry and an identification cannot be made at the place of entry, the plant is usually denied entry on the basis that it may be one of the prohibited genera.

Exclusion of Pests and Pathogens. Most countries prohibit the importation of cultures of pests and pathogens. However, usually exemptions are made for scientific or educational purposes or for zoos and other collections—provided the safeguards are adequate. Exemptions to the exclusion are usually authorized only by permit.

Exclusion of the Host by Added Declaration Requirement on Phytosanitary Certificates. It is possible that a plant which is not officially excluded on a list of prohibited articles may, nevertheless, be excluded in actual practice—if the exporting country cannot certify the "added declaration" required by the importing country on a phytosanitary certificate. Added declarations may require a certification that specified pests and pathogens are not known to occur in an area (e.g., site, county, state, country) where the plants originated. If the former are known to occur in the area, plants from that area are excluded; but plants from other areas are admissible.

Exclusion of Plants if Accompanied by Prohibited Articles. Plants which are not prohibited may, nevertheless, be excluded at ports of entry, if accompanied by prohibited articles. For example, most countries prohibit plants with soil, and many prohibit those established in growing media capable of harboring pests. Plants arriving with these prohibited items may be refused entry. Plants or articles may be excluded if packed with prohibited materials (e.g., hay, straw, forest litter, etc.). Nonprohibited seeds may be excluded if the pulp of the fruit, usually a prohibited item, is present.

Exclusion of Plant Materials Imported for Uses Other Than Propagation. In addition to excluding plants or plant parts intended for propagation, countries may also exclude plant materials imported for consumption, manufacture or industrial utilization. Examples include the exclusion of: 1) fruits imported for consumption when they may harbor fruit fly larvae or other pests and pathogens; 2) by-products of industrial processing of plant materials, if the process does not utilize heat at least as high as pasteurization temperatures; 3) untreated bark either as a horticultural product or on logs, lumber, or dunnage; 4) forest litter for horticultural purposes; 5) untreated hay or straw imported for consumption or as manufactured products, and 6) plant parts intended for consumption which could be readily diverted to planting (e.g., potato tubers that were not treated with sprouting inhibitors, and seeds intended as animal feed); and 7) rarely, fresh cut flowers but not usually dried cut flowers.

Exclusion to Protect the Genetic Purity of local Varieties. Some countries prohibit, for reasons not associated with pest risk, some varieties of a crop while admitting other varieties to prevent local farmers or growers from importing varieties which are genetically inferior to those already grown. Usually, the reason for such prohibition is to prevent the importation of varieties known to be susceptible to local pests and pathogens. Occasionally, the prohibition is to protect the character of the local agricultural products such as the prohibition of certain grape cultivars to protect the quality of the wine in some European countries.

Exclusion of Plants as Weeds, Parasites, or Alternate Hosts. Most countries list certain weeds, parasitic plants or alternate hosts of rust fungi as prohibited. A listing of weeds and parasitic plants mentioned in foreign regulations has been published (Haven, 1979). Among the genera prohibited as alternate hosts of pine rust fungi are *Aster, Callistephus, Cineraria, Coreopsis, Helianthus, Impatiens, Ipomoea, Jacquemontia, Nemesia, Parthenium, Senecio, Solidago, Tropaeolium, Verbena*, and *Wahlenbergia*. Perhaps the most frequently prohibited alternate hosts are those of cereal rusts such as *Berberis, Mahonia* and *Mahoniaberberis* and *Rhamnus*.

Exclusion by Withholding Permits. The most common method for implementing the procedures by which the host as a vector, or pests and pathogens as cultures, are excluded (see Section 7.2) is as follows: 1) a permit is required for the entry of plants, plant parts, and cultures; 2) prohibited articles are listed; 3) permits for prohibited items are not issued (unless exceptions are granted by special permits which stipulate safeguards); and 4) plants or cultures which arrive without permits are denied entry.

7.3 MEASURING ON A WORLDWIDE BASIS THE USE OF EXCLUSION AS A CONTROL

7.3.1 Exclusion by Regulation of the Host as a Vector

The extent to which 125 countries (Table I) prohibit the host because it may be a vector or carrier of significant pests and pathogens was measured by an analysis of quarantine regulations of these countries. The sources for the data were published digests or summaries (USDA, 1933-1979; Ling, 1952, 1954) of national regulations or the documents which are maintained in the files of Plant Protection and Quarantine Programs (PPQ), Animal and Plant Health Inspection Service (APHIS), USDA, as furnished for foreign plant protection services.

The survey included all plants (genera, species, or crops) that were on lists of prohibited articles as described in Section 7.2, "Exclusion by Prohibition of the Host as a Vector." Genera or species were not tabulated when plants were excluded under the following cirucmstances: 1) groups such as nursery stock (four countries), orchids (two countries), ornamentals (three countries), and cacti (nine countries), 2) botanical families such as Bambuseae, Cactaceae and Leguminosae (=Fabaceae), and 3) as noxious weeds, and aquatic, parasitic, or narcotic plants.

The 38 most frequently listed genera or crops are shown in Table II. Citrus, which is the most frequently prohibited crop, is listed by 60% of the 103 countries in which citrus can be grown as a crop. In North America, 62% of the cirtrus-growing countries prohibited citrus, whereas the comparable figure for Africa was 66%.

The survey also included a tabulation of the number of times that crops or genera were prohibited or excluded in regulations which also listed at least one pest or pathogen as the reason for the prohibition (Table III). The tabulations also included whether the prohibition for each of the 38 most frequently cited genera or crops was against plants (including vegetative propagation material), seeds, or both plants and seeds (Table III). For example, 103 countries can grow citrus as a crop, and 62 of these countries prohibit citrus. Of these 62 countries, 55% prohibit plants only, 45% prohibit both plants and seeds, and non prohibit only seeds. For *Coffea*, 49 of the countries list the genus or coffee as prohibited. Of these, 24% prohibit plants only, 18% prohibit seeds only, and 57% prohibit both plants and seeds.

The survey revealed that 246 different genera or crops were named on the lists of prohibited articles published by the 125 countries surveyed. A total of 1,802 "prohibitions" of plants were listed for the 246 crops by the 125 countries (data not shown).

The range of the number of crops prohibited and the average number of crops prohibited by countries in each of six geographic regions and for all 125 countries was calculated (Table IV). For the 125 countries, the range was from 0-132 genera, and the average was 15. The geographic area with the narrowest range was Asia whose 21 countries surveyed had a range of 0-24 genera. The area with the broadest range was Africa whose 32 countries surveyed ranged from 0 to 132 genera that were prohibited. The geographic area whose countries had the highest average, i.e., 26 prohibited genera, was Africa. The areas with the lowest averages, i.e., 7 genera each, were North and South America.

7.3.2 Exclusion of Pests and Pathogens

The extent to which pests and pathogens are singled out by naming in the published quarantine regulations was measured as a means of showing the quarantine significance attached to these organisms by the 125 countries surveyed. The sources of the data were the same as described in Section 7.3.1. The various countries listed pests and pathogens because they prohibited cultures and/or the organism, if encountered on hosts (vectors), or other articles at the time of entry (see Section 7.2, "Exclusion of Pests and Pathogens").

The survey revealed that 1,588 different species of pests and pathogens were named a total of 5,303 times by the 125 countries. (The author estimates that any error due to synonymy or counting would not exceed 1%.) The number of different species named and the total number cited, respectively, were as follows: insects plus mites, 614 and 2,475; nematodes, 46 and 225; fungi, 537 and 1,444; and bacteria, 96 and 514. For viruses (including, for convenience, viroids, mycoplasma-like organisms and spiroplasmas) 292 different names were tabulated with total of 645 citations.

A check list of pests, pathogens, and weeds of quarantine interest to countries and states (of the United States) has been published by the California Department of Agriculture (Haven, 1969). This list is useful because the scientific names are arranged alphabetically and common names are provided. The listing also shows whether the organism has been recorded in California or the group of 48 contiguous states.

In the author's survey, the range of the total number of different pest and pathogen species listed and the average number, listed by countries in each of six geographic regions and for all 125 countries, was calculated (Table IV). For the 125 countries, the range was from 0-275 species, with an average of 35. The geographic area with the narrowest range was South America with 12 countries and a range of from 0-71. The area with the widest range, 0-275, was Africa. The geographic area whose countries had the highest average was Africa whose 32 countries surveyed averaged 46 named species. The area with the lowest average was North America whose 21 countries averaged 12 named species.

The 10 most frequently cited fungi, bacteria, nematodes, viruses and virus-like agents, and insects are listed in Table V which shows the number of countries which cited them and the distribution of these countries in 6 geographic areas. For example, *Synchytrium endobioticum* is listed by 66 countries. In North America, 55% of the countries surveyed listed this species, whereas for the Southwest Pacific, the figure was 25%.

TABLE II. Results of a Survey of Plant Quarantine Regulations of 125 Countries Showing, for the Most Frequently Cited Crops or Genera, the Extent to which they are Prohibited[1]

Crops/Genera	Number of countries in which crops/genera will grow	Percentages of countries in which crops/genera are prohibited	Percentages of Countries in Six Geographic Areas Prohibiting					
			North America	Europe	Southwest Pacific	South America	Africa	Asia
FRUIT CROPS								
Citrus	103	60	62	67	60	54	66	52
Cocos	83	34	24	0	60	9	40	37
Fragaria	68	28	33	12	43	14	67	30
Musa	94	41	43	0	67	54	34	40
Pome Fruits[2]	81	47	40	52	50	40	47	33
Prunus	81	40	40	45	50	20	47	33
Ribes	54	30	67	32	60	20	27	0
Rubus	58	36	0	14	57	43	58	30
Vitis	91	45	18	58	71	31	65	24
VEGETABLE CROPS								
Ipomoea	105	22	14	0	50	8	25	29
Solanum	115	42	36	48	40	15	50	48
FOREST CROPS								
Acer	64	22	20	34	0	0	25	0
Castanea	69	49	50	59	29	12	82	25
Conifers[3]	64	42	60	31	80	12	75	20
Crategus	64	22	0	41	20	12	0	0
Juglans	64	47	20	65	0	12	83	0
Populus	59	54	60	59	23	20	42	0
Quercus	67	54	60	52	33	0	100	80

Salix	64	34	40	34	40	25	42	20
Sorbus	64	37	0	41	40	12	67	20
Ulmus	50	64	40	90	33	20	75	20
OTHER CROPS								
Coffea	79	62	56	0	66	80	74	33
Gossypium	107	48	38	27	67	69	50	43
Helianthus	71	21	0	0	27	7	60	7
Hevea	66	42	9	0	75	12	69	62
Nicotiana	120	23	5	13	60	18	40	24
Palm, oil	83	22	0	0	30	18	36	12
Rice	107	33	9	10	50	46	34	38
Rosa	68	32	40	27	14	20	67	20
Theobroma	81	53	43	0	56	33	70	47
Saccharum	87	45	43	0	60	54	34	40
Tea	36	47	0	0	43	0	75	62

[1] Countries are listed in Table I.

[2] *Chaenomeles, Cydonia, Malus*, and *Pyrus.*

[3] Refers to conifers as a group or specifically to *Picea, Larix, Pinus*, and *Abies.*

TABLE III. Some Particulars of the Quarantine Regulations of 125 Countries Regarding Specific Crops and Plant Genera[1]

Crops/Genera	Number of countries where crops/genera will grow	Number of countries in which crops/genera are prohibited	Percentages of countries which name one or more pests or pathogens	Percentages of countries prohibiting		
				Plants only[2]	Plants and seeds[2]	Seeds only
FRUIT CROPS						
Citrus	103	62	45	55	45	0
Cocos	83	28	32	29	64	7
Fragaria	68	20	55	65	35	0
Musa	94	39	39	54	46	0
Pome Fruits[3]	81	37	68	85	15	0
Prunus	81	37	68	85	15	0
Ribes	58	16	38	69	31	0
Vitis	91	41	41	90	10	0
VEGETABLE CROPS						
Ipomoea	105	23	35	61	39	0
Solanum	115	48	60	77	23	0

FOREST CROPS						
Acer	64	14	43	100	0	0
Castanea	69	34	23	76	24	0
Conifers[4]	64	27	26	100	0	0
Crategus	64	14	86	100	0	0
Juglans	64	21	48	100	0	0
Populus	59	27	44	93	7	0
Quercus	67	25	47	92	8	0
Salix	64	22	45	100	0	0
Sorbus	64	24	58	96	4	0
Ulmus	50	32	47	84	16	0
OTHER CROPS						
Coffea	79	49	31	24	57	18
Gossypium	107	52	23	25	61	14
Helianthus	71	15	40	20	80	0
Hevea	66	28	50	29	71	0
Nicotiana	120	26	31	35	58	7
Palm, oil	83	16	38	56	44	0
Rice	107	33	42	21	61	18
Rosa	68	22	41	100	0	0
Theobroma	81	43	42	19	79	2
Tea	36	20	45	45	55	0
Saccharum	87	40	10	63	37	0

[1] Countries are named in Table I.
[2] Includes plant as well as any parts for vegetative propagation.
[3] *Chaenomeles, Cydonia, Malus*, and *Pyrus*.
[4] Refers to conifers as a group or specifically to *Picea, Larix, Pinus*, and *Abies*.

TABLE IV. Average Number and Numerical Range of Pests, Pathogens and Plant Genera that are Prohibited in the Plant Quarantine Regulations of 125 Countries Surveyed[1]

Geographic region	Number of countries whose regulations were surveyed	Pests and pathogens named to species or common name level[2]		Plant genera prohibited[3]	
		Range	Average	Range	Average
North America	22	0-111	12	0-51	7
Europe	29	0-232	54	0-30	14
Southwest Pacific	10	0-103	25	0-40	22
South America	12	0-71	21	0-27	7
Africa	30	0-275	46	0-132	26
Asia	21	0-129	28	0-24	9
Total All Regions	124	0-275	35	0-132	15

[1] Countries listed in Table I.

[2] Does not include species or common name when all pests and/or pathogens of a plant genus are prohibited, such as all viruses of *Solanum*. In such cases, the number "1" is used in calculations.

[3] Does not include genera when an entire plant family is prohibited or when groups of plants are prohibited, such as annuals, biennials, and forest trees.

7.4 THE BIOLOGICAL BASIS OF EXCLUSION OF HOSTS, PESTS, AND PATHOGENS

A survey (author's unpublished data) of 12 plant pathology or entomology textbooks showed that when quarantine was discussed as a separate topic, generally no more than 5 pages were devoted to from one to four topic areas. These areas were: 1) a listing of regulatory actions and their legal basis, 2) the pros and cons of regulatory actions, 3) the interactions of quarantine with biological, economic, or political factors, and 4) statements of principles or philosophy.

The consensus was that a quarantine action, such as "prohibited," should: 1) be implemented when man rather than nature is the prime mover, 2) be based on biology or pest risk analysis, and 3) constitute the least drastic action which provides the necessary safeguards.

In this section, the author seeks to determine whether exclusion by regulation or inspection as stated in the regulations of 125 countries conforms to these three principles, at least as far as the most frequently prohibited host genera or crops as well as pests and pathogens are concerned.

7.4.1 Man versus Nature as the Prime Mover

Exclusion as a plant quarantine strategy is most effective when man rather than nature acts as the prime mover in relocating pests and pathogens. The argument holds that if pests and pathogens move over long distances, unaided by man, there is not much point in regulating man or his activities to prevent long-distance spread. However, it is significant that such regulation may provide a delaying action that could slow down spread and thus buy time for research on control.

Nature can be considered to be the prime mover of pests and pathogens if they or their efficient vectors (e.g., arthropods) can be transmitted over long distances by air currents, wind, or by their own means of locomotion. Although migratory birds and animals are sometimes vectors, they are regarded in this survey as relatively inefficient, as is the washing ashore of plant materials (e.g., coconuts) harboring pests or pathogens. Long distance spread, however, by natural means can be accomplished by the cumulative effect of a series of short-distance movements over a long period of time.

Man can be considered as an efficient long distance and short distance mover when pests and pathogens can be harbored in or on articles he may transport, including plants or seeds, or these organisms may be hitchhikers on articles or their carriers.

The case of the plum pox virus, which infects many species of *Prunus*, may be used to illustrate this concept. The virus which is known to occur in many countries of Europe is transmitted by aphids, such as *Myzus persicae* (Sulzer), in the nonpersistent manner. The virus is not known to occur in North America, but the aphid vector does occur. If the virus were to be moved from Europe to North America by natural means alone, an aphid vector must feed on an infected host in Europe, acquire the virus, and then feed on a susceptible host in North America within 20 to 30 min—since the aphid loses its ability to transmit after this short period. On the other hand, man may move virus infected plants or parts capable of propagation to North America. Once the infected host is established in North America, the virus could be spread by domestic vectors (natural) or by nursery practices such as budding or grafting (aided by man).

Some pathways along which the 10 fungi, bacteria, and insects listed in Table V may be transported from one region to another are tabulated in Table VI. In each case, the listed species could be readily moved long distances by man on plant materials imported for consumption or propagation. At best, nature can be expected to move long distances only two of the fungi, *Microcylus ulei* and *Peronospore tabacina. M. ulei* does produce air-borne conidia, but the fungus has not yet moved out of the Western Hemisphere. *P. tabacina* has already been moved by man (as an alleged escape from greenhouse inoculations in a case where the fungus was imported as a culture) and by nature. However, further spread by nature may be limited by natural barriers. For example, the Sahara Desert or Sahel area serves as a natural barrier preventing the movement of *viable* air-borne

TABLE V. The 10 Most Frequently Cited Fungi, Bacteria, Nematodes, Insects, Viruses and Viruslike Agents in the Quarantine Regulations of 125 Countries in which at Least One Pest or Pathogen is Listed by Name (of 125 Countries Surveyed[1])

Scientific name	Common name	Number of countries citing organisms	Percentages of countries which cite organisms in each of six geographic areas					
			North America	Europe	Southwest Pacific	South America	Africa	Asia
FUNGI								
Synchytrium endobioticum	potato wart fungus	66	55	93	25	43	42	68
Endothia parasitica	chestnut blight fungus	37	18	62	25	0	43	11
Dothidella ulei (= *Microcyclus ulei*)	South American leaf blight of rubber fungus	25	0	3	50	8	43	67
Ceratocystis fagacearum	oak wilt fungus	23	4	66	13	8	5	3
Ceratocystis ulmi	Dutch elm fungus	19	6	28	25	8	25	0
Phytophthora fragariae	Red stele or core of strawberry fungus	19	12	28	13	15	14	11
Peronospora tabacina	blue mold of tobacco fungus	18	0	14	25	0	32	16
Puccinia horiana	chrysanthemum white rust fungus	18	12	45	0	8	4	5
Urocystis cepulae	onion smut fungus	16	0	7	25	0	29	21
Mycosphaerella ligulicola (=*Aschochyta ligulicola*)	chrysanthemum ray blight fungus	16	0	41	8	0	4	5
BACTERIA								
Erwinia amylovora	fire blight bacterium	44	6	86	13	17	39	21
Corynebacterium sepedonicum	potato bacterial ring rot	42	18	70	38	33	29	32
Rickettsialike organism	Pierce's disease of grapevine agent	30	4	45	38	17	29	16
Pseudomonas solanacearum	tomato bacterial wilt, moko disease of banana, potato brown rot, etc.	26	12	31	0	23	32	25
Xanthomonas citri	citrus canker bacterium	26	12	17	38	17	46	6
Erwinia stewartii	Stewart's wilt of corn bacterium	25	0	28	25	8	32	32

Corynebacterium michiganense	tomato bacterial canker	24	6	21	13	8	39	21
Agrobacterium tumefaciens	crown gall bacterium	20	0	45	0	8	7	21
Pseudomonas pisi	pea bacterial blight	18	0	10	0	17	36	16
Corynebacterium flaccumfaciens	bean bacterial blight	15	0	10	0	17	29	11
NEMATODES								
Globodera rostochiensis (=*Heterodera rostochiensis*)	golden nematode, potato eelworm	51	35	51	38	25	43	37
Ditylenchus dipsaci	bulb and stem nematode	23	4	48	0	0	21	11
Heterodera schachtii	sugarbeet cyst nematode	16	0	21	0	0	25	11
Ditylenchus angustus	rice stem nematode	13	0	28	0	0	7	16
Aphelenchoides fragariae	strawberry spring dwarf nematode	13	0	35	0	0	7	5
Ditylenchus distructor	potato rot nematode	12	0	17	0	0	10	11
Radopholus similis	burrowing nematode	11	0	3	0	0	7	26
Meloidogyne javanica	a root knot nematode	11	0	21	0	0	16	14
Aphelenchoides ritzemabosi	chrysanthemum foliar nematode	11	6	28	0	0	0	11
Aphelenchoides besseyi	rice white tip, strawberry summer dwarf nematode	9	0	21	0	0	11	11
VIRUSES AND VIRUSLIKE AGENTS[2]								
–	plum pox virus	28	18	66	0	0	14	11
–	potato spindle tuber viroid	24	12	45	38	0	29	16
–	apple proliferation agent (MLO)	22	7	66	13	0	5	0
–	peach yellows agent (MLO)	22	7	52	38	0	0	11
–	citrus tristeza virus	19	0	10	13	0	43	16
–	rose wilt agent	18	12	34	0	0	18	5
–	strawberry yellow edge agent	18	0	48	0	0	11	0
–	strawberry witches' broom agent	18	0	48	0	0	11	0
–	strawberry green petal agent (MLO)	18	0	48	0	0	11	0
–	potato witches' broom agent (MLO)	17	0	34	38	8	7	5

TABLE V. (continued)

Scientific name	Common name	Number of countries citing organisms	Percentages of countries which cite organisms in each of six geographic areas					
			North America	Europe	Southwest Pacific	South America	Africa	Asia
INSECTS								
Quadraspidiotus perniciosus	San Jose scale	44	12	86	0	25	36	26
Leptinotarsa decemlineata	Colorado potato beetle	40	18	52	38	17	32	42
Ceratitis capitata	Mediterranean fruit fly	39	47	55	25	25	77	42
Rhagoletis pomonella	apple maggot	33	24	76	13	8	7	16
Dacus dorsalis	oriental fruit fly	28	29	17	50	8	36	26
Popillia japonica	Japanese beetle	28	24	58	0	8	11	16
Anthonomus grandis grandis	boll weevil	21	12	21	13	17	21	21
Anastrepha ludens	Mexican fruit fly	20	18	24	13	8	21	11
Rhagoletis cerasi	cherry maggot	20	6	24	13	0	28	16
Phthorimaea operculella	potato tuberworm	19	0	48	0	0	7	16

[1] Countries listed in Table I.
[2] Includes mycoplasmalike organisms (MLO).

TABLE VI. Factors Relating to Spread and Geographic Distribution of the 10 Most Frequently Prohibited Fungi, Insects, and Bacteria

Species	Reported geographic distribution[1]						Host range	Known vectors in addition to man	Long distance spread[3]		Remarks
	NA	Eur.	SWP	SA	Af.	Asia			Natural	Man's activities[4]	
FUNGI											
Sychytrium endobioticum	+	+++	+	+	+	+	Narrow	Soil	–	Potato tubers, soil	Different races of the fungus exist in different areas. Containment or eradication is practiced
Endothia parasitica	+	++	0	0	0	++	Narrow	Wind, rain birds, insects	–	Nursery stock, logs, lumber, bark, chestnut, fruits	
Microcylus ulei	+	0	0	++	0	0	Narrow	Air-borne conidia, splashing rain	Conidia not a factor	Planting or nursery stock	Races of the fungus exist
Ceratocystis fagacearum	+	0	0	0	0	0	Narrow (Fagaceae)	Root grafts, beetles		Logs, lumber, nursery stock	
Ceratocystis ulmi	+	+++	0	0	0	0	Narrow	Root grafts, beetles		Logs, lumber, nursery stock	
Phytophthora fragaria	+	++	+	0	0	+	Narrow	Soil, drainage, water		Planting stock, soil	
Peronospora tabacina	+++	+++	+	+	+	+	Narrow	Air-borne conidia (seeds?)	Air-borne spores can spread a long distance but spores are short-lived	Planting stock, (seeds?) uncured tobacco	Spores will not stand dessication

TABLE VI. (continued)

Species	Reported geographic distribution[1]						Host range	Known vectors in addition to man	Long distrance spread[3]		
	NA	Eur.	SWP	SA	Af.	Asia			Natural	Man's activities[4]	Remarks
Puccinia horiana	0	++	0	0	+	+	Narrow	Air currents, splashing water	–	–	–
Urocystis cepulae	+++	++	+	+	+	+	Narrow	Persists in soil	–	Planting stock	–
Mycosphaerella legulicola	+	++	+	0	+	+	Narrow	–	–	Planting stock	–
INSECTS											
Quadraspidiotus perniciosus	++	+	+	++	+	+	Wide	Wind, birds	–	Nursery stock, fruits	–
Leptinotarsa decemlineata	++	+++	0	0	0	0	Narrow (Solanceae)	Wind	–	Plants, packaging, hitchhiker, soil	Wind-borne adults account for spread within a region
Ceratitus capitata	+	++	+	++	+++	+	Wide	Winds, flier	–	Fruit, plants, hitchhiker	–
Rhagoletis pomonella	+	0	0	0	0	0	Narrow (Rosaceae)	Flier	–	Fruits, soil around nursery stock, packaging	–
Dacus dorsalis	+	0	+	0	0	++	Wide	Flier	–	Fruits	(Hawaii only)
Popillia japonica	+	0	0	0	0	+	Wide	Flier	–	Nursery stock, hitchhiker	–
Anthonomus grandis grandis	+++	0	0	+	0	0	Narrow (Malvaceae)	Flier, air currents	–	Plants, cotton bolls	–

Anastrepha ludens	+	0	0	0	0	0	–	Wind	–	Fruits, hitchhiker	–
Rhagoletus											
R. cingulata	+	0	0	0	0	0	Narrow				
R. cerasi	0	+++	0	0	0	+(?)	Narrow				
Phthorimaea operculella	+++	++	++	+++	++	++	Narrow (Solanaceae)	–	–	Potato tubers, fruits, tobacco leaves	–
BACTERIA											
Erwinia amylovora	+	++	0	+	+	+	Narrow (Rosaceae)	Insects, rain, birds	Not likely	Nursery stock, fruit, containers	Containment or eradication in Europe
Corynebacterium sepedonicum	++	++	0	+	0	+	Narrow (Solanaceae)	Insects, (volunteer plants)	–	Seed tubers, contaminated equipment, logs	Distribution and prevalence uncertain
Pierce's disease of grapevine agent	++	0	+	+	+	0	Wide	Insects	–	Nursery stock, grafting	–
Pseudomonas solonacearum	+++	++	++	++	+++	+++	Wide	Soil, insects, surface water	–	Planting stock, soil, tools	Species is present but all host specific strains are not
Xanthomonas citri	0	0	+	+	+	+++	Narrow (Rutaceae)	Rain, air currents, flood water, insects, animals	–	Nursery stock, fruit	

TABLE VI. (continued)

Species	Reported geographic distribution[1]						Host range[2]	Known vectors in addition to man	Long distrance spread[3]		
	NA	Eur.	SWP	SA	Af.	Asia			Natural	Man's activities[4]	Remarks
Xanthomonas stewartii	+++	+	0	+	0	+	Narrow	Insects, (seeds = low percentage)	–	Seeds	Insects migrate but probably not between regions. Cold winters reduce overwintering of insect
Corynebacterium michiganense	++	+++	+	+	++	+	Narrow	Seeds, debris, water splash	–	Seeds, soil	–
Agrobacterium tumefaciens	+++	+++	++	++	++	++	Wide	Soil, insects	–	Nursery stock, soil	–
Pseudomonas pisi	++	++	+	+	+	+	Narrow	Seeds, surface water, splashing rain	–	Seeds	–
Corynebacterium flaccumfaciens	+	+	+	0	+	+	Narrow (Fabaceae)	Seeds, soil	–	Seeds	–

[1] 0 = not known to be present; + = report in one or up to about 20 percent of the countries in a region; ++ = reported in 20 to 60% of the region; +++ = reported in 60 to 100% of the region. NA = North America; Eur. = Europe; SWP = Southwest Pacific; SA = South America; Af. = Africa.

[2] Narrow = one genus, or several genera in one botanical family when family name is given; wide = many genera in many families.

[3] Spread between regions.

[4] *Nursery stock* includes spread by nursery practices (grafting) to plants in a nursery which are exported to other regions, sometimes with soil adhering to roots or as baled and burlapped; *planting stock* is similar to nursery stock but includes sets, cuttings, and transplants often with soil; *fruits* refers to import of fruits for consumption.

conidia from North Africa where *P. tabacina* is established to areas south of the Sahara since the spores cannot survive heat and dessication. However, man could move the fungus on infected tobacco plants or as cultures.

Bacteria are not usually moved over long distances by natural means except, perhaps, by migrating birds passing undigested seeds containing bacteria borne internally. While insects move bacteria locally, insects are not involved in long distance spread. However, all 10 of the listed bacteria (Table VI) could easily be moved by man on plants or plant parts, including seeds. Man is much more effective in moving bacteria with seeds than are birds.

The 10 insect species listed are not likely to be moved by natural means between regions although this always remains as a possibility for migratory or wind-blown insects (e.g., *Leptinotarsa decemlineata*, fruit flies, and *Anthonomus grandis grandis*). Man could efficiently move the insects on platns or plant parts and, in some cases, as hitchhikers.

Nematodes, viruses or viruslike agents, including mycoplasma-like organisms, viroids, and spiroplasmas, also are not likely to be moved over long distances by natural means; although insects or mites are often involved in local spread. A few species of nematodes are seed-borne. Others are transmitted by beetles, but the range of the beetles is relatively short. However, man could move these agents or their vectors in plant materials imported for planting or consumption. However, none of the ten most frequently cited nematodes are known to be insect transmitted, whereas two are seed-borne in rice (*Aphelenchoides besseyi* and *Ditylenchus angustus*). Most plant viruses or viruslike agents do not survive in the absence of a living host or vector. Consequently, their spread over long distances is a function of the dissemination of living plant materials or vectors which usually are more effectively transported by man than nature.

7.4.2 The Biology of Exclusion by Prohibition

The biology of exclusionary practices was analyzed for the five most frequently cited fungi, bacteria and insects to determine whether these listings could be attributed to sound biological practices. For this purpose one could assign countries to four categories depending on whether the pest occurred in the country, whether the country listed the pest in its regulations, or both. The categories are as follows:

Category	Pest/Pathogen Reported	Regulation Prohibits Pest/Pathogen
A	yes	no
B	yes	yes
C	no	yes
D	no	no

Table VII shows the number of countries listing 15 pests and pathogens and the percentage of these countries that may be placed in either category A, B, or C. Category D is not included in this comparison because insufficient data were available. The "biological soundness" index was calculated from data in categories A and C, since these two categories represent a proper biological position in considering geographic distribution and exclusion.

Quarantine positions held by countries in category B, unless there are extenuating circumstances (see Section 7.2, "Exclusion of Pests and Pathogens at Ports of Entry"), may be considered as biologically unsound. Of the 15 pests and pathogens listed in Table VII which represent the most frequently listed fungi, bacteria and insects, 12 scored over 80 in the biological soundness index, indicating that at least 80% of the exclusionary regulations were related to whether the pest and pathogens occurred in the country. Three organisms, *Synchytrium endobioticum, Quadraspidiotus pernicious* and *L. decemlineata* scored less than 80 (68, 65 and 78, respectively), but with all three either eradication, suppression, containment, or certification programs were in effect in many countries.

Support to the biological soundness of these exclusionary procedures is given by the fact that most of the 50-most-frequently-cited pests or pathogens (Table V) have narrow host ranges which makes regulatory actions more effective and easier to enforce. For example, all 10 fungi, 5 of the 10 insects, and 7 of the 10 bacteria (Table VI) have narrow host ranges. In addition, all 10 viruses or virus-like agents and 8 of the 10 nematodes (all except *Meloidogyne javanica* and *Radopholus similis*) have narrow host ranges (data not shown).

Additional support for the biological basis of the regulations which prohibit the host because it is a vector may be obtained by comparing 1) the data given above for narrow host range, 2) the listing of most frequently cited prohibited genera (Table II), and 3) the percentages of countries which name at least one pest or pathogen at the time the host is prohibited (Table III). Table III shows that organisms are named in *less* than 50% of prohibitions for 27 of the 38 most frequently prohibited or excluded plant genera or crops. Nevertheless, the hosts for 29 of the 40 pests and pathogens with narrow host ranges are on the list of the 38 most frequently prohibited host genera.

7.4.3 Least Drastic Action

A guiding principle in plant quarantine circles is that a quarantine action should be the least drastic action which provides an adequate safeguard (see Section 7.2, "Exclusion of Pests and Pathogens"). However, prohibition or exclusion is perhaps the most drastic action that can be taken in regulatory matters. The question is when exclusion is utilized by the 125 countries, how often is it the least drastic action?

We can assume that prohibition is the least drastic action if 1) the country has a zero tolerance for the organism, 2) the organism is obscure insofar as detection at ports of entry is concerned, 3) the organism does not occur in the country

TABLE VII. Estimate of the Biological Soundness of Naming 15 Fungi, Insects, and Bacteria as Pests and Pathogens of Quarantine Significance[1]

Pest/pathogen	Number of countries which name Pests/pathogens[2]	Percentage of countries in category[3]			Biological soundness index[4]
		A	B	C	
FUNGI					
Synchytrium endobioticum	66	59	32	9	68
Endothia parasitica	37	67	17	17	84
Microcylus ulei	25	66	0	34	100
Ceratocystis fagacearum	23	91	4	4	95
Ceratocystis ulmi	19	31	19	50	81
BACTERIA					
Erwinia amylovora	44	62	19	19	81
Corynebacterium sepedonicum	42	68	12	20	88
Pseudomonas solanacearum	30	23	13	64	87
Xanthomonas citri	26	42	12	46	88
Xanthomonas stewartii	26	43	11	46	89
INSECTS					
Quadraspidiotus perniciosus	44	35	35	30	65
Leptinotarsa decemlineata	40	53	22	25	78
Ceratitis capitata	39	28	16	56	84
Rhagoletis pomonella	33	97	3	0	97
Dacus dorsalis	28	69	6	25	94

[1] Most frequently cited as listed in Table V.
[2] Number of countries citing pest/pathogen out of 124 listed in Table I.
[3]

Category	Pest/Pathogen known to Occur in Country	Regulations Prohibit Pest or Pathogen
A	yes	no
B	yes	yes
C	no	yes

[4] *Biological soundness index* equals the sum of the percentage figures shown in columns A and C.

(unless there are extenuating circumstances as described in Section 7.2, "Exclusion of Pests and Pathogens at Ports of Entry"), 4) man rather than nature is the prime mover, and 5) the host cannot be freed of the pest or pathogen by treatment.

While it is difficult to provide data to prove or disprove whether each and every prohibition is the least drastic action, one can measure, from data obtained in the survey of 125 countries, the extent to which this prohibition is practiced. If prohibitions are not widely practiced when regulatory actions are taken, one can infer that other actions, necessarily less drastic, are taken.

The survey showed that on a worldwide basis, exclusion by prohibiting the host in regulations because it is a vector is not a widely exercised regulatory option. The 246 plant species or genera, reported herein as named by at lease one of the 125 countries surveyed, constitute only 8% of the 3,000 species listed (USDA, 1977) as economically important in the world.

The extent to which prohibition is used can be determined by calculating the prohibition percentage (PP) from data presented in Section 3 as follows:

$$PP = \frac{\text{Total number of citations in regulations of 125 countries}}{\text{Number of countries} \times \text{number of economic species}} \times 100$$

$$\text{or } PP = \frac{1802}{125 \times 3000} \times 100 = 0.486.$$

Thus, of the total possible prohibitions of 3,000 species by 125 countries, less than one-half of 1% are exercised.

7.5 REFERENCES

Agrios, G. N. (1978). "Plant Pathology," 2nd ed. Academic Press, New York.

Canada Department of Agriculture (1978). "Intercepted Plant Pests, 1976-1977." Ottawa.

European and Mediterranean Plant Protection Organization. (1977). Data sheets on quarantine organisms. EPPO lists A1 and A2.

Gillaspie, A. G. Jr., and McKnew, C. C. (1975). Improved sugar quarantine facilities and procedures at Beltsville, Maryland. *Sugar J.* **38**:40-43.

Gram, E. (1960). *In* "Plant Pathology" (J. G. Horsefall and A. E. Dimond, eds.), Vol. III, pp. 318-355. Academic Press, New York.

Haven, O. (1979). "Plant Pests of Phytosanitary Significance to Importing Countries and States." 1st revision (looseleaf), May 11, 1979. State of California, Department of Food and Agriculture, Sacramento.

Inter-African Phytosanitary Council (1962). A memorandum for phytosanitary procedure in Africa. Publ. No. 82.

Inter-African Phytosanitary Council (1980). Coordinated phytosanitary regulations. In press.

Kahn, R. P. (1970). *In* "Genetic Resources in Plants; Their Exploration and Conservation" (O. H. Frankel and E. Bennett, eds.), pp. 403-411. Blackwell, Oxford.

Kahn, R. P. (1979). A concept of pest risk analysis. *Bull. Eur. Med. Plant Protection Org.* 9:119-130.

Kahn, R. P., Monroe, R. L., Hewitt, W. B., Goheen, A. C., Wallace, J. M., Roistracher, C. N., Nauer, E. M., Ackerman, W. L., Winters, H. F., Seaton, C. A., and Pifer, W. (1967). Incidence of virus detection in vegetatively propagated plant introduction under quarantine in the United States, 1957-1967. *Plant Dis. Reptr.* **51**:715-741.

Kahn, R. P., Waterworth, H. E., Gillaspie, Jr., A. G., Foster, J. A., Goheen, A. C., Monroe, R. L., Povich, W. L., Mock, R. G., Luhn, C. F., Calavan, E. C., and Roistracher, C. N. (1980). Detection of viruses or virus-like agents in vegetatively propagated plant impor-

tations under quarantine in the United States, 1968-1978. *Plant Dis. Reptr.* In press.
Ling, L. (1952). "Digest of Plant Quarantine Regulations." FAO Dev. Paper No. 23, pp. 1-23.
Ling, L. (1954). "Digest of Plant Quarantine Regulations," Suppl. I, pp. 1-100. FAO.
MacGregor, R. C. (1978). *In* "Plant Disease, An Advanced Treatise," Vol. II, How Diseases Develop in Populations" (J. G. Horsefall and E. B. Cowling, eds.), pp. 383-396. Academic Press, New York.
McCubbin, W. A. (1946). Preventing plant disease introduction. *Bot. Rev.* **12**:101-139.
McCubbin, W. A. (1956). "The Plant Quarantine Problem." Ejnar Munksgaard, Copenhagen.
National Academy of Sciences (1968). "Principles of Plant and Animal Pest Control," Vol. I. Plant Disease Development and Control. Publication 1596, National Academy of Sciences, Washington, D. C.
Stakman, E. C., and Harrar, J. G. (1957). "Principles of Plant Pathology." Ronald Press, New York.
USDA (1933-1979). "Export Certification Manual," Vols. 1 and 2, Washington, D. C.
USDA (1977). A checklist of names for 3,000 vascular plants of economic importance. Agriculture Handbook No. 404, Washington, D. C.
USDA (1978a). Code of Federal Regulations; Title 7, Agriculture; Chapter III, Animal and Plant Health Inspection Service, Washington, D. C.
USDA Washington, D. C. (1978b). *Cooperative Plant Pest Rep.* **3**:323.
USDA, Washington, D. C. (1979). List of Intercepted Plant Pests, July 31, 1973 to September 30, 1977.
University of California, Division of Agricultural Sciences (1969). "Plant Quarantines in California, A Committee Report."
Walker, J. C. (1969). "Plant Pathology," 3rd ed. McGraw-Hill, New York.

Chapter 8

PLANT QUARANTINE PROBLEMS IN PREVENTING THE ENTRY INTO THE UNITED STATES OF VECTOR-BORNE PLANT PATHOGENS

J. A. Foster

USDA-APHIS-PPQ
U.S. Plant Introduction Station
Glenn Dale, Maryland

ISBN 0-12-326440-5

8.1 INTRODUCTION

The exclusion of foreign plant pests and pathogens from the United States by regulating the importation of foreign plants and plant parts was first authorized by the passage of the Plant Quarantine Act in 1912. During the following decades, the foreign plant pests and pathogens associated with economic plant genera have been studied with regard to their potential for introduction, establishment and damage in U.S. crops. These pest risk evaluations (Kahn, 1979) have been used to formulate and justify the regulations imposed by the U.S. Department of Agriculture. For example, plant material which may carry pests or pathogens that are easily detectable by visual examination are subject to inspection and treatment when necessary at ports of entry. Importations of plant material that may contain significant pests or pathogens which can not be detected by routine inspection procedures are prohibited to the general public but may be made available under Departmental Permit to scientists only after extensive testing demonstrates the absence of these pests and pathogens from the imported plants or specially treated propagations. The promulgation and implementation of quarantine regulations which affect the health of imported plant material is the responsibility of the Plant Protection and Quarantine (PPQ) program of the Animal and Plant Health Inspection Service (APHIS) in the U.S. Department of Agriculture (USDA).

Foreign plant pests and pathogens which would be subject to quarantine action if detected in imported plant material may be imported legally into the United States as cultures for educational or scientific purposes in accordance with the provisions specified in the Federal Plant Pest Act of 1957. A permit which specifies the safeguards necessary for the safe importation, utilization and final disposal of the foreign pest or pathogen must be obtained from APHIS-PPQ after approval from the appropriate state regulatory officials. Before issuing such a permit, qualified entomologists or plant pathologists analyze the pest risk inherent in the importation of each organism and may refuse permission or impose restrictions or safeguards, when necessary, to minimize the threat to U.S. agriculture. For example, insects not known to be plant pests will be subject to fewer restrictions than foreign insects capable of causing damage to plants and acting as a vector of either foreign or domestic plant pathogens. In general, the safeguards specified in the permit consist of the most practical but least disruptive measures necessary to protect U.S. agriculture in light of our present knowledge.

8.2 ENTRY OF POTENTIAL VECTORS

After studying these plant quarantine procedures, it becomes apparent that a vector carrying a plant pathogen could be introduced into the United States either as a contaminant of imported cargo and baggage or as a culture requested by a scientist for research. In order to evaluate the potential for entry, the number of interceptions at ports of entry and the number of requests recorded by APHIS-PPQ for some families known to contain one or more vectors of plant pathogens are shown in Table I. Since entry of potential vectors probably occurs without the knowledge of APHIS-PPQ or, if detected, is not always reported by APHIS-PPQ, these data underestimate the probability of introduction for some types of vectors and, therefore, must be regarded as only our best estimate of reality. As fascinating as it may be to speculate about these data, it is even more enlightening to examine the quarantine procedures in order to discover why the data appear as they do.

8.2.1 Inspection of Enterable Commodities for Plant Pests and Pathogens

Every day of the year, millions of items of plant material and other cargo are imported into the United States. The inspection and treatment of this foreign material to prevent the introduction of harmful pests or pathogens is the responsibility of PPQ inspectors working at designated ports of entry. Dips, sprays or fumigations with approved chemicals are prescribed on a routine basis for certain materials from areas known to be infested with specific pests or pathogens of quarantine significance to the United States. However, in general, imported material from all areas is routinely inspected for the presence of contaminating organisms.

If such an organism is discovered during inspection, an attempt at identification is made at the port of entry by authorized PPQ personnel qualified by previous experience and training. Specimens of taxonomically difficult plant pests or pathogens are usually sent to specialists for identification. If the identified organism has not previously been reported or is not widely distributed in the United States, the interception is reported, and action is taken at the port of entry. If a treatment to eradicate the contaminating organism is not available, the article may be denied entry. Some of the well-known foreign vectors recent ly intercepted at ports of entry include *Diaphorina citri* Kuwayana from China and India, *Planococcus lilacinus* (Cockerell) from Asia, *Mamestra brassicae* (Linné) and *Pieris brassicae* (Linné) from Europe, *Laodelphax striatellus* (Fallén) from Korea, and *Nephotettix cincticeps* (Uhler), *N. nigropictus* (Stål) and *N. virescens* (Distant) from various countries in Asia (APHIS, 1979).

However, when a pest or pathogen of a species widely distributed in the United States is intercepted, such as *Myzus persicae* (Sulzer), the interception may or may not be reported, and importation without treatment usually occurs. Need-

TABLE I. Number of Interceptions at Ports of Entry and Requests for Cultures Recorded by APHIS-PPQ for Insects, Mites, Fungi and Nematodes in Some Families Containing Known Vectors of Plant Pathogens

Families with Vectors	Number of Interceptions (July 1973-Sept. 1977)[a]	Number of Requests (Sept. 1977-Aug. 1979)[b]
Phylum: Arthropoda[c]		
Class: Acari		
Family: Acaridae	3	1
Family: Ascidae	0	0
Family: Cheyletidae	0	0
Family: Eriophyidae	8	1
Family: Pyemotidae	0	0
Family: Tetranychidae	50	11
Class: Insecta		
Order: Coleoptera		
Family: Anthribidae	194	0
Family: Bostrychidae	2515	0
Family: Brentidae	0	0
Family: Buprestidae	372	9
Family: Cerambycidae	2415	8
Family: Chrysomelidae	616	16
Family: Coccinellidae	12	7
Family: Cucujidae	0	0
Family: Curculionidae	4608	16
Family: Elateridae	173	1
Family: Histeridae	0	0
Family: Meloidae	0	0
Family: Nitidulidae	71	3
Family: Phalacridae	9	0
Family: Scarabaeidae	535	0
Family: Scolytidae	1545	6
Order: Dermaptera		
Family: Forficulidae	0	0
Order: Diptera		
Family: Agromyzidae	318	2
Family: Anthomyiidae	28	1
Family: Bibionidae	3	0
Family: Calliphoridae	0	2
Family: Chloropidae	29	0
Family: Drosophilidae	0	18
Family: Lonchaeidae	71	0
Family: Muscidae	0	0
Family: Sciaridae	0	0
Family: Syrphidae	0	0
Family: Tephritidae	7140	9
Order: Hemiptera		
Sub-Order: Heteroptera		
Family: Anthocoridae	3	0
Family: Coreidae	36	0
Family: Lygaeidae	230	0
Family: Miridae	163	1
Family: Nabidae	0	0
Family: Pentatomidae	157	2
Family: Piesmatidae	0	0
Family: Pyrrhocoridae	48	1
Family: Tingidae	10	0
Sub-Order: Homoptera		
Family: Aleyrodidae	2773	129
Family: Aphididae	122	4
Family: Cercopidae	9	0
Family: Cicadellidae	167	0
Family: Cixiidae	0	0
Family: Delphacidae	11	0
Family: Derbidae	0	0
Family: Flatidae	12	0

Family: Membracidae	22	0
Family: Pseudococcidae	457	20
Family: Psyllidae	55	0
Order: Hymenoptera		
Family: Agaonidae	0	0
Family: Andrenidae	0	1
Family: Apidae	4	5
Family: Formicidae	402	66
Family: Ichneumonidae	0	12
Family: Siricidae	709	0
Family: Tenthredinidae	9	4
Family: Vespidae	2	1
Family: Xylocopidae	0	0
Order: Lepidoptera		
Family: Arctiidae	109	0
Family: Gelechiidae	742	0
Family: Lymantriidae	25	4
Family: Noctuidae	1363	21
Family: Pieridae	35	5
Family: Pyralidae	3249	11
Family: Saturniidae	10	0
Family: Sphingidae	12	0
Order: Orthoptera		
Family: Acrididae	50	22
Family: Gryllidae	191	4
Family: Tettigoniidae	298	6
Order: Thysanoptera		
Family: Heterothripidae	0	0
Family: Phlaeothripidae	77	1
Family: Thripidae	2714	0
Phylum: Nematoda		
Order: Dorylaimida[d]		
Family: Longidoridae	1	0
Family: Trichodoridae	0	0
Order: Tylenchida		
Family: Anguinidae[e]	7	0
Family: Aphelenchoididae[f]	0	0
Division: Eumycota		
Class: Chtyridiomycetes[g]		
Order: Chytridiales		
Family: Olpidiaceae	0	0
Family: Synchytriaceae	0	0
Class: Plasmodiophoromycetes[h]		
Order: Plasmodiophorales		
Family: Plasmodiophoraceae	0	0

[a]Information compiled from data published by the Animal and Plant Health Inspection Service (1979).

[b]In September 1977, APHIS-PPQ began to consistently differentiate requests for foreign cultures from requests for domestic cultures in its computer records.

[c]Vectors known to the author have been classified by specialists in the USDA Systematic Entomology Laboratory.

[d]Vectors known to the author have been classified according to the taxonomic system used by Ferris (1971).

[e]Vectors known to the author have been classified according to the taxonomic system used by Siddiqi (1971).

[f]Vectors known to the author have been classified according to the taxonomic system used by Golden (1971).

[g]Vectors known to the author have been classified according to the taxonomic system used by Sparrow (1973).

[h]Vectors known to the author have been classified according to the taxonomic system used by Waterhouse (1973).

less to say, these foreign pests or pathogens may serve as vectors of foreign plant pathogens with a high potential for damage. From Table I, it is impossible to determine whether insects, mites, fungi and nematodes in certain families known to contain vectors are not being detected at ports of entry, or are not being prevented from entering the United States when detected.

The probability that certain soil-borne vectors could carry plant pathogens into the United States is minimized by the fact that soil is prohibited entry into this country except in circumstances specified by Departmental Permit. Whenever soil is discovered contaminating plant material or other cargo, the PPQ inspector has the option of either denying entry of the cargo or requiring the soil to be washed off before entry. This washing procedure should eliminate many ectoparasitic pests or pathogens and most pathogens causing obvious root infections when the contaminated roots are examined. The presence of soil-borne organisms would usually be reported only if the PPQ inspector had time to analyze the soil and then only if an organism of quarantine significance is discovered.

8.2.2 Requests for Cultures of Potential Vectors

During the last 5 years, a growing interest in importing organisms for biological control purposes is apparent. Either parasite cultures or their infected hosts may be requested. Many requests have also been received for insects to be used in pheromone studies or to be displayed in private or public collections. Most of the foreign plant pathogens are requested for the purpose of comparing them with similar domestic pathogens or assessing their impact on U.S. crops if the foreign pathogen became established. Comparatively few applications list studies in physiology, taxonomy, nutrition, ecology, genetics, behavior, pest potential, or vector capacity as justification for importation.

Recently, APHIS-PPQ has become more concerned about the possibility that pests or pathogens imported as cultures may be carrying foreign plant pathogens. This concern has been evident from an increased effort to collect information on vectors and to incorporate additional safeguards in permits for the importation of known vectors. Some vectors requested by scientists include *Cicadulina mbila* (Naude) from Africa, *Cicadulina pastusae* Ruppel and DeLong from South America, *Epilachna varivestis* Mulsant from Central America, and *Spongospora subterranea* (Wallr.) Lagerh. from South America. Some of the safeguards that have been considered during the preparation of permits for vectors will be analyzed later in this discussion.

8.3 INFORMATION NECESSARY FOR QUARANTINE DECISIONS CONCERNING VECTORS

Before any discussion of this subject, it is important to define the vector-pathogen relationships of significance to regulatory officials. For the purposes of plant quarantine decisions, the term "vector" should include only those organisms which are capable of acquiring and transmitting plant pathogens. Organisms which are observed in the vicinity of pathogens in nature or which aid the invasion of pathogens by wounding the host are not considered vectors for the purposes of this discussion unless these organisms have been shown experimentally to transmit a pathogen to a healthy host.

In the case of vectors of fungi and true bacteria, evidence that the pathogen can be isolated from vectors collected in nature is considered as incriminating as actual transmission studies in assessing an organism's capabilities as a vector. Most of these pathogens require for infection a natural opening or a wound in the epidermis of a susceptible host. Therefore, the assumption is that there is a high probability that fungal spores or bacterial ooze from a contaminated vector will be deposited in the vicinity of a wound or natural opening during the normal activity of the vector.

However, ingestion of viruses, mycoplasmalike organisms (MLOs), spiroplasmas or rickettsialike bacteria (RLBs) by an organism does not insure that the pathogen will be deposited in susceptible tissues during the normal interaction of the vector with its host. Some viruses and all plant pathogenic MLOs, RLBs and spiroplasmas must be injected into the phloem or xylem to insure infection of the susceptible host. In order to be deposited in these tissues, the vector-borne pathogen must contaminate the salivary glands of the vector so that the pathogen will be injected into the susceptible plant tissues during the feeding process of the vector. In these cases, transmission studies with organisms either collected in nature or reared in captivity are the only valid evidence of vector capability.

8.3.1 Geographic Distribution of Vector-Borne Pathogens

Once a vector-pathogen combination has been established and confirmed in the scientific literature, the distribution of the vector-borne pathogen becomes of critical importance to regulatory officials. Movement of vectors of pathogens widely distributed in the United States may not be as severely regulated by APHIS-PPQ as movement of vectors of pathogens not known to occur in this country. The importation of potential vectors from locations distant from the known occurrence of a foreign pathogen represents a lower risk to U.S. agriculture than importation of potential vectors from countries reporting outbreaks of the foreign pathogen the vector is reported to transmit. In order to make these evaluations, regulatory officials must have access to the most recent information on the distribution of each pathogen transmitted by a vector.

8.3.2 Distribution of the Viable Pathogen on or in the Vector

An artificial classification of pathogens transmitted by vectors can be made on the basis of whether the pathogen is reported on external surfaces of vectors or whether the pathogen can survive in the internal organs of the vector. Examples of pathogenic fungi and bacteria located on every possible external surface of vectors have been reported (Carter, 1973). Some of these pathogens can also be ingested, passed through the intestinal tract and excreted in a viable condition. There are even examples of fungi and bacteria spreading to other organs of their vectors. However, despite the fact many fungi and bacteria are only reported as external contaminants on their vectors, there are few available data to suggest that these bacteria and fungi can only be carried on external surfaces and would not retain their viability if ingested. Recent data suggest that some bacterial and fungal pathogens are transmitted to plants by homopterous and heteropterous vectors via an ingestion-egestion mechanism (Harris, 1979, 1980; Harris *et al.*, 1980). Vectors that can only carry plant pathogens on external surfaces would be of particular interest, since effective surface decontamination would eliminate the pathogens from the vector.

One of the criteria often used to classify submicroscopic pathogens is the mechanism of transmission by vectors. Traditionally, certain viruses were thought to be carried on the surface of the mouthparts or stylets of arthropod vectors. Even though these viruses may be ingested and remain infectious in their vector, there was no evidence to suggest that ingested particles could be used by the vector to infect a plant. More recent data, however, indicate that vector ingestion-egestion behavior plays an important, if not essential, role in "stylet-borne" and semipersistent, noncirculative virus transmission by aphids and leafhopper (Harris, 1977, 1978, 1979; Harris and Childress, 1980).

In contrast, other viruses and plant pathogenic MLOs, RLBs and spiroplasmas circulate in the digestive and circulatory systems of their vectors before being injected by the vector into susceptible host tissues during salivation and feeding (Harris, 1979). In these cases, the evidence indicates the entity causing infection survives and sometimes multiplies within the vector but would not be a threat to a plant as a contaminant on any external surface of the vector.

8.3.3 Retention Period of Pathogens by Vectors

For regulatory purposes, a distinction can be made between vectors which retain their pathogens for life and vectors which lose the ability to transmit acquired pathogens long before death. Vectors which retain viable components of pathogens for only a few hours or a few days can be freed of the pathogen in different ways and, therefore, have the potential for a lower pest risk evaluation. Beetles, grasshoppers, mealybugs (Table II), eriophyid mites, and certain aphids retain plant pathogenic viruses for a short period of time not approaching their life span. Unfortunately, most aphids that carry *certain* viruses for short periods of time (nonpersistently or semipersistently) also are capable of transmitting *other* viruses for their life span (persistently). Nematodes and fungi as well as most insects in the families Aleyrodidae, Cicadellidae, Cixiidae, Delphacidae (Table III), Miridae Piesmatidae, Psyllidae, and Thripidae, which transmit viruses,

MLOs, RLBs and spiroplasmas, may retain their pathogens for periods approaching their life span.

Unfortunately, there is less data available on the retention of fungi and true bacteria by vectors, but it seems safe to assume that the retention period is closely associated with the viability of the different fungal spores or bacterial cells on external surfaces or in the internal organs of vectors. Some fungi and bacteria are ingested by their vectors, passed through the gut and excreted after various periods of time in a viable condition in feces or during deposition of eggs or the offspring. The retention period of fungal spores contaminating external portions of a vector obviously will also depend on how easily the spores are dislodged. It would be interesting to determine how long bacteria will remain viable on external parts of a vector. Ark and Thomas (1936) have demonstrated that *Erwinia amylovora* (Burr.) Winslow *et al.* contaminating the heads of honeybees, *Apis mellifera* L., does not remain viable longer than 24 hr.

8.3.4 Transmission of Pathogens to the Progeny of Vectors

Of those pathogens retained for long periods, most are not passed to the progeny of the vector. Regulatory officials must be aware of those vectors that can be imported as eggs free of known plant pathogens since such importations would considerably lower the risk to U.S. agriculture. For example, the importation of nematode eggs reliably certified as to species and produced in pure culture would eliminate the risk that viruses normally transmitted by nematodes would be imported with the vector, because these viruses are not transmitted to the progeny of viruliferous adults through the eggs.

In some cases involving fungal and bacterial pathogens, the pathogen is ingested by the vector and contaminates the surface of the egg or progeny at the time of discharge from the adult. In these cases, the pathogen may be sufficiently exposed to be inactivated by physical or chemical treatments. However, there are pathogens which are transmitted internally in the egg or resting spore to the vector's progeny (Table IV). In these cases, disinfestation of external surfaces would be ineffective in eliminating the pathogen from the vector.

8.4 APPROACHES TO PREVENT THE INTRODUCTION OF POTENTIAL VECTORS WITH IMPORTED CARGO

When a pest or pathogen is intercepted and subsequently identified, the decision to treat the contaminated material should be based not only on the organism's damage potential or its absence from the United States, but also its capacity to transmit a plant pathogen. Theoretically, PPQ inspectors at ports of entry should be supplied with a listing of vectors in a form similar to Tables II and III so that they will be able to recognize potential vectors and, therefore, prevent the introduction of vector-borne plant pathogens.

TABLE II. Vectors in the Family Pseudococcidae (Order: Hemiptera)

Vector[a]	Pathogen	Pathogen Distribution	References[b]
Allococcus sp. near *proteae* (Hall)	cacao swollen shoot virus[c]	Ghana, Indonesia, Ivory Coast, Nigeria, Sabah (Malaysia), Sierra Leone, Sri Lanka, Togo	Thorold, 1975
Delococcus tafoensis (Strickland)	cacao swollen shoot virus	Ghana, Indonesia, Ivory Coast, Nigeria, Sabah (Malaysia), Sierre Leone, Sri Lanka, Togo	Roivainen, 1980
Dysmicoccus brevipes (Cockerell)	cacao mottle leaf virus[c] (=cacao swollen shoot virus)[d]	Ghana, Nigeria, Togo	Thorold, 1975
	cacao swollen shoot virus	Ghana, Indonesia, Ivory Coast, Nigeria, Sabah (Malaysia), Sierra Leone, Sri Lanka, Togo	Thorold, 1975
	Trinidad cacao disease agent[e]	Trinidad	Thorold, 1975
D. pseudobrevipes (Mamet) (see *Dysmicoccus brevipes*)			Williams, 1970
D. sp. near *brevipes* (Cockerell)	Trinidad cacao disease agent	Trinidad	Thorold, 1975
Ferrisia virgata (Cockerell)	cacao swollen shoot virus	Ghana, Indonesia, Ivory Coast, Nigeria, Sabah (Malaysia), Sierra Leone, Sri Lanka, Togo	Thorold, 1975
	Calonectria rigidiuscula (Berk. & Br.) Sacc.[f]	worldwide	Booth & Waterston, 1964
	Trinidad cacao disease agent	Trinidad	Thorold, 1975
Ferrisiana virgata (Cockerell) (see *Ferrisia virgata*)			Williams, 1970
Formicococcus tafoensis Strickland (see *Delococcus tafoensis*)			Williams, 1958b
Maconellicoccus ugandae (Laing)	cacao swollen shoot virus	Ghana, Indonesia, Ivory Coast, Nigeria, Sabah (Malaysia), Sierra Leone, Sri Lanka, Togo	Thorold, 1975
Paracoccus sp. near *proteae* (Hall) (see *Allococcus* sp. near *proteae*)			De Lotto, 1977

Paraputo anomalus (Newstead)	cacao swollen shoot virus	Ghana, Indonesia, Ivory Coast, Nigeria, Sabah (Malaysia), Sierra Leone, Sri Lanka, Togo	Thorold, 1975
P. multispinosa James (see *Paraputo anomalus*)			De Lotto, 1964
P. ritchiei Laing (see *Paraputo anomalus*)			De Lotto, 1964
Phenacoccus hargreavesi (Laing)	cacao mottle leaf virus (=cacao swollen shoot virus)	Ghana, Nigeria, Togo	Thorold, 1975
	cacao swollen shoot virus	Ghana, Indonesia, Ivory Coast, Nigeria, Sabah (Malaysia), Sierra Leone, Sri Lanka, Togo	Thorold, 1975
P. saccharifolii (Green)	sugarcane spike disease agent[f]	India	Ali, 1963
P. sp. H5418 (see *Maconellicoccus ugandae*)			Thorold, 1975
Planococcoides njalensis (Laing)	cacao mottle leaf virus (=cacao swollen shoot virus)	Ghana, Nigeria, Togo	Thorold, 1975
	cacao swollen shoot virus	Ghana, Indonesia, Ivory Coast, Nigeria, Sabah (Malaysia), Sierra Leone, Sri Lanka, Togo	Thorold, 1975
	Calonectria rigidiuscula (Berk. & Br.) Sacc.	worldwide	Booth & Waterston, 1964
Planococcus citri (Risso)	cacao mottle leaf virus (=cacao swollen shoot virus)	Ghana, Nigeria, Togo	Thorold, 1975
	cacao swollen shoot virus	Ghana, Indonesia, Ivory Coast, Nigeria, Sabah (Malaysia), Sierra Leone, Sri Lanka, Togo	Thorold, 1975
	small bacilliform taro virus[f]	Southwest Pacific	Gollifer *et al.*, 1977
	Trinidad cacao disease agent	Trinidad	Thorold, 1975
P. kenyae (Le Pelley)	cacao mottle leaf virus (=cacao swollen shoot virus)	Ghana, Nigeria, Togo	Thorold, 1975
	cacao swollen shoot virus	Ghana, Indonesia, Ivory Coast, Nigeria, Sabah (Malaysia), Sierra Leone, Sri Lanka, Togo	Thorold, 1975

TABLE II. (continued)

Vector[a]	Pathogen	Pathogen Distribution	References[b]
P. lilacinus (Cockerell)	cacao swollen shoot virus	Ghana, Indonesia, Ivory Coast, Nigeria, Sabah (Malaysia), Sierra Leone, Sri Lanka, Togo	Thorold, 1975
P. sp. near *celtis* (Strickland)	cacao mottle leaf virus (=cacao swollen shoot virus)	Ghana, Nigeria, Togo	Thorold, 1975
	cacao swollen shoot virus	Ghana, Indonesia, Ivory Coast, Nigeria, Sabah (Malaysia), Sierra Leone, Sri Lanka, Togo	Thorold, 1975
Pseudococcus adonidum (L.) (see *Pseudococcus longispinus*)			Williams, 1970
P. brevipes (Cockerell) (see *Dysmicoccus brevipes*)			Williams, 1958a
P. bromeliae (Bouche) (see *Dysmicoccus brevipes*)			Williams, 1970
P. bukobensis Laing (see *Phenacoccus hargreavesi*)			Williams, 1970
P. calceolariae minor (Maskell) (see *Planococcus citri*)			Williams, 1958a
P. cannae Green (see *Dysmicoccus brevipes*)			Williams, 1958a
P. citri (Risso) (see *Planococcus citri*)			Williams, 1970
P. comstocki (Kuwana)	Trinidad cacao disease agent	Trinidad	Thorold, 1975
P. concavocerarii James	cacao mottle leaf virus (=cacao swollen shoot virus)	Ghana, Nigeria, Togo	Thorold, 1975
	cacao swollen shoot virus	Ghana, Indonesia, Ivory Coast, Nigeria, Sabah (Malaysia), Sierra Leone, Sri Lanka, Togo	Thorold, 1975
P. crotonis Green (see *Planococcus lilacinus*)			Ezzat & McConnell, 1956
P. exitiabilis Laing (see *Planococcoides njalensis*)			De Lotto, 1964
P. hargreavesi Laing (see *Phenacoccus hargreavesi*)			Williams, 1970

P. kenyae Le Pelley (see *Planococcus kenyae*)			De Lotto, 1964
P. kenyaensis Betrem (see *Planococcus kenyae*)			De Lotto, 1964
P. lilacinus Cockerell (see *Planococcus lilacinus*)			Williams, 1958a
P. longirostralis James (see *Dysmicoccus brevipes*)			De Lotto, 1964
P. longispinus (Targioni-Tozzetti)	cacao mottle leaf virus (=cacao swollen shoot virus)	Ghana, Nigeria, Togo	Thorold, 1975
	cacao swollen shoot virus	Ghana, Indonesia, Ivory Coast, Nigeria, Sabah (Malaysia), Sierra Leone, Sri Lanka, Togo	Thorold, 1975
	small bacilliform taro virus	Southwest Pacific	Gollifer *et al.*, 1977
P. njalensis Laing (see *Planococcoides njalensis*)			De Lotto, 1964
P. pseudobrevipes Mamet (see *Dysmicoccus brevipes*)			Williams, 1970
P. saccharifolii (Green) (see *Phenacoccus saccharifolii*)			Williams, 1970
P. ugandae Laing (see *Maconellicoccus ugandae*)			De Lotto, 1964
P. virgatus (Cockerell) (see *Ferrisia virgata*)			Williams, 1970
P. sp near *brevipes* (Cockerell) (see *Dysmicoccus* sp. near *brevipes*)			Willaims, 1970
P. sp. near *calceolariae* (Maskell)	cacao mottle leaf virus (=cacao swollen shoot virus)	Ghana, Nigeria, Togo	Thorold, 1975
	cacao swollen shoot virus	Ghana, Indonesia, Ivory Coast, Nigeria, Sabah (Malaysia), Sierra Leone, Sri Lanka, Togo	Thorold, 1975

TABLE II. (continued)

Vector[a]	Pathogen	Pathogen Distribution	References[b]
P. sp. near *celtis* Strickland (see *Planococcus* sp. near *celtis*)			Williams, 1958b
P. sp. near *fragilis* Brain (see *Pseudococcus* sp. near *calceolariae*)			Williams & De Boer, 1973
P. sp. near *gahani* Green (see *Pseudococcus* sp. near *calceolariae*)			Williams & De Boer, 1973
P. sp. near *masakensis* James	cacao swollen shoot virus	Ghana, Indonesia, Ivory Coast, Nigeria, Sabah (Malaysia), Sierra Leone, Sri Lanka, Togo	Thorold, 1975
P. sp. near *proteae* Hall (see *Allococcus* sp. near *proteae*)			De Lotto, 1977
Tylococcus mauritiensis Mamet (see *Planococcus lilacinus*)			Williams, 1958a
T. westwoodi Strickland	cacao swollen shoot virus	Ghana, Indonesia, Ivory Coast, Nigeria, Sabah (Malaysia), Sierra Leone, Sri Lanka, Togo	Thorold, 1975

[a]Plant pathogens are listed opposite the currently accepted scientific name for each vector species. Synonyms commonly used during the last 50 years are included to help the reader recognize vectors when they are referred to by other names.

[b]The listed references are provided only to aid the reader to verify synonymies or vector-pathogen combinations and may or may not be appropriate for each individual reader's interests.

[c]Viruses may be retained by the vectors up to 4 days (Roivainen, 1980).

[d]Serological relationship between cacao mottle leaf virus and cacao swollen shoot virus demonstrated by Kenten and Legg (1971).

[e]Pathogen may be retained by the vectors up to 22½ h (Kirkpatrick, 1950).

[f]Retention period of this pathogen in its vector(s) is unknown at the present time.

Note: Evidence for mealybug transmission of tobacco mosaic virus or certain poorly defined pathogens is inconclusive (Roivainen, 1980).

TABLE III. Vectors in the Family Delphacidae (Order: Hemiptera)[a]

Vector[b]	Pathogen	Pathogen Distribution	References[c]
Asiraca striata (Fabricius) (see *Javesella pellucida*)			Metcalf, 1943; Nast, 1972
Callidelphax striatella (Fallen) (see *Laodelphax striatellus*)			Ling, 1973
Calligypona marginata (Fabricius) (see *Javesella pellucida*)			Harpaz, 1972
C. obscurella (Boheman) (see *Javesella obscurella*)			Mochida & Okada, 1971
C. pellucida (Fabricius) (see *Javesella pellucida*)			Mochida & Okada, 1971
C. propinqua (Fieber) (see *Toya propinqua*)			Harpaz, 1972; Nast, 1972
C. sordidula (Stål) (see *Megadelphax sordidulus*)			Raatikainen, 1960
C. striatella (Fallén) (see *Laodelphax striatellus*)			Mochida & Okada, 1971
C. vibix (Haupt) (see *Sogatella suezensis*)			Harpaz, 1972; Nast, 1972
Chloriona furcifera (Horvath) (see *Sogatella furcifera*)			Mochida & Okada, 1971
C. kolophon (Kirkaldy) (see *Sogatella kolophon*)			Mochida & Okada, 1971
Delphacinus hamatus (Boheman) (see *Dicranotropis hamata*)			Metcalf, 1943
Delphacodes akashiensis (Matsumura) (see *Laodelphax striatellus*)			Ling, 1973

TABLE III. (continued)

Vector[b]	Pathogen	Pathogen Distribution	References[c]
D. albifascia (Matsumura) (see *Unkanodes albifascia*)			Ling, 1973
D. albolineosa (Fowler) (see *Sogatella furcifera*)			Metcalf, 1943; Nast, 1972
D. albovittata (Matsumura) (see *Terthron albovittatus*)			Mochida & Okada, 1971
D. colocasiae (Matsumura) (see *Tarophagus proserpina*)			Ishihara, 1949
D. concolor (Fieber) (see *Javesella pellucida*)			Metcalf, 1943; Nast, 1972
D. difficilis (Edwards) (see *Javesella dubia*)			Metcalf, 1943; Nast, 1972
D. discolor (Boheman) (see *Javesella discolor*)			Metcalf, 1943; Nast, 1972
D. discreta (Edwards) (see *Javesella obscurella*)			Metcalf, 1943; Nast, 1972
D. fairmairei (Perris) (see *Muellerianella fairmairei*)			Mochida & Okada, 1971
D. flavipennis (Sahlberg) (see *Javesella pellucida*)			Metcalf, 1943; Nast, 1972
D. frontalis (Kirschbaum) (see *Javesella pellucida*)			Metcalf, 1943; Nast, 1972
D. furcifera (Horvath) (see *Sogatella furcifera*)			Metcalf, 1943; Nast, 1972
D. giffuensis (Matsumura) (see *Laodelphax striatellus*)			Ling, 1973
D. herrichii (Kirschbaum) (see *Javesella dubia* or *J. pellucida*)			Metcalf, 1943; Nast, 1972
D. longifurcifera Esaki & Ishihara (see *Sogatella longifurcifera*)			Mochida & Okada, 1971
D. maikoensis (Matsumura) (see *Laodelphax striatellus*)			Ling, 1973
D. marginata (Fabricius) (see *Javesella pellucida*)			Harpaz, 1972

D. minoensis (Matsumura) (see *Laodelphax striatellus*)			Ling, 1973
D. nipponica (Matsumura) (see *Laodelphax striatellus*)			Ling, 1973
D. nitidipennis (Kirschbaum) (see *Javesella dubia*)			Metcalf, 1943; Nast, 1972
D. obscurella (Boheman) (see *Javesella obscurella*)			Mochida & Okada, 1971
D. oryzae (Matsumura) (see *Nilaparvata lugens*)			Ling, 1973
D. pargasensis (Reuter) (see *Javesella dubia*)			Metcalf, 1943; Nast, 1972
D. pellucida (Fabricius) (see *Javesella pellucida*)			Mochida & Okada, 1971
D. propinqua (Fieber) (see *Toya propinqua*)			Metcalf, 1943; Nast, 1972
D. sapporona (Matsumura) (see *Unkanodes sapporonus*)			Ling, 1973
D. shirozui Ishihara (see *Toya propinqua*)			Mochida & Okada, 1971; Nast, 1972
D. similis (Kirschbaum) (see *Javesella discolor*)			Metcalf, 1943; Nast, 1972
D. sordidula (Stål) (see *Megadelphax sordidulus*)			Metcalf, 1943; Nast, 1972
D. striatella (Fallén) (see *Laodelphax striatellus*)			Ling, 1973
D. vastatrix (Breddin) (see *Perkinsiella vastatrix*)			Metcalf, 1943
Delphax (old generic name)			Metcalf, 1943
Dicranotropis albovittata Matsumura (see *Terthron albovittatus*)			Mochida & Okada, 1971
D. hamata (Boheman)	Arrhenatherum blue dwarf virus (=oat sterile dwarf virus)[d]	GDR, GFR, ?Czechoslovakia	Milne & Lovisolo, 1977
	cereal tillering disease virus	Sweden	Milne & Lovisolo, 1977
	oat sterile dwarf virus	Europe	Milne & Lovisolo, 1977

TABLE III. (continued)

Vector[b]	Pathogen	Pathogen Distribution	References[c]
D. vastatrix Breddin (see *Perkinsiella vastatrix*)			Metcalf, 1943
Hikona formosana Matsumura (see *Nilaparvata lugens*)			Ling, 1973
Javesella discolor (Boheman)	Arrhenatherum blue dwarf virus (=oat sterile dwarf virus)	GDR, GFR, ?Czechoslovakia	Milne & Lovisolo, 1977
	oat sterile dwarf virus	Europe	Milne & Lovisolo, 1977
J. dubia (Kirschbaum)	Arrhenatherum blue dwarf virus (=oat sterile dwarf virus)	GDR, GFR, ?Czechoslovakia	Milne & Lovisolo, 1977
	Dubia disease agent	GDR	Brcak, 1979
	European wheat striate mosaic virus	Europe, ?Turkey, ?Egypt	Bremer & Raatikainen, 1975
J. obscurella (Boheman)	European wheat striate mosaic virus	Europe, ?Turkey, ?Egypt	Bremer & Raatikainen, 1975
	oat sterile dwarf virus	Europe	Milne & Lovisolo, 1977
J. pellucida (Fabricius)	Arrhenatherum blue dwarf virus (=oat sterile dwarf virus)	GDR, GFR, ?Czechoslovakia	Milne & Lovisolo, 1977
	barley yellow striate mosaic virus	Italy	Conti, 1980
	Dubia disease agent	GDR	Brcak, 1979
	European aster yellows MLO[e]	Europe, ?Turkey	Mochida & Okada, 1971
	*European wheat striate mosaic virus	Europe, ?Turkey	Bremer & Raatikainen, 1975
	maize rough dwarf virus	Europe, Israel	Milne & Lovisolo, 1977
	*oat sterile dwarf virus	Europe	Milne & Lovisolo, 1977
	stolbur MLO	Europe, ?Middle East	Kostiw, 1976
Laodelphax striatellus (Fallén)	*barley yellow striate mosaic virus	Italy	Conti, 1980;Wiese, 1977
	cereal tillering disease virus	Sweden	Milne & Lovisolo, 1977
	*maize rough dwarf virus	Europe, Israel	Milne & Lovisolo, 1977
	northern cereal mosaic virus	Japan, Korea	Lee & Shikata, 1977
	oat pseudorosette agents (rhabdovirus & MLO)	U.S.S.R.	Milne & Lovisolo, 1977
	rice black-streaked dwarf virus	China, Japan, Korea	Ling, 1979
	*rice stripe virus	China, Japan, Korea, Taiwan, U.S.S.R.	Ling, 1979
	Spiroplasma citri Saglio *et al.*	worldwide in citrus	Bove *et al.*, 1979
	*wheat chlorotic streak virus	France	Wiese, 1977

Liburnia albicollis Haupt (see *Toya propinqua*)	Metcalf, 1943; Nast, 1972
L. albolineata Matsumura (see *Terthron albovittatus*)	Ishihara, 1949; Nast, 1972
L. albovittata (Matsumura) (see *Terthron albovittatus*)	Mochida & Okada, 1971
L. brazilensis (Muir) (see *Sogatodes orizicola*)	Ling, 1973
L. colocasiae Matsumura (see *Tarophagus proserpina*)	Ishihara, 1949
L. concolor (Fieber) (see *Javesella pellucida*)	Metcalf, 1943; Nast, 1972
L. cubana (Crawford) (see *Sogatodes cubanus*)	Ling, 1973
L. difficilis Edwards (see *Javesella dubia*)	Metcalf, 1943; Nast, 1972
L. discolor (Boheman) (see *Javesella discolor*)	Metcalf, 1943; Nast, 1972
L. discreta Edwards (see *Javesella obscurella*)	Metcalf, 1943; Nast, 1972
L. fairmairei (Perris) (see *Muellerianella fairmairei*)	Metcalf, 1943; Nast 1972
L. flavipennis Sahlberg (see *Javesella pellucida*)	Metcalf, 1943; Nast 1972
L. frontalis (Kirschbaum) (see *Javesella pellucida*)	Metcalf, 1943; Nast 1972
L. furcifera (Horvath) (see *Sogatella furcifera*)	Mochida & Okada, 1971
L. furcifera kolophon (Kirkaldy) (see *Sogatella kolophon*)	Metcalf, 1943; Mochida & Okada, 1971
L. Haupti Lindberg (see *Laodelphax striatellus*)	Ling, 1973
L. herrichii (Kirschbaum) (see *Javesella dubia* or *J. pellucida*)	Metcalf, 1943; Nast, 1972
L. marginata Haupt (see *Laodelphax striatellus*)	Ling, 1973
L. niveopicta Haupt (see *Laodelphax striatellus*)	Nast, 1972

TABLE III. (continued)

Vector[b]	Pathogen	Pathogen Distribution	References[c]
L. obscurella (Boheman) (see *Javesella obscurella*)			Mochida & Okada, 1971
L. orizicola (Muir) (see *Sogatodes orizicola*)			Ling, 1973
L. oryzae (Matsumura) (see *Nilaparvata lugens*)			Ling, 1973
L. pellucida (Fabricius) (see *Javesella pellucida*)			Mochida & Okada, 1971
L. propinqua (Fieber) (see *Toya propinqua*)			Metcalf, 1943; Nast, 1972
L. sordidula (Stål) (see *Megadelphax sordidulus*)			Metcalf, 1943; Nast, 1972
L. striatella (Fallén) (see *Laodelphax striatellus*)			Metcalf, 1943; Nast, 1972
L. suezensis (Matsumura) (see *Sogatella suezensis*)			Metcalf, 1943; Nast, 1972
L. venosa (Germar) (see *Javesella discolor*)			Metcalf, 1943; Nast, 1972
L. vibix Haupt (see *Sogatella suezensis*)			Nast, 1972
Malaxodes farinosus Fennah	maize tassel abortion agent	East Africa	Guthrie, 1977
	molasses dwarf disease agent	East Africa	Guthrie, 1977
Megadelphax sordidulus (Stål)	*Phleum* green stripe MLO	Finland	Bremer & Raatikainen, 1975
Megamelus proserpina Kirkaldy (see *Tarophagus proserpina*)			Ishihara, 1949
M. venosus (Germar) (see *Javesella discolor*)			Metcalf, 1943; Nast, 1972
Metadelphax propinqua (Fieber) (see *Toya propinqua*)			Nast, 1972

Muellerianella fairmairei (Perris)	northern cereal mosaic virus	Japan, Korea	Lee & Shikata, 1977
Nilaparvata greeni Distant (see *Nilaparvata lugens*)			Metcalf, 1943
N. lugens (Stål)	rice grassy stunt agent	India, Indonesia, Malaysia, Philippines, Sri Lanka, Taiwan, Thailand	Ling, 1979
	rice ragged stunt virus	India, Indonesia, Malaysia, Philippines, Sri Lanka, Thailand	Ling, 1979
N. oryzae (Matsumura) (see *Nilaparvata lugens*)			Ling, 1973
N. sordescens (Motschulsky) (see *Nilaparvata lugens*)			Ishihara, 1949
Peregrinus maidis (Ashmead)	*Digitaria* striate virus	Australia	Greber, 1979a
	maize line virus	East Africa, ?Reunion	Guthrie, 1977
	maize mosaic virus (rhabdovirus)	?Australia, Caribbean Islands, East Africa, Hawaii, India, Madagascar, Mauritius, South America	Shurtleff, 1973
	maize stripe virus	Africa, Peru, U.S., Venezuela, ?India, ?Philippines	Gingery *et al.*, 1979
	sorghum chlorosis virus	India	Raychaudhuri *et al.*, 1977
	B37 stunt virus	Australia	Greber, 1979b
Perkinsiella saccharicida Kirkaldy	*sugarcane Fiji disease virus	Australia, Madagascar, South Pacific Islands, Southeast Asia	Milne & Lovisolo, 1977
P. vastatrix (Breddin)	sugarcane Fiji disease virus	Australia, Madagascar, South Pacific Islands, Southeast Asia	Milne & Lovisolo, 1977
P. vitiensis Kirkaldy	sugarcane Fiji disease virus	Australia, Madagascar, South Pacific Islands, Southeast Asia	Milne & Lovisolo, 1977
Pundaluoya simplicia Distant (see *Peregrinus maidis*)			Metcalf, 1943

TABLE III. (continued)

Vector[b]	Pathogen	Pathogen Distribution	References[c]
Ribautodelphax albifascia (Matsumura) (see *Unkanodes albifascia*)			Ling, 1973
Sogata albovittata (Matsumura) (see *Terthron albovittatus*)			Mochida & Okada, 1971
S. cubana (Crawford) (see *Sogatodes cubanus*)			Ling, 1973
S. furcifera (Horvath) (see *Sogatella furcifera*)			Mochida & Okada, 1971
S. furcifera kolophon (Kirkaldy) (see *Sogatella kolophon*)			Mochida & Okada, 1971
S. longifurcifera (Esaki & Ishihara) (see *Sogatella longifurcifera*)			Mochida & Okada, 1971
S. oryzicola Muir (see *Sogatodes orizicola*)			Ling, 1973
S. pallescens Distant (see *Sogatella furcifera*)			Metcalf, 1943; Nast, 1972
Sogatella furcifera (Horvath)	pangola stunt virus	Fiji, South America, Taiwan	Milne & Lovisolo, 1977
S. kolophon (Kirkaldy)	*Digitaria* striate virus	Australia	Greber, 1979a
S. longifurcifera (Esaki & Ishihara)	ragi mosaic virus (rhabdovirus)	India	Maramorosch *et al.*, 1977
S. suezensis (Matsumura)	maize rough dwarf virus	Europe, Israel	Harpaz, 1972
S. vibix (Haupt) (see *Sogatella suezensis*)			Nast, 1972
Sogatodes cubanus (Crawford)	rice hoja blanca virus	Central & South America, Caribbean Islands, Mexico, U.S.	Ling, 1979
S. brazilensis (Muir) (see *Sogatodes orizicola*)			Ling, 1973

S. orizicola (Muir)	*rice hoja blanca virus	Central & South America, Caribbean Islands, Mexico, U.S.	Ling, 1979
Tarophagus proserpina (Kirkaldy)	large bacilliform taro virus (=bobone)	Solomon Islands	Gollifer *et al.*, 1977
Terthron albovittatus (Matsumura)	northern cereal mosaic virus	Japan, Korea	Mochida & Okada, 1971
	rice stripe virus	China, Japan, Korea, Taiwan, U.S.S.R.	Ling, 1979
Toya catilina (Fennah)	African cereal streak virus	Kenya, ?Ethiopia	Wiese, 1977
T. propinqua (Fieber)	maize rough dwarf virus (transmitted after injection only)	Europe, Israel	Harpaz, 1972
	Spiroplasma citri Saglio *et al.* (detected in insect only)	worldwide in citrus	Bove *et al.*, 1979
Unkana sapporona Matsumura (see *Unkanodes sapporonus*)			Ling, 1973
Unkanella sapporona (Matsumura) (see *Unkanodes sapporonus*)			Ling, 1973
Unkanodes albifascia (Matsumura)	northern cereal mosaic virus	Japan, Korea	Lee & Shikata, 1977
	rice black-streaked dwarf virus	China, Japan, Korea	Ling, 1979
	rice stripe virus	China, Japan, Korea, Taiwan, U.S.S.R.	Ling, 1979
U. sapporonus (Matsumura)	northern cereal mosaic virus	Japan, Korea	Lee & Shikata, 1977
	rice black-streaked dwarf virus	China, Japan, Korea	Ling, 1979
	rice stripe virus	China, Japan, Korea, Taiwan, U.S.S.R.	Ling, 1979
Weidnerianella (see *Javesella*)			Le Quesne, 1964

*Pathogen carried within the egg of the vector.

[a]Vectors in this family usually transmit plant pathogens until death.

[b]Plant pathogens are listed opposite the currently accepted scientific name for each vector species. Synonyms used during the last 50 years are included to help the reader recognize vectors when they are referred to by other names.

[c]The listed references are provided only to aid the reader to verify synonymies or vector-pathogen combinations and may or may not be appropriate for each individual reader's interests.

[d]Milne and Lesemann (1978) consider *Arrhenatherum* blue dwarf virus to be a form of oat sterile dwarf virus.

[e]MLO = mycoplasmalike organism.

TABLE IV. Vectors that Carry Pathogens within the Egg (Arthropods) or Resting Spore (Fungi)

Vector	Pathogen	Pathogen Distribution	References
Phylum: Arthropoda[a]			
Order: Diptera			
Family: Tephritidae			
Dacus oleae Rossi	*Pseudomonas savastonoi* (Sm.) Stev. (olive knot pathogen)	Argentina, Brazil, Iran, Mediterranean region, U.S.	Carter, 1973
Order: Hemiptera			
Family: Aphididae			
Myzus persicae (Sulzer)	potato leafroll virus	worldwide in potato	Carter, 1973
Nasonovia lactucae (L.)	sowthistle yellow vein virus	England, U.S. France, the Netherlands	Carter, 1973
Family: Cicadellidae			
Agallia constricta Van Duzee	potato yellow dwarf virus	Canada, U.S.	Nielson, 1968; Carter, 1973
	wound tumor virus	U.S.	Nielson, 1968
Agalliopsis novella (Say)	clover club-leaf RLB[b]	U.S.	Nielson, 1968; Carter, 1973
	wound tumor virus	U.S.	Nielson, 1968; Carter, 1973
Austroagallia torrida Evans	clover rugose leaf-curl RLB	Australia	Nielson, 1968; Carter, 1973
Cicadulina bipunctata bimaculata Evans	maize wallaby ear virus	Australia	Grylls, 1975
C. pastusae Ruppel & DeLong	cereal enanismo agents	Colombia, Ecuador	Wiese, 1977
Euscelis plebejus (Fallén)	clover phyllody MLO[c] (?)	Canada, Europe, U.S.	Carter, 1973
	clover witches-broom MLO (?)	England	Carter, 1973
Nephotettix cincticeps (Uhler)	rice dwarf virus	China, Japan, Korea	Nielson, 1968; Carter, 1973
Nephotettix nigropictus (Stål)	rice dwarf virus	China, Japan, Korea	Nielson, 1968
Psammotettix striatus (Linné)	Russian winter wheat mosaic virus	Eastern Europe, U.S.S.R.	Carter, 1973; Wiese, 1977
Recilia dorsalis (Motschulsky)	rice dwarf virus	China, Japan, Korea	Nielson, 1968; Carter, 1973
Family: Delphacidae			
Javesella pellucida (Fabricius)	Dubia disease agent	GDR, ?Egypt	Brcak, 1979
	European wheat striate mosaic virus		Carter, 1973; Ammar, 1975
	oat sterile dwarf virus	Europe	Milne & Lovisolo, 1977
Laodelphax striatellus (Fallén)	barley yellow striate mosaic virus	Italy	Conti, 1980

	maize rough dwarf virus	Europe, Israel	Milne & Lovisolo, 1977
	rice stripe virus	China, Japan, Korea, Taiwan, U.S.S.R.	Carter, 1973
	wheat chlorotic streak mosaic virus	France	Wiese, 1977
Perkinsiella saccharicida Kirkaldy	sugarcane Fiji disease virus	Australia, Madagascar, South Pacific Islands, Southeast Asia	Chang, 1977
Sogatodes orizicola (Muir)	rice hoja blanca virus	Central & South America, Caribbean Islands, Mexico, U.S.	Carter, 1973
Division: Eumycota			
Class: Chytridiomycetes[d]			
Order: Chytridiales			
Family: Olpidiaceae			
Olpidium brassicae (Wor.) Dang.	lettuce big vein agent	Australia, Canada, Colombia, Europe, Israel, Japan, New Zealand, U.S.	Campbell & Fry, 1966
	tobacco stunt virus	Japan	Hiruki, 1965
Class: Plasmodiophoromycetes[e]			
Order: Plasmodiophorales			
Family: Plasmodiophoraceae			
Polymyxa betae Keskin	beet necrotic yellow vein virus	Japan, Europe	Tamada, 1975
Polymyxa graminis Ledingham	barley yellow mosaic virus	Japan, GFR	Inouye & Saito, 1975
	oat mosaic virus	U.K., U.S., ?New Zealand	Hebert & Panizo, 1975
	rice necrosis mosaic virus	Japan	Inouye & Fujii, 1977
	soil-borne wheat mosaic virus	Argentina, Brazil, Egypt, Italy, Japan, U.S.	Brakke, 1971; Wiese, 1977
	wheat spindle streak mosaic virus	Canada, Japan, U.S., ?France, ?India	Slykhuis, 1976
Spongospora subterranea (Wallr.) Lagerh.	potato mop-top virus	Bolivia, Peru, Western Europe	Harrison, 1974

[a] Vectors were classified by specialists in the USDA Systematic Entomology Laboratory.
[b] RLB = rickettsia-like bacteria
[c] MLO = mycoplasma-like organism
[d] Vector was classified according to the taxonomic system used by Sparrow (1973).
[e] Vectors were classified according to the taxonomic system used by Waterhouse (1973).

However, before such a program could be implemented, some difficult decisions must be made concerning which vectors and pathogens would be regulated. Would it be wise to prevent entry of all vectors, including those of domestic pathogens, or would it be more practical to regulate only against vectors of foreign pathogens? Some may argue that the introduction of a new biotype of a vector known to transmit a domestic pathogen or the introduction of a new foreign strain of a domestic pathogen in an imported vector may trigger a new epidemic in the United States. Would only vectors from countries reporting the vector-borne pathogen be eradicated or would vectors from nearby countries not yet reporting the pathogen also be treated? How would foreign plant material exposed to the potential vector be handled? Another difficult question involves whether regulatory officials have the legal right to treat material contaminated with a potential vector known to transmit a pathogen from the country of origin, even though the presence of that pathogen on or in the vector has not been demonstrated. If evidence of the pathogen is required, the detection of most viruses, MLOs and RLBs would have to be done by a specialist. How important is the destination of the shipment? The treatment of mealybugs found on plant material from Africa destined for Minnesota hardly seems justifiable since cacao is not grown in our northern States. The answers to these and other questions will be quite difficult to integrate into a simple but workable program.

8.5 APPROACHES FOR THE SAFE IMPORTATION OF POTENTIAL VECTORS AS CULTURES

Since the objective of plant quarantine activities at ports of entry is to insure that imported cargo is free of exotic pests and pathogens, the basic decision facing each inspector who detects a potential vector is whether to eradicate it or not. However, when a pest or pathogen is imported by a scientist as a culture, more subtle measures must be employed, because in this case the objective is to keep the potential vector alive while eliminating any plant pathogens it may be carrying. Therefore, knowledge of the vector-pathogen relationship is necessary in order to manipulate the vector in a way that selectively eliminates the pathogen. For the regulatory official, the decision becomes a matter of choosing which vector-pathogen combinations to regulate and then which of the following safeguards, based on the vector-pathogen relationship, to incorporate into the permit in order to minimize the threat to U.S. agriculture.

8.5.1 Importation of Vectors Collected in Nature from Symptomless Plants

When collecting insects, mites, fungi or nematodes in the field, special care can be taken to reduce the possibility that the organisms collected will be carrying a plant pathogen. Since most vectors are plant pests or pathogens, these organisms must be collected either in the vicinity or on the surface of living plants in nature. Visual observations of the plants on which the organism is collected will often detect symptoms of diseases caused by pathogens that may be carried either internally or externally by the collected organism. For example, many insects collected from *Pyrus* species showing symptoms of fireblight caused by *E. amylovora* may be carrying this pathogen (Van der Zwet and Keil, 1979).

Unfortunately, not all infected plants exhibit symptoms throughout the growing season. Species of *Xiphinema* and *Longidorus* may be carrying one or more nepoviruses, even though the plant species nearby exhibit no symptoms of disease. In this case, only laboratory or greenhouse tests would detect these viruses in the plants or nematodes. Consequently, the absence of symptoms in the plant serving as a food source is no guarantee that the feeding organism is free of plant pathogens.

Even if the plant on which a vector is collected is healthy, the vector may be carrying a pathogen acquired as a result of its previous activity on an infected plant. Bacterial ooze or fungal spores may retain their viability for long periods of time either as external contaminants or after ingestion by the vector, and many viruses, MLOs, RLBs, and spiroplasmas, circulate for weeks through the gut, blood, hemolymph, and salivary glands of their vectors. Consequently, a requirement on a permit to collect potential vectors only from healthy plants in the field would not, by itself, be very effective at excluding pathogens in most instances.

8.5.2 Importation of Vectors Previously Reared in Captivity

Importation of organisms collected in nature and cultured on plants in a greenhouse or growth chamber by an entomologist in a foreign country, for whatever reason, represents less of a threat to U.S. agriculture than importation of field-collected potential vectors. In many cases, any pathogen carried by a field-collected vector will be transmitted to the host used as a food source, and the developing symptoms would alert the observer to a potential problem. However, if the rearing conditions are not conducive to infection and symptom development, the pathogen will remain undetected during the rearing of the vector.

The most likely pathogens to remain undetected are the fungi, bacteria, MLOs, RLBs, spiroplasmas and viruses which are carried internally for periods of time approaching the life of the vector. If the pathogen is only retained for a few hours or days by the vector, the pathogen can be eliminated by rearing the vector on a host immune to the pathogen. In this case, the observer may be unaware of the fact the vectors collected in nature were contaminated. However, if symptoms develop in the host, the observer must assume the vector was and probably still is carrying a plant pathogen.

In addition to foreign entomologists, foreign scientists who conduct research in vector relationships may also be good sources of uncontaminated potential vectors. In order to demonstrate transmission of a pathogen by a vector, a specialist in vector research usually maintains a vector colony free of the pathogen so that "clean" individuals from this colony can be exposed to the pathogen during experiments designed to characterize vector transmission. Individuals from the "clean" colony are more likely to be pathogen-free than similar individuals collected in nature. However, the possibility that the vectors in a foreign scientist's colony are contaminated, unknown to him, with a pathogen of quarantine significance to the United States must be considered. Therefore, further tests on vectors from foreign scientists may be necessary upon arrival in the United States, at least until the reliability of the foreign scientist is determined.

8.5.3 Packaging Vectors for Shipment

Since the organisms capable of carrying plant pathogens are either plant pests or pathogens and, therefore, require plant material as a food source, plant tissue is often included in the package to sustain the organism during shipment to the United States. Undoubtedly, the most hazardous plant material to use for the food source would be plant material collected in the field. If the plant tissue is infected or contaminated with a pathogen and packaged as a food source, the organism may acquire and later transmit the pathogen. Therefore, the risk to U.S. agriculture of importing this organism is increased by including field-collected plant tissue in the package. On the other hand, apparently healthy plant material grown from seed in a greenhouse or growth chamber would be preferable to field-collected plant tissue as a food source.

The species or genetic character of the plant material included with the organism during shipment may greatly affect the possibility that the organism is carrying a pathogen upon arrival in the United States. If a quarantine significant pathogen is retained for a short period of time by the vector and the host material used as a food source is immune to the pathogen, the pathogen may lose viability in the vector during shipment, and the threat to U.S. agriculture will be eliminated. For example, if the mealybug *Planococcus citri* (Risso) from Africa is packaged with tissue from plants in orders other than Malvales and Tiliales (Thorold, 1975), cacao swollen shoot virus should not be transmissible by the mealybugs after 4 days (Roivainen, 1980).

In recent years, important advances have been made in the development of artificial media for the culture of insects. Where possible, the substitution of an artificial medium for plant material in a package may be advantageous, especially in the case of vectors that retain plant pathogens for short periods. Of the vectors which transmit viruses in a nonpersistent manner, aphids (Pirone, 1977) and mealybugs (Okusanya, 1969) are capable of acquiring and later transmitting certain viruses from artificial media. However, it would be interesting to determine whether viruliferous aphids or mealybugs released on uncontaminated media would lose their ability to transmit nonpersistent viruses or would be able to reacquire these viruses from the media they had contaminated during the feeding process. Such studies should reveal the potential for using artificial media to safely import certain potential vectors.

8.5.4 Visual Inspection of Vectors for Plant Pathogens at Ports of Entry

This quarantine action would successfully detect only fungal spores or bacterial ooze carried on the external portions of the vector. Pathogens that are too small to observe under a light microscope or that are ingested by the vector would not be detected by visual inspection at ports of entry.

8.5.5 Decontamination of Imported Vectors

Very little research has been done to assess whether physical or chemical disinfectants would be useful on a routine basis in eliminating plant pathogens from living vectors. Fungal spores or bacterial ooze, viruses carried on external surfaces of vectors would seem to be most susceptible to treatment. Treatment of the exposed stylet of infective aphids with chemicals or ultraviolet radiation (Sylvester & Richardson, 1964; Bradley, 1966), may prevent transmission of some noncirculative viruses by inactivating virus or, and perhaps more likely, by temporarily altering the probing activity (sap sampling or ingestion-egestion behavior) responsible for transmission or both (Harris, 1977, 1979). In addition, tobacco necrosis virus has been inactivated by specific antisera or acid treatments of infectious *Olpidium brassicae* (Wor.) Dang. (Campbell and Fry, 1966). Some research on exposure of infectious leafhoppers to antibiotics has demonstrated reduced transmission of MLOs, but in most cases multiplication of pathogens in the vector resumes normally after a few weeks (Freitag and Smith, 1969; Whitcomb and Davis, 1970; Amin and Jensen, 1971). Future research with fungicides, bacteriocides, ultraviolet radiation or antibiotics may reveal instances when decontamination of imported vectors may be routinely possible.

8.5.6 Testing of Vectors for the Presence of Plant Pathogens

When adequate quarantine facilities are available, specific tests of potential vectors for specific pathogens may be possible. These tests usually involve caging suspected vectors on indicator plants chosen to detect the hazardous pathogen. In order for these tests to be accurate, the indicator plant must truely indicate the presence of the pathogen under the conditions used and the potential vector must be reared for a long enough period for transmission to occur. Of the tests available, transmission studies with suspected vectors are most likely to detect the widest variety of known and unknown pathogens.

Very sensitive serological tests, such as ELISA (Gera *et al.*, 1978; Clarke *et al.*, 1980) or fluorescent antibody techniques (Sinha and Paliwal, 1977), show potential for detecting pathogens in individual vectors. Fungi, bacteria and spiroplasmas could also be detected by culturing vectors on special media. Unfortunately, serological and cultural methods of detection usually require sacrificing imported organisms for the test, and, therefore, only a sample of the imported organisms could be tested. A low percentage of vectors could be infectious and be missed by these tests. Another disadvantage of these tests is that in the best of circumstances only pathogens that are being tested for are detected.

8.5.7 Importation of Vectors After Molting

The success of this procedure will depend on the location of the infectious entity in or on the vector. During a molt, a vector sheds its exoskeleton, including portions of its mouthparts and the intima lining its foregut. If the pathogen is located only on external surfaces or is carried in the fore portion of the alimentary canal, success at eliminating the pathogen from the vector may be achieved. Pathogens that circulate internally will not be affected during a molt.

Unfortunately, even if the pathogen is located on an external surface and is shed with the old skin, recontamination of the emerging vector may occur. The probability of recontamination is greater with fungal or bacterial pathogens than with viruses. An unfortunate brush against the discarded contaminated skins could recontaminate a vector with fungal spores or bacterial ooze. On the other hand, a nonpersistent virus must be reacquired by the vector during probing and feeding (sap sampling). For this reason, molting may be useful in the importation of aphid vectors known to transmit viruses exclusively in this manner.

8.5.8 Importation of the Progeny of Vectors

At present, the best method for assuring imported potential vectors are free of plant pathogens is to obtain the progeny of the organisms collected in the foreign country. Of the pathogens known to be transmitted by vectors, relatively few pathogens are transmitted by infectious adult vectors to their progeny. Functionally, this procedure involves removing the offspring immediately after birth or immediately after hatching from the egg and placing them on healthy plant tissue before they can feed on the parental host plant. In a permit, this procedure could be required of a foreign scientist before shipment or of a U.S. scientist immediately after importation into the United States. Importation of eggs or progeny certified as to species would prevent the introduction of most pathogens carried by vectors but would not eliminate pathogens carried either internally or externally by the egg or resting spore (fungi).

If the technology were available, the eggs or recently born or hatched progeny could be disinfected as a further precaution designed to reduce the risk of importation. This procedure would eliminate most fungi and bacteria which are ingested by the vector and eventually contaminate the surface of eggs during oviposition or the surface of progeny during birth. In Table IV, pathogens that are transovarially transmitted to their vector's progeny within the egg and which, therefore, are not susceptible to a surface decontamination procedure are listed. If one of these vectors were to be imported, specific tests to detect the listed pathogens must be undertaken, if feasible, to further protect U.S. agriculture.

8.6 CURRENT APPROACH TO THE EXCLUSION OF VECTORS

From the proceeding discussion, it should be obvious that there is no single procedure presently known that will detect and eliminate all pathogens from all known vectors. In view of the quantities of cargo imported into the United States daily, it is also practically impossible to detect and eliminate all known vectors from all cargo. In addition, we can also assume that not all vectors of plant pathogens are presently known. Such uncertainties may lead some quarantine officials to advise against the importations of any living insects, mites, fungi or nematodes.

At the present time, U.S. plant quarantine officials believe that a complete prohibition of the importation of plant pests and pathogens would impede commerce and cripple valuable scientific research. Instead, APHIS-PPQ reviews each individual situation on the basis of its risks and benefits. Such a policy will hopefully avoid the obvious danger of unrestricted importations but not restrict beneficial commercial or scientific activities through complete prohibition.

The policy of reviewing each application on its merits requires up-to-date information on known vectors and on procedures for safely importing potential vectors without plant pathogens. Therefore, APHIS-PPQ is developing a listing of known vectors in the form of tables similar to Tables II and III. With these lists, PPQ entomologists and plant pathologists can look up the pest or pathogen which is intercepted or requested, find any pathogens reported to be transmitted by this organism, and then determine by looking at the pathogen's distribution if the vector could be carrying a pathogen of quarantine significance to the United States. As in Tables II and III, synonyms used in the taxonomic and vector literature during the last 50 years will also be listed so that a potential vector by any valid name will be recognized as such. Only with such information can regulatory officials assess risks and then formulate decisions concerning vectors.

However, the assumption that all vectors are now known and will be included in our lists is unrealistic. New vectors are being discovered every year. Obviously, it would be impossible to test all requested plant pests or pathogens for all plant pathogens they might transmit; just as it would be impossible to inspect every item of cargo for potential vectors. Therefore, some future thought should be given to the development of standard importation procedures that would minimize the threat of vector-borne pathogens and also be simple enough to perform routinely on all imported specimens. At present, the best procedure for insuring unreported vectors are free of plant pathogens is to obtain and then decontaminate the eggs or progeny of the requested organisms originally collected in foreign countries. For ports of entry, the exclusion of vectors could involve statistical sampling of cargos for inspection and then treatment of all cargos found contaminated with any pest or pathogen.

8.7 ACKNOWLEDGMENTS

The author would like to recognize the valuable assistance provided by Dr. A. A. Brunt, Glasshouse Crops Research Institute, England on vectors of cacao viruses; Dr. J. P. Kramer, USDA Systematic Entomology Laboratory, on taxonomy of the vectors in the family *Delphacidae*; Dr. K. C. Ling, The International Rice Research Institute, Philippines on vectors of rice diseases; and Dr. D. R. Miller, USDA Systematic Entomology Laboratory, on taxonomy of the mealybug vectors. The Plant Importation and Technical Support Staff of the USDA-APHIS-PPQ must also be thanked for the data used in Table I and for their critical review of this manuscript. If the reader finds any statements in error, he or she should assume that the author has misinterpreted the information published in the scientific literature or provided by the experts he has consulted.

8.8 REFERENCES

Ali, S. M. (1963). Some experimental studies on transmission of sugarcane spike disease by a coccid, *Pseudococcus saccharifolii* (Green). *Ind. J. Sug. Res. Dev.* 7:189-191.

Amin, P. W., and Jensen, D. D. (1971). Effects of tetracycline on the transmission and pathogenicity of the western X disease agent in its insect and plant hosts. *Phytopathology* **61**:696-702.

Ammar, E. D. (1975). Effect of European wheat striate mosaic, acquired transovarially, on the biology of its planthopper vector *Javesella pellucida*. *Ann. Appl. Biol.* **79**:203-213.

Animal and Plant Health Inspection Service. (1979). List of Intercepted Plant Pests (Pests Recorded from July 1, 1973 through September 30, 1977). APHIS Publication 82-5.

Ark, P. A., and Thomas, H. E. (1936). Persistence of *Erwinia amylovora* in certain insects. *Phytopathology* **26**:375-381.

Booth, C., and Waterston, J. M. (1964). *Calonectria rigidiuscula. C.M.I. Descriptions of Pathogenic Fungi and Bacteria*, No. 21.

Bove, J. M., Moutous, G., Saillard, C., Fos, A., Bonfils, J., Vignault, J., Nhami, A., Abassi, M., Kabbage, K., Hafidi, B., Mouches, C., and Viennot-Bourgin, G. (1979). Mise en evidence de *Spiroplasma citri*, l'agent causal de la maladie du "stubborn" des argumes dans 7 cicadelles du Maroc. *C.R. Acad. Sci. Paris, Ser. D*, **288**:335-338.

Bradley, R. H. E. (1966). Which of an aphid's stylets carry transmissible virus? *Virology* **29**: 396-401.

Brakke, M. K. (1971). Soil-borne wheat mosaic virus. *C.M.I./A.A.B. Descriptions of Plant Viruses*, No. 77.

Breák, J. (1979). Leafhopper and planthopper vectors of plant disease agents in central and southern Europe. *In* "Leafhopper Vectors and Plant Disease Agents" (K. Maramorosch and K. F. Harris, eds.), pp. 97-154. Academic Press, New York and London.

Bremer, K., and Raatikainen, M. (1975). Cereal diseases transmitted or caused by aphids and leafhoppers in Turkey. *Ann. Acad. Sci. Fenn. A, IV Biologica* **203**:1-14.

Campbell, R. N., and Fry, P. R. (1966). The nature of the associations between *Olpidium brassicae* and lettuce big-vein and tobacco necrosis viruses. *Virology* **29**:222-233.

Carter, W. (1973). "Insects in Relation to Plant Disease," 2nd ed. Wiley (Interscience), New York.

Chang, V. C. S. (1977). Transovarial transmission of the Fiji disease virus in *Perkinsiella saccharicida* Kirk. *Sugarcane Pathol. News.* **18**:22-23.

Clarke, R. G., Converse, R. H., and Kojima, M. (1980). Enzyme-linked immunosorbent assay to detect potato leafroll virus in potato tubers and viruliferous aphids. *Plant Disease* **64**:43-45.

Conti, M. (1980). Vector relationships and other characteristics of barley yellow striate mosaic virus (BYSMV). *Ann. Appl. Biol.* **95**:83-92.

De Lotto, G. (1964). Observations on African mealy bugs (Hemiptera: Coccoidea). *Bull. Brit. Mus. (Nat. Hist.) Ent.* **14**:341-397.

De Lotto, G. (1977). On some African mealybugs (Homoptera: Coccoidea: Pseudococcidae). *J. Ent. Soc. South. Afr.* **40**:13-36.

Ezzat, Y. M., and McConnell, H. S. (1956). A classification of the mealybug tribe Planococcini (Pseudococcidae, Homoptera). *Univ. Md. Agr. Exp. Sta. Bull.* **A-84.**

Ferris, V. R. (1971). Taxonomy of the Dorylaimida. *In* "Plant Parasitic Nematodes," Vol. 1, Morphology, Anatomy, Taxonomy, and Ecology" (B. M. Zuckerman, W. F. Mai, and R. A. Rohde, eds.), pp. 163-189. Academic Press, New York.

Freitag, J. H., and Smith, S. H. (1969). Effects of tetracyclines on symptom expression and leafhopper transmission of aster yellows. *Phytopathology* **59**:1820-1823.

Gera, A., Loebenstein, G., and Raccah, B. (1978). Detection of cucumber mosaic virus in viruliferous aphids by enzyme-linked immunosorbent assay. *Virology* **86**:542-545.

Gingery, R. E., Nault, L. R., Tsai, J. H., and Lastra, R. J. (1979). Occurrence of maize stripe virus in the United States and Venezuela. *Plant Dis. Rep.* **63**:341-343.

Golden, M. A. (1971). Classification of the genera and higher categories of the order Tylenchida (Nematoda). *In* "Plant Parasitic Nematodes," Vol. 1, Morphology, Anatomy, Taxonomy, and Ecology (B. M. Zuckerman, W. F. Mai, and R. A. Rohde, eds.), pp. 191-232. Academic Press, New York.

Gollifer, D. E., Jackson, G. V. H., Dabek, A. J., Plumb, R. T., and May, Y. Y. (1977). The occurrence and transmission of viruses of edible aroids in the Solomon Islands and the Southwest Pacific. *PANS* **23**:171-177.

Greber, R. S. (1979a). *Digitaria* striate virus—a rhabdovirus of grasses transmitted by *Sogatella kolophon* (Kirk.). *Aust. J. Agr. Res.* **30**:43-51.

Greber, R. S. (1979b). Cereal chlorotic mottle virus—a rhabdovirus of Gramineae in Australia transmitted by *Nesoclutha pallida* (Evans). *Aust. J. Agric. Res.* **30**:433-443.

Grylls, N. E. (1975). Leafhopper transmission of a virus causing maize wallaby ear disease. *Ann. Appl. Biol.* **79**:283-296.

Guthrie, E. J. (1977). Virus diseases of maize in East Africa. *Proc. Int. Maize Virus Dis. Colloq. Workshop, Aug. 16-19, 1976* (L. E. Williams, D. T. Gordon, and L. R. Nault, eds.), pp. 62-66. Ohio Agricultural Research and Development Center, Wooster.

Harpaz, I. (1972). "Maize Rough Dwarf. A Planthopper Virus Disease Affecting Maize, Rice, Small Grains and Grasses." Israel Univ. Press, Jerusalem.

Harris, K. F. (1977). An ingestion-egestion hypothesis of noncirculative virus transmission. *In* "Aphids as Virus Vectors" (K. F. Harris and K. Maramorosch, eds.), pp. 165-220. Academic Press, New York and London.

Harris, K. F. (1978). Aphid-borne viruses: ecological and environmental aspects. *In* "Viruses and Environment" (E. Kurstak and K. Maramorosch, eds.), pp. 311-337. Academic Press, New York and London.

Harris, K. F. (1979). Leafhoppers and aphids as biological vectors: Vector-virus relationships. *In* "Leafhopper Vectors and Plant Disease Agents" (K. Maramorosch and K. F. Harris, eds.), pp. 217-308. Academic Press, New York and London.

Harris, K. F., and Childress, S. A. (1980). Fate of maize chlorotic dwarf virus (MCDV) in its black-faced leafhopper vector, *Graminella nigrifrons. XVI Int. Congr. Entomol. Abstracts,* Kyoto, Japan, 3-9 Aug., 1980. *Daily Bull. No. 4* (Abst. 8R-15-47).

Harris, K. F. (1980). Aphids, leafhoppers and planthoppers. *In* "Vectors of Plant Pathogens" (K. F. Harris and K. Maramorosch, eds.), pp. 1-13. Academic Press, New York and London.

Harris, K. F., Treur, B., Tsai, J., and Toler, R. (1981). Observations on leafhopper ingestion-egestion behavior: its likely role in the transmission of semipersistent viruses and other plant pathogens. In manuscript.

Harrison, B. D. (1974). Potato mop-top virus. *C.M.I./A.A.B. Descriptions of Plant Viruses* **No. 138.**

Hebert, T. T., and Panizo, C. H. (1975). Oat mosaic virus. *C.M.I./A.A.B. Descriptions of Plant Viruses* **No. 145.**

Hiruki, C. (1965). Transmission of tobacco stunt virus by *Olpidium brassicae. Virology* **25**: 541-549.

Inouye, T., and Fujii, S. (1977). Rice necrosis mosaic virus. *C.M.I./A.A.B. Descriptions of Plant Viruses* **No. 172.**

Inouye, T., and Saito, Y. (1975). Barley yellow mosaic virus. *C.M.I./A.A.B. Descriptions of Plant Viruses* **No. 143.**

Ishihara, T. (1949). Revision of the Araeopidae of Japan, Ryukyu Islands and Formosa. *Sci. Rep. Matsuyama Agr. College* **No. 2.**

Kahn, R. P. (1979). A concept of pest risk analysis. *EPPO Bull.* **9**:119-130.

Kenten, R. H., and Legg, J. T. (1971). Serological relationships of some viruses from cocoa (*Theobroma cacao* L.) in Ghana. *Ann. Appl. Biol.* **67**:195-200.

Kirkpatrick, T. W. (1950). Insect transmission of cacao virus disease in Trinidad. *Bull. Ent. Res.* **41**:99-118.

Kostiw, M. (1976). Vectors of potato viruses and mycoplasma-like bodies. *Biul. Inst. Ziemniaka* **17**:109-124.

Lee, S. H., and Shikata, E. (1977). Occurrence of northern cereal mosaic virus in Korea. *Kor. J. Plant Prot.* **16**:87-92.

Le Quesne, W. J. (1964). Some taxonomic observations on the British Delphacidae (Hemiptera). *Proc. Roy. Ent. Soc. London (B)* **33**:56-58.

Ling, K. C. (1973). Synonymies of insect vectors of rice viruses. International Rice Research Institute, Los Banos, Philippines.

Ling, K. C. (1979). Rice virus and virus-like diseases. International Rice Research Institute, Los Banos, Philippines.

Maramorosch, K., Govindu, H. C., and Kondo, F. (1977). Rhabdovirus particles associated with a mosaic disease of naturally infected *Eleusine coracana* (finger millet) in Karnataka State (Mysore), South India. *Plant Dis. Rep.* **61**:1029-1031.

Metcalf, Z. P. (1943). "General Catalogue of the Hemiptera. Fascicle IV. Fulgoroidea," Part 3. Araeopidae (Delphacidae). (W. E. China and H. M. Parshley, eds.). Smith College Press, Northampton, Massachusetts.

Milne, R. G., and Lesemann, D. E. (1978). An immunoelectron microscopic investigation of oat sterile dwarf and related viruses. *Virology* **90**:299-304.

Milne, R. G., and Lovisolo, O. (1977). Maize rough dwarf and related viruses. *Adv. Virus Res.* **21**:267-341.

Mochida, O., and Okada, T. (1971). A list of the Delphacidae (Homoptera) in Japan with special reference to host plants, transmission of plant diseases and natural enemies. *Bull. Kyushu Agr. Exp. Sta.* **15**:737-843.

Nast, J. (1972). "Palaearctic Auchenorrhyncha (Homoptera), an Annotated Check List." Polish Academy of Sciences Institute of Zoology, Polish Scientific Publishers, Warsaw.

Nielson, M. W. (1968). The leafhopper vectors of phytopathogenic viruses (Homoptera, Cicadellidae) taxonomy, biology, and virus transmission. *U. S. Dep. Agr. Tech. Bull.* **No. 1382.**

Okusanya, B. A. M. (1969). Cocoa swollen shoot virus. *Ann. Rep. Cocoa Res. Inst. Nigeria* **1967-1968**: 88-94.

Pirone, T. P. (1977). Accessory factors in nonpersistent virus transmission. *In* "Aphids As Virus Vectors" (K. F. Harris and K. Maramorosch, eds.), pp. 221-235. Academic Press, New York and London.

Raatikainen, M. (1960). The biology of *Calligypona sordidula* (Stal) (Hom., Auchenorrhyncha). *Ann. Ent. Fenn.* **26**:229-242.

Raychaudhuri, S. P., Seth, M. L., Renfro, B. L., and Varma, A. (1977). Principal maize virus diseases in India. *Proc. Int. Maize Virus Dis. Coll. Workshop, Aug. 16-19, 1976* (L. E. Williams, D. T. Gordon, and L. R. Nault, eds.), pp. 69-76. Ohio Agricultural Research and Development Center, Wooster.

Roivainen, O. (1980). Mealybugs. *In* "Vectors of Plant Pathogens" (K. F. Harris and K. Maramorosch, eds.), pp. 15-38. Academic Press, New York and London.

Shurtleff, M. C. (1973). "A Compendium of Corn Diseases." The American Phytopathological Society, St. Paul, Minnesota.

Siddiqi, M. R. (1971). Structure of the oesophagus in the classification of the superfamily Tylenchoidea (Nematoda). *Indian J. Nematol.* **1**:25-43.

Sinha, R. C., and Paliwal, Y. C. (1977). Detection of wheat streak mosaic virus antigens in vector mites with fluorescent antibodies. *Phytopathology* **67**:570-572.

Slykhuis, J. T. (1976). Wheat spindle streak mosaic virus. *C.M.I./A.A.B. Descriptions of Plant Viruses* **No. 167.**

Sparrow, F. K. (1973). Chytridiomycetes, Hyphochytridiomycetes. *In* "The Fungi, An Advanced Treatise," Vol. IVB. A Taxonomic Review with Keys: Basidiomycetes and Lower Fungi" (G. C. Ainsworth, F. K. Sparrow, and A. S. Sussman, eds.), pp. 85-110. Academic Press, New York.

Sylvester, E. S., and Richardson, J. (1964). Transmission of cabbage mosaic virus by green peach aphids–stylet transmission studies. *Virology* **22**:520-538.

Tamada, T. (1975). Beet necrotic yellow vein virus. *C.M.I./A.A.B. Descriptions of Plant Viruses* **No. 144.**

Thorold, C. A. (1975). "Diseases of Cocoa." Clarendon Press, Oxford.

Van der Zwet, T., and Keil, H. L. (1979). Fire blight, a bacterial disease of rosaceous plants. *U. S. Dep. Agr. Handbook* **No. 510.**

Waterhouse, G. M. (1973). Plasmodiophoromycetes. *In* "The Fungi, An Advanced Treatise," Vol. IVB. A Taxonomic Review with Keys: Basidiomycetes and Lower Fungi (G. C. Ainsworth, F. K. Sparrow, and A. S. Sussman, eds.), pp. 61-73. Academic Press, New York.

Whitcomb, R. F., and Davis, R. E. (1970). Evidence on possible mycoplasma etiology of aster yellows disease. II. Suppression of aster yellows in insect vectors. *Infect. Immun.* **2**: 209-215.

Wiese, M. V. (1977). "Compendium of Wheat Diseases." The American Phytopathological Society, St. Paul, Minnesota.

Williams, D. J. (1958a). The mealy-bugs (Pseudococcidae: Homoptera) described by W. M. Maskell, R. Newstead, T. D. A. Cockerell, and E. E. Green from the Ethiopian region. *Bull. Brit. Mus. (Nat. Hist.) Entomol.* **6**:205-236.

Williams, D. J. (1958b). The mealy bugs (Pseudococcidae: Homoptera) described by W. J. Hall, F. Laing and A. H. Strickland from the Ethiopian region. *Bull. Brit. Mus. (Nat. Hist.) Entomol.* **7**:1-37.

Williams, D. J. (1970). The mealybugs (Homoptera, Coccoidea, Pseudococcidae) of sugarcane, rice and sorghum. *Bull. Entomol. Res.* **60**:109-188.

Williams, D. J., and De Boer, J. A. (1973). The taxonomy of some New Zealand Pseudococcidae (Homoptera: Coccoidea). *Trans. Roy. Entomol. Soc. Lond.* **125**:227-252.

Chapter 9

NATURE OF INHERITED NEMATODE RESISTANCE IN PLANTS

Sven Bingefors

Department of Plant and Forest Protection
Swedish University of Agricultural Sciences
Uppsala, Sweden

PATHOGENS, VECTORS, AND PLANT DISEASES

ISBN 0-12-326440-5

9.1 INTRODUCTION

A number of plant parasitic nematodes cause great economic losses to cultivated crops. Their control by chemical means is expensive because the treatment often must be repeated at short intervals. Crop rotation to prevent nematode damage often requires long intervals between susceptible crops. The use of nematode resistant cultivars of susceptible crops seems to be a good way of controlling nematode damage. Therefore, research on resistance in plants to nematode attacks has been carried out to an increasing extent by plant breeders, geneticists, and nematologists, also in cooperation with biochemists.

There are often a number of hosts to each nematode species or to each host race within a species. This number may vary from a few to many but plant susceptibility is as unusual to parasitic nematodes as to other pathogens. Inherited nematode resistance is clearly limited to "host resistance" (Day, 1974), which depends on a genetic change in an otherwise susceptible host. Therefore, this review will consider varietal resistance, developed from variation within a plant species or transferred from related species, more or less intercrossable. There is little experimental background to discuss "nonhost resistance" in relation to varietal resistance to nematodes as well as to other pathogens.

Resistance against nematodes is normally demonstrated in their ability to feed, develop, and reproduce on a plant (Howard and Cotten, 1978). Nematodes reproduce poorly on a resistant plant. If the nematodes cause little injury the plant is tolerant (Dropkin and Nelson, 1960). Tolerance is desirable in a resistant cultivar as the nematodes may be able to cause considerable damage although they are not able to reproduce to a large extent. A tolerant, susceptible cultivar may support a large number of nematodes, sometimes with only limited injury.

Observations of resistance in plants to nematodes are available from the beginning of this century, but nematode-resistant rootstocks were recommended for citrus, coffee, etc., even earlier (Malo, 1964). Although basic knowledge on plant parasitic nematodes, including taxonomy, was limited, important advances in the development of resistant cultivars were made. These early investigations concerned *Meloidogyne* spp., at that time grouped into one "species," with resistant cultivars in cowpeas in the United States (Webber and Orton, 1902), *Hetero-*

dera spp. with resistant barley cultivars in Sweden (Nilsson-Ehle, 1908), *Ditylenchus dipsaci* with resistant rye in the Netherlands (Koeslag, 1931), alfalfa in Argentina (Burkart, 1937), and red clover in Sweden (Sylvén, 1937).

Resistance in plants to nematodes has been described mainly for sedentary endoparasites and for some migratory species. Here, some examples of important nematode resistance will be given. A number of earlier reviews describe resistant cultivars in crops or in some groups of crops. This chapter will mainly review investigations on structural and biochemical differences between resistant and susceptible hosts and discuss such information in relation to inherited nematode resistance in plants.

9.2 SOME EXAMPLES OF INHERITED NEMATODE RESISTANCE IN PLANTS

Heterodera. The cereal cyst nematode (*H. avenae*) attacks cereal crops and grasses. Resistance in barley was investigated early, and the interest was renewed in the 1950s (Andersen, 1961) in Western Europe, Australia, and India. Sources of resistance have been found in the cereal species–barley, oats, wheat, and rye–either in available cultivars, in more primitive local varieties, or in related species. Resistant barley varieties are more tolerant to nematode attacks than resistant oat varieties. Pathotypes have been clearly differentiated in barley (Cook and Williams, 1972). A working group has proposed a set of common differential cultivars for international investigations on pathotypes in this species (Holm Nielsen, 1972). Some cyst nematode populations on cereals have been described as separate, allied species (*H. hordecalis, H. iri*, and *H. mani*).

Nilsson-Ehle (1920) first described regular inheritance of resistance to nematodes in barley to *H. avenae*. The genetics of resistance in different cereal species has continued to be of great interest (Cook, 1974). One dominant gene, derived from barley No. 191, was found to confer resistance against two pathotypes. This and another dominant gene, independent or loosely linked to the first one but giving resistance to only one of the pathotypes and found in several barley cultivars, were described in extensive investigations in Denmark (Andersen, 1961; Andersen and Andersen, 1968, 1970). Another dominant gene for resistance to both pathotypes, closely linked to the one from No. 191, has been found in barley No. 14. Additional genes from other sources, probably also on chromosome 2, have been described. The occurrence of at least six genes was predicted on theoretical grounds (Hayes and Cotten, 1971). In an Australian investigation (O'Brien *et al.*, 1979) at least three, and possibly five, different genes were involved in barley resistance in the cultivars Athinais, Morocci, Nile, CI 8147, and Marocaine 079.

Resistance in oats to *H. avenae* has been derived from different *Avena* species, including *A. sativa, A. sterilis*, and *A. byzantina*. In *A. sativa*, the same locus, possibly with different alleles (Cook, 1974), confers dominant resistance to several cultivars. In the cultivar Mortgage Lifter, resistance was controlled by two recessive genes. Resistance in *A. sterilis* was found to be controlled by two dominant genes, while in *A. byzantina* resistance was dominant and monogenic (Cotten and Hayes, 1972). As Cook (1974) pointed out, the hexaploid constitution of cultivated oats complicates the interpretation of experimental results.

Resistance in wheat to *H. avenae* is controlled by a dominant gene in chromosome 2 B in the cultivar Loros (Holm Nielsen, 1966; Slootmaker *et al.*, 1974).

Useful resistance to *H. schachtii* in beets has not been found but transmission of resistance from *Beta procumbens* is being attempted. The resistance is controlled by one dominant gene or by three complementary genes (Yu, 1978). The transmission rate is very low (Savitsky, 1975), although more promising rates may have been reached recently (Nakamura and Tsuchiya, 1978). A low degree of resistance in *B. maritima* was believed (Heijbroek, 1977) to be recessive and polyfactorial.

Steele (1977) found in tomatoes that resistance to *H. schachtii* was "conferred by a single dominant gene, or that susceptibility is inherited as an epistatic combination of single recessive and single dominant genes."

H. glycines is a widely spread parasite, able to cause serious losses in soybeans. Resistant cultivars are available although there have been some difficulties in combining resistance and commercial quality and also in developing good cultivars resistant to new pathotypes (Riggs, 1977). Four or more pathotypes ("races") have been found, and they are able to hybridize (Price *et al.*, 1978). Triantaphyllou (1975) showed that the known races are populations differing from each other in the frequencies of three groups of genes for parasitism, acting quantitatively and controlling the reproduction on three differential cultivars.

Caldwell *et al.* (1960) considered resistance to *H. glycines* in soybeans to be controlled by three independent recessive genes. A dominant gene, closely linked to the I locus for seed-coat color, was later found to be required (Matson and Williams, 1965). Thomas *et al.* (1975) found that resistance to a pathotype 4 appeared to be conditioned by one dominant and two recessive genes.

Globodera. Breeding for resistance in potatoes to cyst nematodes *G. rostochiensis* and *G. pallida* has become a very important goal in most potato-breeding institutions. A number of cultivars resistant to *G. rostochiensis* have been released and are widely used. Tolerance (Huijsman *et al.*, 1969; Evans and Franco, 1979) but no useful resistance has been found in *Solanum tuberosum* spp. *tuberosum*. Ellenby (1952), after a ten-year search in *Solanum* collections, found a few resistant clones in *S. tuberosum* ssp. *andigena*, which were immediately used in breeding programs with European potatoes (Toxopeus and Huijsman, 1952). Potato cyst nematodes, originally considered as belonging to one species (*Heterodera rostochiensis*), were very soon found to be differentiated by the resistant clones into a number of pathotypes, and some of them could be separated as a species (*G. pallida*). Sources of resistance to both species and to their different known pathotypes in Europe and the United States have been found (Howard and Cotten, 1978). However, investigations indicate that "European populations of potato cyst nematodes contain only a fraction of the genetic variability found in South American populations" (Franco and Evans, 1978). Sources of resistance to some of these pathotypes have also been found (Mayer de Scurrah, 1978). International schemes for identifying and classifying pathotypes of potato cyst nematodes have been worked out (Kort *et al.*, 1977; Canto Saenz and Mayer de Scurrah, 1977).

Resistance to *G. rostochiensis* and *G. pallida* has been transferred from different cultivated and wild *Solanum* spp. The dominant gene H_1 from *S. tuberosum* spp. *andigena* gives resistance to the most common pathotype of *G. rostochiensis* and has been widely used in producing commercial cultivars. Resistance to other pathotypes of both species has been transferred, especially from *S. vernei*, which may confer resistance, probably polygenically inherited, to several pathotypes. *S. multidissectum* has a dominant gene (H_2) for resistance to one pathotype of *G. pallida* (Howard and Cotten, 1978). Some clones of *S. tuberosum* ssp. *andigena* also confer resistance to *G. pallida* (Fuller and Howard, 1974) and so do *S. spegazzinii* and *S. oplocense* (Bouwman and Ross, 1972). De Maine (1978) found it possible to produce dihaploids of potatoes that were more resistant to one pathotype (Pa 3) of *G. pallida* than their tetraploid parents.

In tomatoes, resistance to potato cyst nematodes was conferred by a dominant gene from *Lycopersicon pimpinellifolium* (Ellis and Maxon Smith, 1971).

Rotylenchulus. Resistance to *R. reniformis* often—but not always—occurs together with resistance to cyst nematodes in soybeans (Rebois *et al.*, 1970; Caviness and Riggs, 1976). In tomatoes, some 20 cultivars resistant to *R. reniformis* were also resistant to *Heterodera schachtii* (Rebois *et al.*, 1977), but in potatoes resistance was independent of resistance to potato cyst nematodes (Rebois and Webb, 1979).

Meloidogyne. A dominant, or incompletely dominant, gene—Mi in linkage group IV—was considered to confer resistance in tomatoes to the root-knot species *M. incognita incognita, M. incognita acrita, M. javanica*, and *M. arenaria*. Resistance was originally introduced from *Lycopersicon peruvianum* (Gilbert and McGuire, 1956; Barham and Winsted, 1957). This source of resistance has been widely used, and resistant varieties are cultivated in most parts of the world (Laterrot, 1973). Pathotypes, able to develop on resistant cultivars, were soon differentiated (Riggs and Winsted, 1959). Resistant tomato cultivars improve the yield very much in infested fields but pathotypes able to overcome the resistance constitute a threat in some areas, especially when resistant tomatoes are grown repeatedly. To prevent the development of such pathotypes, integrated control systems have been recommended (Netscher, 1976).

Sidhu and Webster (1973) found that two cultivars (Nematex and Small Fry) resistant to *M. incognita* each had a dominant resistance gene, whereas a third cultivar (Cold Set) had a recessive gene for resistance. They labelled their genes tentatively as $LMiR_1$ (from Nematex), $LMiR_2$ (from Small Fry), and $LMir_3$ (from Cold Set). Further investigations (Sidhu and Webster, 1975) showed that $LMiR_1$ and $LMiR_2$ were closely linked and that $LMiR_1$ and Mi were identical or allelic. It was not possible to demonstrate relationships to $LMir_3$. Fatunla and Salu (1977) considered resistance against *M. incognita acrita* in the cultivars

Nematex and Rossol to be controlled by separate, incompletely dominant, single genes, which were nonallelic.

A number of soybean cultivars are resistant to *M. incognita*, but most of them are susceptible to other *Meloidogyne* spp. (*M. javanica, M. arenaria*, and *M. hapla*) (Caviness and Riggs, 1976). There are rather few facts on the inheritance; there is often doubt about the number of genes involved, although there are speculations about one to three major genes and additional minor genes (Caviness and Riggs, 1976). Boquet *et al.* (1976) found that susceptibility to one pathotype of *M. incognita* was partially dominant–resistance was "a qualitative character conditioned by one major gene with at least one modifying gene."

Resistance in tobacco to *M. incognita* was found to be conferred by a dominant gene (Malo, 1964). Slana and Stavely (1978a) showed, using monosomics in their crosses, that the gene in *Nicotiana tabacum* responsible for resistance against *M. incognita* was located on chromosome G.

In cotton, root-knot resistance is important also in the control of other diseases, for instance, cotton wilt caused by a *Fusarium* species. Hare (1956) referred to results indicating that two recessive genes controlled resistance to *M. incognita acrita* in crosses between resistant and susceptible *Gossypium barbadense* breeding stocks. Shepherd (1974) developed, by repeated selection, F_{10} lines of *G. hirsutum*, which were transgressive segregates for resistance to *M. incognita acrita*. Resistance was supposed to be controlled by multiple genes.

Cultivars resistant against *M. incognita* and also against *M. javanica* and *M. arenaria* are important in a number of other crops (Malo, 1964; Hare, 1965; Bingefors, 1969, 1971; Cook, 1974; Howard and Cotten, 1978). Pathotypes often occur in *M. incognita* and other species, and populations within a species may vary widely in their host ranges (Sturhan, 1966). Viglierchio (1978) demonstrated very great variability in aggressiveness among Californian populations of *M. incognita*.

In alfalfa, resistance to *M. hapla* was conditioned in two clones by one dominant gene, tetrasomically inherited, while no definite inheritance pattern was found in two other clones (Goplen and Stanford, 1960). The resistance gene was closely linked to another dominant gene, also tetrasomically inherited, which gave resistance to *M. javanica*. Hunt *et al.* (1969) confirmed that one dominant gene with tetrasomic inheritance gave resistance to *M. hapla*. Sullivan and Christie (1979) shoed, however, that resistance was controlled by two loci that segregated independently. Resistance was dominant: a plant must have at least two dominant alleles at both loci to be resistant.

Anguina and Ditylenchus radicicolus. Very little search for resistance against *A. tritici* has been done, but investigations in East Europe and in Asia have demonstrated variety differences in susceptibility (Cook, 1974). Resistance to *D. radicicolus* has been found in barley and rye cultivars in Norway (Bumbulucz, 1970).

Ditylenchus dipsaci. *D. dipsaci* consists of a great number of "host races," studied especially by Seinhorst (1956) and by Sturhan (1969). Each can develop on several plants but some are very specialized in their host requirements. Natural selection and breeding has developed resistant cultivars in some of their main hosts. Thus, resistance is found in several crops, including cereals (Cook, 1974) and forages (Bingefors, 1969). Little is known about pathotypes in the oat race (Goodey and Hooper, 1962) but they are considered unimportant (Clamot, 1972). Griffiths *et al.* (1957) found that resistance against *D. dipsaci* in the oat cultivar Grey Winter was dependent on one dominant gene. This opinion was supported by Cook in his review (1974). Resistance in Grey Winter means inhibited nematode reproduction. In another cultivar, Early Miller, there is also a single independent gene that gives a type of tolerance that may be modified by the genetic background. Crosses between these two types of cultivars gave progenies that reacted as *Avena ludoviciana*. Resistance in this species is considered to be controlled by two independent genes.

It is possible to find stem nematode resistant plants in most plant populations or cultivars of alfalfa and red clover, although their number may be very small. Selection may result in an increase from close to zero up to 50% "resistant" plants in two or three generations (Lundin, 1969). Complete resistance has not been obtained in any cultivar. There do not seem to be any clear indications of pathotypes occurring either in alfalfa (Elgin *et al.*, 1977) or in red clover (Frandsen, 1966; Bingefors and Eriksson, 1968) although there are some differences between nematode populations in their aggressiveness.

Resistance in alfalfa to *D. dipsaci* was first described in Argentina by Burkart (1937), who found it impossible to explain the inheritance on a factorial basis. Later attempts have given varying results. For one or two sources of resistance a major factor with tetrasomic inheritance has been considered to control the resistance. In other cultivars one or more minor factors were considered responsible and, in still another, it was not possible to explain resistance on a factorial basis (Grundbacher and Stanford, 1962). Elgin (1979) explained inheritance, even in earlier reports, assuming two complementary, dominant, tetrasomically inherited genes. Resistance in red clover was not explained on a factorial basis, although the selection effect was rapid, as in alfalfa. Resistance was considered to depend on a small number of major genes (Bingefors, 1956). An attempt has been made to explain the inheritance as depending on two dominant genes, one of them linked to an incompatibility factor (Nordenskiöld, 1971).

Aphelenchoides. Resistance against *A. besseyi* is found and utilized in local rice varieties. As summarized by Cook (1974) there are some divergences in the rating of resistance in different parts of the world. Resistance to other *Aphelenchoides* spp. has been described in chrysanthemum (*A. ritzemabosi*; Hesling and Wallace, 1960) and in strawberries (*A. fragariae* and *A. ritzemabosi*; Szczygiel and Danek, 1975).

Tylenchulus. Root-stocks resistant to *T. semipenetrans* are used in citrus-growing areas. Pathotypes attacking some resistant root-stocks occur (O'Bannon *et al.*, 1977). Resistance was highly dominant in the F_1 (Hare, 1965).

Pratylenchus. A degree of resistance to *Pratylenchus* spp. has been demonstrated in different crops. *Tagetes* spp. suppressed especially *P. penetrans* in the soil, and a number of other members of Compositae have also been shown to produce nematicidal substances active against this species (Gommers and Voor in't Holt, 1976). Resistance in alfalfa clones to *P. penetrans* is used in plant breeding (Townshend and Baenziger, 1976). Peanut cultivars showed differences in resistance to *P. brachyurus* (Smith *et al.*, 1978).

9.3 ATTRACTIVENESS TO RESISTANT AND SUSCEPTIBLE PLANTS

Hatching. Hatching of the larvae in *Heterodera, Globodera*, and *Meloidogyne* is often stimulated by root exudates. This is especially pronounced in some species of *Heterodera* and *Globodera* in which hatching is rather low in the absence of host or other hatching-stimulating plants. However, "there is no correlation between resistance to female development and the production of active root exudates" (Williams, 1978). Potato cyst nematodes are known to require stimulants for intensive hatching and cultivars carrying resistance from *Solanum tuberosum* spp. *andigena* also stimulate hatching. Reports (Forrest, 1978) indicate that there are clones both in *S. vernei* × *S. tuberosum* and in *S. tuberosum* differing in their hatching activity. *S. juzepezukii* was found to suppress hatching in *G. rostochiensis*, but most in *G. pallida*, whereas *G. pallida* hatched more than *G. rostochiensis* in exudates of a cultivar of *S. tuberosum* ssp. *andigena* (Jatala *et al.*, 1977). Evans and Franco (1979), studying factors increasing tolerance to potato cyst-nematode attacks, considered reduced hatching stimulation as a possibility.

Attraction. Nematodes are often attracted to plant roots in the soil. There are different opinions about the reasons but there are few investigations on differences among cultivars in this respect. Larvae of *Meloidogyne hapla* were attracted to both resistant and susceptible alfalfa plants but more were found at the susceptibles when the larvae could choose (Griffin and Waite, 1971). Peacock (1959) demonstrated that *M. incognita* larvae were attracted to excised root tips of root-knot resistant *Lycopersicon peruvianum* but not so strongly as to tomatoes. Roy (1975) found no difference in this respect for *M. graminicola* in resistant and susceptible rice. In a sweet potato cultivar (Nemagold) the production of exudates toxic or repellent to the nematode larvae was considered a probable cause of its resistance to *M. incognita* (Jatala and Russell, 1972). The exudates were believed to preclude or reduce the contact between larvae and roots.

Blake (1962a), working with *Ditylenchus dipsaci*, did not find much difference in attraction to roots of resistant and susceptible oats. Stem nematodes were normally equally attracted to resistant and susceptible alfalfa seedlings (Griffin and Waite, 1971).

Penetration. Penetration of nematode larvae is seldom much reduced in resistant compared with susceptible cultivars. *Heterodera avenae* invades roots of both resistant and susceptible plants of barley (Wålstedt, 1936), oats (Cook, 1974), and wheat (O'Brien and Fisher, 1977). However, roots of a resistant wheat cultivar were not penetrated when the inoculation was repeated on the same roots. This resistance to penetration was not transferred to other roots on the plant (O'Brien and Fisher, 1978).

Riggs and Winsted (1959) reported that root-knot nematode larvae penetrated resistant as well as susceptible tomato plants, but Taylor (1975) found that the resistant cultivar Rossol was less penetrated by nematode larvae than susceptible cultivars. Peacock (1959) also demonstrated penetration by lesser numbers in excised root tips of *Lycopersicon peruvianum* than in *L. esculentum*. Susceptible cowpea was entered by more *Meloidogyne incognita* larvae than a resistant cultivar (Sethi and Sharma, 1976); however, equal penetration in resistant and susceptible cultivars has also often been observed (Barrons, 1939).

Dropkin and Webb (1967) found that fewer larvae of *M. hapla* entered resistant than susceptible tomato plants. There is a report on resistance to penetration by *M. hapla* in a clone of alfalfa (Goplen and Stanford, 1959), but later investigations showed (Griffin and Elgin, 1977) that larvae penetrated initially to the same extent, as in susceptible seedlings, although the number soon decreased.

Ditylenchus dipsaci penetrates both resistant and susceptible plants of red clover (Bingefors, 1951) and alfalfa (Hanna and Hawn, 1965). However, very soon there were fewer nematodes per plant in resistant than in susceptible cultivars. Also fewer plants contained nematodes in resistant than in susceptible cultivars and complete infestation could be obtained neither in resistant red clover nor in alfalfa.

9.4 DIFFERENCES IN STRUCTURAL CHANGES BETWEEN RESISTANT AND SUSCEPTIBLE PLANTS

9.4.1 Cyst Nematodes

The formation of syncytia in susceptible plants invaded by *Heterodera* or *Globodera* nematodes has been studied in different plant species as reviewed by Endo (1971). Investigations of ultrastructural changes confirmed that the multinucleate syncytia of *H. glycines* in soybean were formed by dissolution of cell walls (Gipson *et al.*, 1971). Jones and Northcote (1972a) described the syncytia in potatoes induced by *G. rostochiensis* as transfer cells for nutrients provided by the xylem sap and with the nematode as the nutrient sink. The nematode larva "controls the differentiation of unspecialized cells to cells with a specific physiological function."

It has been pointed out that even highly resistant potato cultivars are hosts for the potato cyst nematode and many of the invading larvae are able to initiate syncytia. The cause of the death of many of these larvae is considered to be necrosis of cells around the syncytia, which become isolated and die. This process is slow and allows many males, but only seldom any females, to become adults (Huijsman, 1974). The internal response has been investigated in potato clones carrying resistance from different sources (Hoopes *et al.*, 1978). A cultivar carrying the gene H_1 from *S. tuberosum* ssp. *andigena* had a few cells with granular cytoplasm in the early stage of infection. Vacuolation of cytoplasm, walling off and collapse of cells, and necrosis allowed maximum development only to the third stage. Males were seen. In a clone based on *S. vernei*, some cell wall dissolution and changes in the cytoplasm typical of the syncytium were initiated. Maximal development was to the fourth stage, but vacuolation of the cytoplasm, walling off, and necrosis again limited the syncytium development. One of the clones, deriving its resistance from *S. spegazzinii*, also allowed an initial development of the syncytium and maximal development of the larvae to the third stage. Vacuolation of the cytoplasm occurred and the syncytium "appeared to collapse from within." In the fourth clone, carrying resistance from *S. multi-dissectum*, the syncytium was initiated but it was limited to the cortical region by the formation of a heavily thickened wall around it. Maximum larval development was to the third stage. Such differences in resistance reactions are of great interest from both practical and theoretical points of view.

Larvae of *H. avenae* stimulated the initiation of syncytia in the resistant barley cultivar Sabarlis; the cytoplasm became sparse and vacuolated and only males developed. Invading larvae in the roots of resistant oats caused hyperplasia and necrosis, and syncytia were not initiated (Cook, 1974). In an oat cultivar, Silva, with a lower degree of resistance (Harmuth, 1977), some of the penetrating larvae induced syncytia. The cysts were smaller and had fewer eggs than cysts on a susceptible cultivar. Females were underrepresented, whereas males occurred in near-normal numbers in the resistant cultivar.

Endo (1965) described the resistance reaction in soybeans where nematodes (*H. glycines*) seldom grew beyond the second larval stage. Small syncytia sometimes developed, allowing larval growth and development in some cultivars, mainly of males. Ross (1958) had considered the resistance to be caused by a hypersensitive reaction, but Endo (1965) traced necrotic regions to deteriorated syncytia—when the larvae stopped feeding there was necrosis and collapse of the syncytia. He found the presence of syncytia and developing larvae in resistant plants to "deemphasize" the hypersensitivity theory. The conclusion of an investigation of ultrastructural changes in a resistant soybean cultivar (Riggs *et al.*, 1973) was that "deposition of secondary wall material to wall off the diseased area may be the mechanism of resistance to *H. glycines*." Syncytia were observed within 42 hr after inoculation, and they did not begin to deteriorate until four days after the inoculation. Webster (1975) thought it more likely that the cytoplasmic degeneration of giant cells beginning after four days inhibited selective transport and amino acid synthesis enough to prevent nematode development before the increase in cell wall thickening occurred.

9.4.2 Reniform Nematodes

A study of ultrastructural changes in soybean cultivars resistant and susceptible to *Rotylenchulus reniformis* showed that infection and development in the first initial phase were alike (Rebois *et al*., 1975). Syncytia were initiated by partial cell wall lysis and cell separation. The lysis continued and was accelerated in the resistant cultivar, whereas developing syncytia went into an anabolic phase in the susceptible cultivar. A hypersensitive reaction was pronounced four days after the inoculation. *R. reniformis* was also studied in susceptible and resistant cotton (Carter, 1974). The initial development was equal. However, after six days, further development in resistant plants became restricted and cells near the nematode began to deteriorate. Complete necrosis occurred in cells around the nematode head.

9.4.3 Root-Knot Nematodes

When *Meloidogyne* spp. have penetrated a susceptible plant, a gall is usually formed in one or two days (Dropkin, 1972) and the larvae induce the development of giant cells in the differentiating vascular tissue behind the root tip. A few giant cells develop in the gall for each establishing larva. The fine structure of these giant cells was first described by Bird (1961) and the information collected in several important studies was reviewed by Endo (1971) and Webster (1975). Giant cells are multinucleate cells, formed by repeated mitosis without cytokinesis (Huang and Maggenti, 1969); cell wall breakdown does not occur or is not important. Thus, their formation differs from that of syncytia induced by *Heterodera* spp., which are formed by cell wall breakdown. Their function, however, is similar: they are transfer cells, transporting nutrients to the developing nematode female (Jones and Northcote, 1972b).

The induction of giant cells was considered by Jones and Dropkin (1976) to depend on (1) a stimulation of mitosis and blockage of cytokinesis, at the same time as cells enlarge, and (2) the existence of nematode larvae as a nutrient sink causing the cell metabolism to maintain homeostatis, which stresses the transfer function of the giant cell. A synthesis of wall polysaccharides also occurs in giant cells.

Tomatoes, resistant to *M. incognita*, were studied extensively by Paulson and Webster (1972; Webster and Paulson, 1972; Paulson, 1976). There was a hypersensitive reaction in cells adjacent to the larvae, which suppressed giant cell initiation. There were indications of cellular activity rather than of cellular breakdown in the beginning of the hypersensitive reaction. Later, there was a breakdown of cell membranes and organelles. The hypersensitivity reaction occurred within 8-12 hr and was considered rapid enough to prevent the formation of giant cells, which requires 24-48 hr in susceptible plants. Nematode larvae that were removed from hypersensitive roots two days after inoculation were able to infect susceptible roots. Later studies (Sawhney and Webster, 1979) made it less convincing that the hypersensitivity reaction is the "only" mechanism of root-knot resistance in the investigated tomato cultivar (Nematex). When using chemicals to inhibit protein synthesis, no hypersensitive reaction was obtained in resistant roots; the cells surrounding nematodes did not die, but the plants did not become susceptible because of the inhibition of the hypersensitive response.

Structural changes in both susceptible and resistant plants have not been studied so extensively in any other interaction between root-knot nematodes and hosts as in tomatoes infected by *M. incognita*. However, necrosis has been observed in the resistant cultivars and mostly the resistance has been considered due to hypersensitivity. For example, giant cells began to develop in resistant as well as in susceptible cultivars of beans (*Phaseolus vulgaris*) within four days after inoculation with *M. incognita*. Giant cells in the resistant cultivar were small, and necrosis was evident within four days. Resistance was considered due to "absence of adequate giant cell development and to hypersensitive reaction" (Fassuliotis *et al.*, 1970).

Nevertheless, there are also examples of root-knot resistance that have not been explained by necrosis or hypersensitivity. A study of two *Cucumis* spp., resistant to *M. incognita acrita* (Fassuliotis, 1970), showed that giant cells developed in resistant as well as in susceptible species but some of them became small in the resistant plants. Few females but many males developed in these plants. In peach rootstocks, resistance to *M. javanica* was considered due to an extensive walling-off of the giant cells, which later broke down (Malo, 1967). In resistant cotton (McClure *et al.*, 1974a, 1974b) resistance was not associated with increased root necrosis, but nematode development was greatly retarded. In alfalfa, a "symptomless" resistance reaction has been reported (Reynolds *et al.*, 1970). *M. incognita acrita* larvae entered the roots in large numbers, but three to four days after inoculation their numbers in the roots dropped rapidly. They left the roots, and there was no evidence of a hypersensitive reaction.

9.4.4 Stem Nematodes

Ditylenchus dipsaci, attacking oats, caused mesophyll cells around the nematode to enlarge and to separate in a susceptible cultivar (Blake, 1962b). Wide air spaces developed between the cells, which became spherical when they enlarged. Effects could be seen up to four cells away from the nematode. In a resistant cultivar, only cells in contact with the nematode separated and the space between them was very narrow. Cell walls around the nematodes collapsed when most of the cytoplasm had been removed and cavities were formed in both cultivars. No hyperplasia or necrosis was observed in the resistant cultivar. There was little or no hypertrophy. Thus, the nematodes became isolated in local lesions (Cook, 1974).

Seinhorst (1956) studied the reactions of several plant species, nonhosts or susceptible and resistant hosts, to a number of stem nematode races. He considered resistance in red clover plants to the red clover race of stem nematodes to be caused by hypersensitivity resulting in necrosis. Stem nematodes in resistant alfalfa plants were found in cavities (Bingefors, 1961). Cell walls adjacent to these cavities were stained by saffranin; cellular structure was very little disrupted around the cavity two weeks after inoculation. In susceptible plants the nematodes occurred in less distinct cavities in a disrupted cellular structure with rounded cells. Susceptible and resistant red clover were considered to react principally in the same way as alfalfa.

Chang *et al.* (1973, 1975) investigated ultrastructural changes in alfalfa resistant and susceptible to stem nematodes. There were no morphological changes after one day but there were more lipid bodies per cell in infected than in noninfected plants and more in the susceptible cultivar than in the resistant. After three days, nearly all cells of the resistant cultivar contained normally developing chloroplasts, cytoplasm and nuclei while cells in the susceptible cultivar had swollen chloroplasts, sometimes swollen nuclei, and also changes in the cytoplasm. Very few cells in the resistant cultivar had similar changes. After seven days most chloroplasts in cells of the resistant cultivar were still normal–in the susceptible cultivar they were partially or completely disrupted. Although there were heavily infected areas in the resistant cultivar with some similar changes in nuclei and cytoplasm as in the susceptible cultivar, cell walls remained normal both after three and after seven days. In the susceptible cultivar cell walls were broken down after seven days, allowing the cytoplasm of adjacent cells to mix.

Both resistant and susceptible cultivars of alfalfa had lignified cell walls surrounding the cavities formed by *D. dipsaci* (Reed *et al.*, 1979) but the cavities were smaller and stained more heavily in the resistant cultivar. In the resistant cultivar chromatin content increased less than in the susceptible cultivar. Cellular disruption was not distinctly different but cell function was lost earlier in the resistant than in the susceptible cultivar.

9.4.5 Hypersensitivity and Resistance

Although other reasons have been reported for the resistance reaction in plants to nematode attacks it is obvious that a hypersensitive reaction in the plants is most commonly referred to as the cause of the resistance or at least as one factor in the resistant response. Christie, as early as 1949, discussed "the attractive idea" that the nematodes die because "they kill the tissues on which they depend for food" but he did not find it quite adequate. Rohde, in his review (1965), points to the fact that the connection between products formed in browning reactions and hypersensitivity is not clear. Necrosis is very often associated with resistance. Seinhorst (1961) refers to necrosis becoming apparent a few days after the penetration by *Meloidogyne* larvae as a quick indication for a poor host in a breeding program. Webster (1969) characterizes hypersensitivity as "a mild form of necrosis," and Dropkin (1969b) concludes that "certain, but not all types of genetic resistance" depend upon necrotic reactions to the nematodes. Resistance to root-knot and cyst nematodes is most commonly of this type. Rohde (1972) considers the detrimental reaction to endoparasitic nematodes typically to be necrosis.

Probably, resistance against nematodes is too often characterized as being due to hypersensitivity or to necrosis. An example is stem nematodes in alfalfa, where Chang *et al.* (1973, 1975) failed to demonstrate a hypersensitive reaction. Also in a plant with so typical a hypersensitive response as root-knot resistant tomato, Sawhney and Webster (1979) made it clear that this reaction is not alone responsible for the resistance. When "hypersensitivity" sometimes is used to characterize a slow process preventing the complete development of all the nematode larvae that have penetrated a plant and finally resulting in some necrosis, the term seems especially misused. Probably, necrosis in plant nematology is often a result of the resistant response of a plant to an incompatible parasite: it may be a consequence, not the cause (Király *et al.*, 1972).

9.5 SOME BIOCHEMICAL DIFFERENCES BETWEEN RESISTANT AND SUSCEPTIBLE PLANTS

In tissues infected by *Meloidogyne* spp. there is a considerable increase in the levels of free amino acids, organic acids, nucleotides, protein, RNA, DNA, lipids, hemicellulose, and minerals but a decrease in carbohydrates, pectins, cellulose, and lignin (Owens and Novotny Specht, 1966). There is also an increase in the activity of oxidoreductive, hydrolytic, and oxidative enzymes (Veech and Endo, 1970). Here only differences between resistant and susceptible plants will be discussed. Some chemical compounds–phenolics (Section 9.6), growth hormones (Section 9.7), phytoalexins, and toxins (Section 9.8)–have been treated under separate headings.

Amino acids. Investigations often concern changes in proteins and amino acids. Lewis and McClure (1975) studied the occurrence of 17 free amino acids in cotton cultivars susceptible and resistant to *M. incognita*. Maximum differences were obtained six days after infection. At that time, total free amino acids was greatest in the resistant cultivar, but several amino acids increased more in the susceptible cultivar. The most apparent difference concerns proline, which in ten days increased 1000-fold and more in the susceptible cultivar against only a small increase in the resistant cultivar.

In rice, Jena and Rao (1977) found resistance to *M. graminicola* associated with high aspartic acid and alanine content but low valine, tryptophane, and methionine content in the roots.

A study of the changes of aspartic acid, glutamic acid, and glutamine in *Beta vulgaris* (susceptible) and *B. patellaris* (resistant) indicated significant increases in *B. vulgaris* after infection with *Heterodera schachtii* but no change in *B. patel-*

laris (Doney *et al.*, 1970). The ratio of protein proline to hydroxyproline was found to increase in susceptible but to decrease in resistant potatoes after infection with *Globodera rostochiensis* (Giebel and Stobiecka, 1974). Giebel and Krenz (1975) considered these differences indicative of a role for hydroxyproline in resistance, but only as one among other factors.

Energy. There is an investigation on the effect of infection by *M. incognita* in resistant and susceptible cotton on the concentration of adenylates and energy charge values (Misaghi *et al.*, 1975). There was a drastic increase in adenylate levels both in susceptible and in resistant hosts, but the values for energy charge remained nearly constant, 0.75-0.80. This was thought to indicate that the nematode infection did not upset the energy metabolism either in susceptible or resistant plants.

Giebel *et al.* (1977) observed an increase of phosphatase activity in resistant potato plants after infection with *G. rostochiensis*. They thought the high activity in resistant plants could lead to a decomposition of high-energy compounds and reduce cell metabolism.

9.6 PHENOLS AND PHENOLIC COMPOUNDS IN NEMATODE RESISTANCE

In plant pathology, phenolics are considered to be the most important group of chemical compounds in disease resistance, although there is seldom a correlation between resistance and the level of total or special phenolics (Levin, 1976). Many of them act as phytoalexins, accumulating in plant tissues after the infection. Phenols normally occur in the plants as glycosides, which are nontoxic. They may be decomposed to free phenols by the action of a glycosidase, available in the plant cells or in the nematodes' secretion. Phenolics may be hydrolyzed to aglycones or oxidized—by polyphenol oxidases and peroxidases—to quinones.

9.6.1 Cyst Nematodes

Phenolics in relation to resistance in potatoes to cyst nematodes (*Globodera* spp.) have been extensively studied by Polish researchers, who interpreted their results to indicate that resistance "is dependent on the presence of phenolic glucosides in the root cells of resistant potatoes which are hydrolized by β-glucosidase secreted by the larvae." A difference in activity of the enzyme in different nematode pathotypes was postulated (Wilski and Giebel, 1966). Further research led to a hypothesis indicating that aglucones, released by β-glucosidase, interact with indole acetic acid oxidase in infected tissues and more (in susceptible plants) or less (in resistant plants) inhibits its auxin-destructing effect, thus regulating the development of giant cells (see Section 9.7). Investigations showed that

β-glucosidase activity was localized to cells lying close to the nematodes (Giebel and Wilski, 1970; Giebel *et al.*, 1971). In a more recent paper (Giebel, 1976) the importance is stressed of glycosidases "as enzymes liberating the primary factors of susceptible/resistant reactions, i.e., phenolic aglycons." While there was no significant difference in the activity of β-glucosidase in uninoculated plants, this activity increased much more in susceptible than in resistant cultivars as a result of the infection.

Studies of enzymes in potatoes in Estonia (USSR) demonstrated a greater increase of peroxidase activity in resistant than in susceptible potatoes. The activation in resistant potatoes was in the middle isozymes 6, 7, and 8 (Haberman, 1972). There was a greater increase of phenolic compounds in resistant than in susceptible cultivars (Padu, 1977).

When glycosidase activity (β-glucosidase and β-galactosidase) was investigated in roots of barley and oats (Torp and Hedeby Jensen, 1978), no direct relationships with the resistance were found in three susceptible cultivars and in four cultivars resistant to *H. avenae.*

9.6.2 Root-Knot Nematodes

There are several reports on high levels of phenolic compounds in tomatoes resistant to *Meloidogyne* spp. Chlorogenic acid, being the main phenolic compound, had a higher concentration in uninfected roots of the resistant cultivar Nemared than in susceptible and moderately resistant cultivars. "Localized accumulation of chlorogenic acid and its oxidized products in host root cells" was considered to be an important resistance mechanism, preventing the nematode larvae from penetrating into the endodermis (Hung and Rohde, 1973). Also Singh and Choudhury (1973) demonstrated a relationship between phenolic content and resistance to *Meloidogyne* spp. in tomato cultivars (resistant cultivar Nematex) and the resistant species *Lycopersicon peruvianum* when phenolics were determined in uninfested plants. When phenolic compounds in leaves of resistant and susceptible tomatoes were determined (Bajaj and Mahajan, 1977), the concentrations in resistant plants of chlorogenic, neochlorogenic, and caffeic acids were found to be higher. High levels of phenolic compounds produced in the leaves and translocated to the roots were supposed to confer resistance to nematodes.

Exogenous application of catechol, cinnamic acid, or salicylic acid by spraying or root treatment of tomato plants susceptible to *M. javanica* made them partially resistant, reducing larval penetration and the number of galls. The nematodes lost some of their egg-laying capability when exposed to the chemical (Sitaramaiah and Pathak, 1979).

Milne *et al.* (1965) demonstrated a higher chlorogenic acid content and a higher activity of chlorogenic acid-oxidase in *Nicotiana repanda*, resistant to *M. javanica*, then in *N. tabacum*, susceptible to this nematode. They suspected that oxidation products of chlorogenic acid were involved in the hypersensitive reaction toward the nematode.

In cotton, resistant and susceptible to *M. incognita*, several forms of peroxidase and 6-phosphogluconate dehydrogenase were isolated (Noel and McClure, 1978). Infection increased the activity of the enzymes in both resistant and susceptible cultivars. Peroxidase activity in uninfested tissues was greater in resistant than in susceptible plants. Resistance in cotton is considered to be partly due to phytoalexins (see Section 9.8).

9.6.3 Lesion and Burrowing Nematodes

When studying phenol levels in tomato cultivars that were good and poor hosts to *Pratylenchus penetrans*, Friedman and Rohde (1977) found only small differences, which were not considered as major factors in the disease syndrome. There were higher levels of bound phenols in the leaves of nematode-infected plants than in healthy plants.

Bound and free phenolics in the roots and leaves of citrus cultivars, susceptible to *Radopholus similis*, decreased after infection, while they increased, sometimes very much, in tolerant cultivars. Attempts to increase the tolerance by application of phenolics did not eliminate the nematodes from the plants, although it sometimes modified the phenolic content in the roots (Feldman and Hanks, 1971).

9.6.4 Summary

Thus, phenolic compounds in plants commonly increase in response to nematode infection as well as to infections by other plant parasites. They accumulate in the tissues, especially in resistant plants. Phenols are important in the resistance response to nematode infections but "it now remains to determine whether the role of phenols is a primary or a secondary response," as Webster (1975) stated in his excellent review.

9.7 GROWTH HORMONES IN RELATION TO RESISTANCE

The formation of swelling and different types of galls caused by many nematodes is influenced by growth hormones, especially auxins (Viglierchio, 1971). The importance of the plant growth hormones for differences observed between susceptible and resistant plants is obvious because they influence the synthesis of DNA, RNA, and protein as well as cytokinesis and cell hypertrophy (Webster, 1975). Growth hormones involved in the reaction of attacked plants may originate in the nematodes or in the host cells, stimulated by the nematodes. *Meloidogyne hapla* and *Ditylenchus dipsaci* were found also to produce auxin inactivators (Viglierchio and Yu, 1965). And a number of nematode species, investigated by Viglierchio and Mjuge (1975), produced auxin inactivators, but none was observed to produce auxins. Larvae of *Globodera* spp. did not secrete auxins but they secreted inactivating substances (Hanschke *et al.*, 1975).

Some comparisons of the activity of growth hormones in resistant and susceptible plant cultivars deserve attention. In addition to auxins, this also concerns cytokinins, gibberellic acid, and ethylene. Most of the investigations have been done with cyst or root-knot nematodes. The great requirements of these nematodes with respect to adaptation of infected plant tissues make them especially interesting to study.

9.7.1 Cyst Nematodes

As discussed earlier (see (Section 9.6) biochemical factors involved in resistance against potato cyst nematodes have been extensively investigated by Giebel and his co-workers. Their results were summarized in a hypothesis putting very great emphasis on localized occurrence and activity of auxins, especially indol acetic acid, in the induction and development of giant cells in susceptible plants. Giebel in his review (1974) admits, however, that there is meager evidence behind his assumption "that the pathogenic changes which occur in susceptible plant tissues after nematode invasion are due to auxins secreted by the nematode or released from complexes by the action of nematode enzymes such as glycosidases and proteases." Results and hypotheses from the work of Giebel and his co-workers have been reviewed also by Webster (1975). Both these reviews describe the intricate problems connected with the activity of different growth hormones and their inhibitors, but the role of auxins in susceptible and resistant potatoes is far from clear.

Giebel and Jackowiak (1976) also demonstrated that infection doubled the activity of tryptophan decarboxylase in susceptible but caused no change in resistant potatoes, thus probably increasing the biosynthesis of auxin in susceptible plants.

9.7.2 Root-Knot Nematodes

In tomatoes (Dropkin *et al.*, 1969) exogenously supplied cytokinins (kinetin and three others) highly increased the susceptibility of a resistant cultivar to *M. incognita*. No such effect was obtained using indole acetic acid, gibberellic acid, or other substances tested. Sawhney and Webster (1975) used 1-naphthaleneacetic acid and kinetin in similar experiments. The susceptibility, measured in the number of larvae present in the roots and in the amount of galling, increased only if both were used in combination, but only a few larvae developed to maturity in the treated, resistant cultivar. The conclusion reached was that auxins and kinins were not alone determining susceptibility or resistance in tomatoes to *M. incognita*. Van Staden and Dimalla (1977) found higher levels of endogenous cytokinins in susceptible than in resistant tomatoes, and nematode infestation increased the level in both.

A growth promoter was found in a susceptible cotton cultivar, whereas there was a growth inhibitor and also an auxin-degrading enzyme in the cultivar Clevewilt, resistant to *M. incognita* (Lewis, 1973).

In peach rootstocks, resistant to *M. javanica*, nematode larvae did not develop to maturity but when 1-naphthylacetic acid was exogenously applied, roots developed vigorously, galls were formed, and nematodes developed to maturity (Kochba and Samish, 1971). Kinetin applied in the same way increased gall formation and female development, although plant development was not affected. The added growth substances were considered to allow syncytia in resistant plants to function long enough for complete nematode development. Endogenous cytokinin activity was lower in resistant than in susceptible rootstocks; however, exogenously added growth substances increased the activity (Kochba and Samish, 1972).

9.7.3 Summary

In spite of some interesting investigations during the last few years the conclusion drawn by Viglierchio and Mjuge (1975) seems still to be valid: "the mechanism and characteristics of the host-parasite interactions, modifying the normal plant hormonal balance, is as yet largely a matter of speculation."

9.8 PHYTOALEXINS AND TOXINS IN NEMATODE RESISTANCE

9.8.1 Phytoalexins

Phytoalexins are defined in different ways. One definition referred to by Drysdale (1978) describes them as "substances which inhibit the growth of certain micro-organisms and which are produced in higher plants in response to a number of chemical, physical and biological stimuli." Phytoalexins have been investigated in very few host-nematode relationships.

Gossypol and related terpenoids have been correlated with insect and fungus resistance in cotton cultivars. They have also been considered important in resistance to *Meloidogyne incognita* (Veech and McClure, 1977). A correlation between the change in concentration of gossypol and related terpenoid aldehydes and host resistance was found (Veech, 1978). Terpenoid aldehydes were located in the epidermis of healthy plants, but within four days after the infection they accumulated in the endodermis and outer stele where they were effective against the sedentary nematodes (Veech, 1979).

Glyceollin has been reported to accumulate in a soybean cultivar resistant to *M. incognita*; the accumulation was correlated with resistance (Kaplan *et al.*, 1978). Glyceollin accumulated mainly in the stele of the resistant cultivar and began within 48 hr of inoculation.

Phaseollin was isolated from bean seedlings (*Phaseolus vulgaris*) infected with *Pratylenchus penetrans* (Abawi *et al.*, 1971). Lima bean (*Ph. lunatus*) accumulated at least four coumestans when exhibiting a hypersensitive response to *P. schribneri*. They were active against the nematode but no fungicidal phytoalexins were detected after nematode infection (Rich *et al.*, 1977).

Negative results were reported from the search for phytoalexins in potatoes infected with *M. hapla* and *M. incognita acrita* (Schindler *et al.*, 1977) and for ipomeamarone in sweet potatoes infected by *M. incognita* (Martin *et al.*, 1976). Jatala and Russell (1972) considered postinfection production of "inhibitory chemicals" a possible resistance factor in sweet potatoes, cultivar Nemagold.

9.8.2 Preexisting Toxins

There are examples of preinfectional chemical substances in plants that are toxic to plant parasitic namatodes. They may reduce or prevent the attack of one or several nematode species in a plant. For most plants investigated no varietal differences have been described; they probably have not been used in breeding programs as sources of resistance. Examples of varietal differences are cucurbitacins (tetracyclic triterpenoids), causing bitterness in cucumbers, which also reduce the attraction and reproduction of *Meloidogyne incognita* in the roots. Nematodes build up faster in nonbitter than in bitter cultivars. Cucurbitacins are also known to be effective against some other pests (Haynes and Jones, 1976).

9.9 NATURE OF NEMATODE RESISTANCE

9.9.1 Mode of Inheritance

In Table I, the mode of inheritance of nematode resistance in different crops to parasitic nematode species has been summarized. It is easily seen that the relationships most often have been explained as monogenic dominant resistance. This is characteristic for all *Heterodera* and *Globodera* species with the exception of *H. glycines*. Certain complications in oats and beets may be due to hexaploidy (oats) or difficulties in the transfer of resistance genes from related species. Resistance to *Meloidogyne incognita* (often including also *M. javanica* and *M. arenaria* but not *M. hapla*) is also normally monogenic and dominant although sometimes considered temperature sensitive (Dropkin, 1969a; Slana and Stavely, 1978b). Resistance to *M. hapla* is reported monogenic, oligogenic, or

TABLE I. Inheritance of Nematode Resistant in Plant Species

Nematode/host species	Monogenic Dominant	Monogenic Recessive	Oligogenic Dominant	Oligogenic Recessive	Polygenic
Heterodera avenae					
Barley	x	(x)			
Oats	x		x	x	
Wheat	x				
Heterodera schachtii					
Beets	x		x?		(x)
Tomatoes	x				
Heterodera glycines					
Soybeans			x	x	
Globodera rostochiensis					
Potatoes	x				x
Tomatoes	x				
Globodera pallida					
Potatoes	x				x
Rotylenchulus reniformis					
Soybeans	x?				
Tomatoes	x?				
Meloidogyne incognita					
Tomatoes	x	(x)			
Soybeans			x?		x?
Tobacco	x				
Cotton (*Gossypium barbadense*)				x	
(*G. hirsutum*)					x
Peach[1]	x?				
Beans (*Ph. vulgaris*)[2]				x	
(*Ph. lunatus*)[2]					x?
Pepper[2]	x				
Cowpea[2]	x				
Grape (*Vitis* spp)[2]	x?				
Meloidogyne javanica					
Tomatoes	x				
Peach[1]	x?				
Alfalfa	x		(x)		
Meloidogyne hapla					
Alfalfa	x		x		x?
Ditylenchus dipsaci					
Oats	x		x		
Alfalfa	x?		x?		x?
Red clover			x?		

[1] Malo, 1964.
[2] Hare, 1965.

possibly polygenic. Oats in relation to *Ditylenchus dipsaci* seem to depend on two independent genes, giving different types of resistance. For alfalfa and red clover there are indications of two or more genes behind the resistance, although for some cultivars dominant monogenic inheritance has been reported.

9.9.2 Pathotypes

Pathotypes have been recognized in several crop-nematode relationships when resistant cultivars have been produced and utilized. Resistance has been overcome by such pathotypes in, for example, barley resistant to *Heterodera avenae*, soybeans resistant to *H. glycines*, potatoes resistant to *Globodera rostochiensis* and to *G. pallida*, tomatoes and soybeans resistant to *Meloidogyne incognita*, and also several other crops resistant to this or other *Meloidogyne* species. Pathotypes have also been described in *Tylenchulus semipenetrans* in its relation to citrus rootstocks. In the complicated picture offered by *Ditylenchus dipsaci*, with at least 20 host races ("forma speciales") described, there is very little evidence in each of them on pathotypes seriously attacking resistant cultivars.

9.9.3 Gene-for-Gene Relationships

The type of inheritance in many host-nematode relationships and the common occurrence of pathotypes able to attack cultivars carrying certain resistance genes may indicate "vertical" resistance as discussed by Rohde (1972) in his review. There are well-known difficulties to obtain durability in this type of resistance. However, nematodes in the soil have reduced possibilities for a rapid spread compared with some parasites dispersed through the air. Their reproduction rate is also often low compared with most fungal parasites. This reduces the danger that the resistance might be overcome by new pathotypes but it is still a very real problem in *Globodera, Heterodera*, and *Meloidogyne* spp.

A gene-for-gene relationship was first suggested by Jones and Parrott (1965) for potato cyst nematodes. Their attempts to prove the theory were hampered because some pathotypes did not interbreed–it was later shown that they represented two different species, *Globodera rostochiensis* and *G. pallida* (Jones, 1974a). Inheritance of virulence in interbreeding pathotypes of each species appeared to be recessive (Parrott and Berry, cited in Jones, 1974b). Recent investigations by Parrott (Jones, 1979) are considered to have confirmed the gene-for-gene theory for potato cyst nematodes. Females reproducing on resistant roots are double recessive, whereas males may have any consitution.

Hayes and Cotten (1971) applied the gene-for-gene hypothesis on the resistance in barley to *Heterodera avenae* and calculated a "predicted minimum number of host resistance genes" of six with a corresponding number of genes in the nematodes. According to Andersen (1965) virulence in *H. avenae* is dominant, thus deviating from the common pattern in gene-for-gene relationships. Clearly, our knowledge of the inheritance of virulence in the nematodes is very limited even in these species with sexual propagation: most populations of *Meloidogyne* spp. reproduce by parthenogenesis (Triantaphyllou, 1971). Crosses between races in *Ditylenchus dipsaci* have concerned different host races and have not been related to resistance genes in the hosts.

9.9.4 Horizontal Resistance

Much interest in resistance breeding and research in nematology has been devoted to vertical resistance, but there are also examples of useful "horizontal" resistance. Resistance against stem nematodes in at least clover and alfalfa appears to be examples of such a resistance. If this is true, it provides a nice illustration of Robinson's (1979) very optimistic views on the practical use of this type of resistance. Although complete resistance has not been obtained, selections for three generations in very susceptible cultivars have improved the resistance in both red clover and alfalfa to a level where the nematode infestation in the soil is not increased and where the economic damage is largely reduced. A red clover cultivar, Merkur, released in 1937 in Sweden has been grown on a large scale but has not lost its resistance.

A few other examples of apparent horizontal resistance have been given in this chapter. For several relationships between hosts and nematodes, as well as many other parasites, there is now a growing interest for the possibilities offered by breeding for durable resistance. There are indications of such resistance against cyst nematodes in potatoes, although breeding up to now largely has been concentrated on vertical resistance (Darsow, 1979) due to the difficulties for selection in a polygenic system (Huijsman, 1974). Partial resistance in potatoes, due to reduced hatching stimulation, was reviewed by Evans and Franco (1979), and reduced invasion by larvae or a reduced development of females was also considered. Reduced occurrence of females developing into adults, commonly found in highly resistant cultivars (Triantaphyllou, 1973), was also described from *Beta* spp. susceptible or partially resistant to *Heterodera schachtii* (Heijbroek, 1977).

9.9.5 Hypotheses on the Nature of Resistance

The last decade has provided vast amounts of information on the reaction of plant tissues to a nematode attack. Information on the influence of cyst and root-knot nematodes on their host plants was collected earlier by using light microscopy, but the information now obtained on the ultrastructure by electron microscopy and on biochemical events by new analytical methods means a revolution in our knowledge, although very much remains to be investigated. Clarifying the role of syncytia and giant cells as transfer cells (Jones, 1976), developed in different ways but serving the same function, seems to be an important progress; and the amount of information gathered on enzymatic function will contribute to our knowledge of the interactions between nematodes and their resistant hosts. Some recent results indicating that the hypersensitive response is not alone responsible for nematode resistance in a plant may indicate a more differentiated view on the resistance mechanism—along with general discussion among plant pathologists.

Naturally, hypotheses and theories produced on nematode resistance in plants have closely followed research on resistance against other, especially fungal, parasites. Results available for nematodes in relation to their resistant hosts are, however, very limited. As all hypotheses up to now seem to have failed to explain the pathogen specificity in a commonly accepted way (Wheeler, 1975), it is not surprising that the biochemical background (gene products eliciting the host response) is not known. Toxins (phytoalexins) seem to take an active part in root-knot resistance in some crops; however, there are no indications of either pathotoxins, preformed toxins in the plants, or phytoalexins being generally linked to resistance. Similarly, investigations on several enzymes and hormones have not provided convincing evidence about their primary role in the resistance mechanism. References to the common antigen hypothesis have been made in nematology as well as in plant pathology and shared antigens of nematodes and their plant hosts have been demonstrated (McClure *et al.*, 1973). However, convincing evidence for a common application of this theory is not available.

Vanderplank (1978) suggested that susceptibility involves copolymerization of protein from the host with protein from the pathogen. Thus, host and pathogen should recognize each other by their proteins in gene-for-gene diseases. He considered proteins to be able to store the very great variability offered by a large number of genes for resistance and genes for virulence. Other substances with large enough molecules were ruled out for different reasons. The conclusion that susceptibility, and not resistance, involves protein copolymerization was based mainly on the fact that this endothermic process proceeds further at high temperatures. There are several examples of increased nematode susceptibility in a resistant cultivar grown at high temperatures (Dropkin, 1969a; Chang *et al.*, 1975). Paulson and Webster (1972) found that the hypersensitive reaction in tomatoes, carrying the resistance gene Mi, was prevented at 33°C and that giant cells were formed. Webster (1975) thought it possible that the heat treatment "may prevent the interaction of the membrane proteins with the nematode secretions." The "protein-for-protein hypothesis" (Vanderplank, 1978) and other theories may stimulate further fruitful research also in nematode-host interactions and contribute to a better understanding of the nature of the resistance.

9.10 ACKNOWLEDGMENTS

I wish to thank Professor V. Umaerus, Professor K. Bengt Eriksson, and Dr. S. Brishammer for reading the manuscript and for making many useful suggestions. I am also grateful to Mr. Nigel Rollison for linguistic revision and to Mrs. Gunnel Reimers for her help in preparing the manuscript.

9.11 REFERENCES

Abawi, G. S., VanEtten, H. D., and Mai, W. F. (1971). Phaseollin production induced by *Pratylenchus penetrans* in *Phaseolus vulgaris*. *J. Nematol.* **3**:301.

Andersen, K., and Andersen, S. (1968). Inheritance of resistance to *Heterodera avenae* in barley. *Nematologica* **14**:128-130.

Andersen, S. (1961). Resistens mod havreål *Heterodera avenae. Den kgl Veterinaer-og Landbohøjskoles Afdeling for Landbrugets Plantekultur, medd nr 68.* Dansk Videnskabs Forlag, Copenhagen.

Andersen, S. (1965). Heredity of race 1 or race 2 in *Heterodera avenae. Nematologica* **11**: 121-124.

Andersen, S., and Andersen, K. (1970). Sources of genes which promote resistance to races of *Heterodera avenae* Woll. *Eppo Publ. Ser. A* **54**:29-36.

Bajaj, K. L., and Mahajan, R. (1977). Phenolic compounds in tomato susceptible and resistant to *Meloidogyne incognita* (Kofoid et White) Chitwood. *Nematol. Medit.* **5**:329-333.

Barham, W. S., and Winsted, N. N. (1957). Inheritance of resistance to rootknot nematodes in tomatoes. *Proc. Am. Soc. Hort. Sci.* **69**:372-377.

Barrons, K. C. (1939). Studies of the nature of root knot resistance. *J. Agr. Res.* **58**:263-271.

Bingefors, S. (1951). The nature of resistance to stem nematodes, *Ditylenchus dipsaci* (Kühn) Filipjev, in red clover, *Trifolium pratense* L. *Acta Agr. Scand.* **1**:180-189.

Bingefors, S. (1956). Inheritance of resistance to stem nematodes in red clover. *Nematologica* **1**:102-108.

Bingefors, S. (1961). Stem nematode in lucerne in Sweden. II. Resistance in lucerne against stem nematode. *K. Lantbrukshögskolans Ann.* **27**:385-398.

Bingefors, S. (1969). The use of nematode-resistant varieties of grasses and legumes. *Herbage Abstr.* **39**:107-111.

Bingefors, S. (1971). Resistance to nematodes and the possible value of induced mutations. *In* "Mutation Breeding for Disease Resistance," pp. 209-235. International Atomic Energy Agency, Vienna.

Bingefors, S., and Eriksson, K. B. (1968). Some problems connected with resistance breeding against stem nematodes in Sweden. *Z. Pflanzenzücht.* **59**:359-375.

Bird, A. F. (1961). The ultrastructure and histochemistry of a nematode induced giant cell. *J. Biophys. Biochem. Cytol.* **11**:701-715.

Blake, C. D. (1962a). Some observations on the orientation of *Ditylenchus dipsaci* and invasion of oat seedlings. *Nematologica* **8**:177-192.

Blake, C. D. (1962b). The etiology of tulip-root disease in susceptible and in resistant varieties of oats infected by the stem nematode, *Ditylenchus dipsaci* (Kühn) Filipjev. II Histopathology of tulip-root and development of the nematode. *Ann. Appl. Biol.* **50**:713-722.

Boquet, D., Williams, C., and Birchfield, W. (1976). Inheritance of resistance to the Wartelle race of root-knot nematode in soybeans. *Crop Sci.* **16**:783-785.

Bouwman, L. A., and Ross, H. (1972). Differentiation between *Heterodera rostochiensis* and an undescribed allied species by female colour, morphometrics and pathogenicity. *Nematologica* **18**:265-269.

Bumbulucz, L. (1970). Kroknematode-"byggål," Race-og resistensundersøkelser. *Jord Avling* **2**:20-23.

Burkart, A. (1937). La selección de alfalfa immune al nemátode del tallo *(Anguillulina dipsaci). Rev. Arg. Agron.* **4**:171-196.

Caldwell, B. E., Brim, C. A., and Ross, J. P. (1960). Inheritance of resistance of soybeans to the cyst nematode, *Heterodera glycines. Agron. J.* **52**:635-636.

Canto Saenz, M., and Mayer de Scurrah, M. (1977). Races of the potato cyst nematode in the Andean region and a new system of classification. *Nematologica* **23**:340-349.

Carter, W. W. (1974). Histological responses of resistant and susceptible *Gossypium arboreum* to *Rotylenchulus reniformis. J. Nematol.* **6**:138.

Caviness, C. E., and Riggs, R. D. (1976). Breeding for nematode resistance. *Proc. World Soybean Res. Conf. 1975*, pp. 594-601.

Chang, D. C. N., Campbell, W. F., and Griffin, G. D. (1973). Ultrastructure changes induced by stem nematodes in hypocotyl tissue of alfalfa. *J. Nematol.* **5**:165-173.

Chang, D. C. N., Campbell, W. F., and Griffin, G. D. (1975). Effects of temperature on the fine-structural responses in the hypototyl region of alfalfa lines to *Ditylenchus dipsaci*. *J. Nematol.* 7:239-246.

Christie, J. R. (1949). Host-parasite relationships of the root-knot nematodes, *Meloidogyne* spp. III. The nature of resistance in plants to root knot. *Proc. Helminthol. Soc. Wash.* **16**:104-108.

Clamot, G. (1972). Amélioration de la résistance de l'avoine au nématode de la tige, *Ditylenchus dipsaci* (Kühn) Filipjev. *Bulle. Rech. Agron. Gembloux N.S.* 7:16-40.

Cook, R. (1974). Nature and inheritance of nematode resistance in cereals. *J. Nematol.* **6**: 165-174.

Cook, R., and Williams, T. D. (1972). Pathotypes of *Heterodera avenae*. *Ann. Appl. Biol.* **71**:267-271.

Cotten, J., and Hayes, J. D. (1972). Genetic studies of resistance to the cereal cyst nematode *(Heterodera avenae)* in oats *(Avena* spp.). *Euphytica* **21**:538-542.

Darsow, U. (1979). Vertikale and horizontale Resistenz in der Kartoffelzüchtung, *Arch. Züchtungsforsch. Berlin* **9**:121-131.

Day, P. R. (1974). "Genetics of Host-Parasite Interaction." W. H. Freeman, San Francisco.

De Maine, M. J. (1978). Field resistance to late blight and potato root eelworm in group *Tuberosum* dihaploids. *Euphytica* **27**:305-315.

Doney, D. L., Fife, J. M., and Whitney, E. D. (1970). The effect of the sugarbeet nematode *Heterodera schachtii* on the free amino acids in resistant and susceptible *Beta* species. *Phytopathology* **60**:1727-1729.

Dropkin, V. H. (1969a). The necrotic reaction of tomatoes and other hosts resistant to *Meloidogyne*: reversal by temperature. *Phytopathology* **59**:1632-1637.

Dropkin, V. H. (1969b). Cellular responses of plants to nematode infections. *Annu. Rev. Phytopathol.* 7:101-122.

Dropkin, V. H. (1972). Pathology of *Meloidogyne*–galling, giant cell formation, effects on host physiology. *OEPP/EPPO Bull.* **6**:23-32.

Dropkin, V. H., and Nelson, P. E. (1960). The histopathology of root-knot nematode infections in soybeans. *Phytopathology* **50**:442-447.

Dropkin, V. H., and Webb, R. E. (1967). Resistance of axenic tomato seedlings to *Meloidogyne incognita acrita* and to *M. hapla. Phytopathology* **57**:584-587.

Dropkin, V. H., Helgeson, J. P., and Upper, C. D. (1969). The hypersensitivity reaction of tomatoes resistant to *Meloidogyne incognita*: reversal by cytokinins. *J. Nematol.* **1**:55-61.

Drysdale, R. B. (1978). The elicitation of phytoalexin production. *Ann. Appl. Biol.* **89**:340-343.

Elgin, J. H., Jr. (1979). Inheritance of stem-nematode resistance in alfalfa. *Crop Sci.* **19**:352-354.

Elgin, J. H., Jr., Evans, D. W., and Faulkner, L. R. (1977). Response of resistant and susceptible alfalfa cultivars to regional isolates of stem nematode. *Crop Sci.* **17**:957-959.

Ellenby, C. (1952). Resistance to the potato root eelworm, *Heterodera rostochiensis* Wollenweber. *Nature* **170**:1016.

Ellis, P. R., and Maxon Smith, J. W. (1971). Inheritance of resistance to potato cyst-eelworm (*Heterodera rostochiensis* Woll.) in the genus *Lycopersicon*. *Euphytica* **20**:93-101.

Endo, B. Y. (1965). Histological responses of resistant and susceptible soybean varieties, and backcross progeny to entry and development of *Heterodera glycines. Phytopathology* **55**:375-381.

Endo, B. Y. (1971). Nematode-induced syncytia (giant cells). Host-parasite relationships of Heteroderidae. *In* "Plant Parasitic Nematodes" (B. M. Zuckerman, W. F. Mai, and R. A. Rohde, eds.), Vol. 2, pp. 91-117. Academic Press, New York.

Evans, K., and Franco, J. (1979). Tolerance to cyst-nematode attack in commercial potato cultivars and some possible mechanisms for its operation. *Nematologica* **25**:153-162.

Fassuliotis, G. (1970). Resistance of *Cucumis* spp. to the root-knot nematode, *Meloidogyne incognita acrita*. *J. Nematol.* **2**:174-178.

Fassuliotis, G., Deakin, J. R., and Hoffman, J. C. (1970). Root-knot nematode resistance in snap beans: breeding and nature of resistance. *J. Am. Soc. Hort. Sci.* **95**:640-645.

Fatunla, T., and Salu, A. (1977). Breeding for resistance to root-knot nematodes in tomatoes. *J. Agr. Sci., Cambridge* **88**:187-191.

Feldman, A. W., and Hanks, R. W. (1971). Attempts to increase tolerance of grapefruit seedlings to the burrowing nematode (*Radopholus similis*) by application of phenolics. *Phytochemistry* **10**:701-709.

Forrest, J. M. S. (1978). Aspects of potato cyst eelworm biology. *Scot. Plant Breeding Station 57th Annu. Rep.* **1977-78**:54-55.

Franco, J., and Evans, K. (1978). Multiplication of some South American and European populations of potato cyst nematodes on potatoes possessing the resistance genes H_1, H_2, and H_3. *Plant Pathology* **27**:1-6.

Frandsen, K. J. (1966). Observations on the attack by populations of *Ditylenchus dipsaci* on strains of red clover. *Acta Agr. Fennica* **107**:18-29.

Friedman, P., and Rohde, R. A. (1977). Phenol levels in leaves of tomato cultivars infected with *Pratylenchus penetrans*. *J. Nematol.* **9**:257-259.

Fuller, J. M., and Howard, H. W. (1974). Breeding for resistance to the white potato cyst-nematode, *Heterodera pallida*. *Ann. Appl. Biol.* **77**:121-128.

Giebel, J. (1974). Biochemical mechanisms of plant resistance to nematodes: a review. *J. Nematol.* **6**:175-184.

Giebel, J. (1976). β-Glucosidase activity in potato roots and its possible role in plant tissue response to *Heterodera rostochiensis*. *Bull. Acad. Pol. Sci. Ser. Sci. Biol. Cl II* **24**:37-42.

Giebel, J., and Jackowiak, N. (1976). Tryptophan decarboxylase in resistant and susceptible potato roots infected with *Heterodera rostochiensis*. *Nematologica* **22**:462-466.

Giebel, J., and Krenz, J. (1975). Role of amino acids in plant tissue response to *Heterodera rostochiensis*. II. Effect of proline and hydroxyproline. *Nematol. Medit.* **3**:49-53.

Giebel, J., and Stobiecka, M. (1974). Role of amino acids in plant tissue response to *Heterodera rostochiensis*. I. Protein-proline and hydroxyproline content in roots of susceptible and resistant solanaceous plants. *Nematologica* **20**:407-414.

Giebel, J., and Wilski, A. (1970). The role of IAA-oxidase in potato resistance to *Heterodera rostochiensis* Woll. *Zeszyty Prob. Postepow Nauk Rolniczych* **92**:239-253.

Giebel, J., Krenz, J., and Wilski, A. (1971). Localization of some enzymes in roots of susceptible and resistant potatoes infected with *Heterodera rostochiensis*. *Nematologica* **17**:29-33.

Giebel, J., Jackowiak, N., and Krenz, J. (1977). Relationship between phosphatase activity and *Heterodera rostochiensis* infection of resistant and susceptible potato roots. *Nematologica* **23**:51-56.

Gilbert, J. C., and McGuire, D. C. (1956). Inheritance of resistance to severe root-knot from *Meloidogyne incognita* in commercial type tomatoes. *Proc. Am. Soc. Hort. Sci.* **68**:437-442.

Gipson, I., Kim, K. S., and Riggs, R. D. (1971). An ultrastructural study of syncytium development in soybean roots infected with *Heterodera glycines*. *Phytopathology* **61**:347-353.

Gommers, F. J., and Voor int́ Holt, D. J. M. (1976). Chemotaxonomy of Compositae related to their host suitability for *Pratylenchus penetrans*. *Neth. J. Plant Pathol.* **82**:1-8.

Goodey, J. B., and Hooper, D. J. (1962). Observations on the attack by *Ditylenchus dipsaci* on varieties of oats. *Nematologica* **8**:33-38.

Goplen, B. P. and Stanford, E. H. (1959). Studies on the nature of resistance in alfalfa to two species of root-knot nematodes. *Agron. J.* **51**:486-488.

Goplen, B. P., and Stanford, E. H. (1960). Autotetraploidy and linkage in alfalfa–a study of resistance to two species of root-knot nematodes. *Agron. J.* **52**:337-342.

Griffin, G. D., and Elgin, J. H., Jr. (1977). Penetration and development of *Meloidogyne hapla* in resistant and susceptible alfalfa under differing temperatures. *J. Nematol* **9**:51-56.

Griffin, G. D., and Waite, W. W. (1971). Attraction of *Ditylenchus dipsaci* and *Meloidogyne hapla* by resistant and susceptible alfalfa seedlings. *J. Nematol.* **3**:215-219.

Griffiths, D. J., Holden, J. H. W., and Jones, J. M. (1957). Investigations on resistance of oats to stem eelworm, *Ditylenchus dipsaci* Kühn. *Ann. Appl. Biol.* **45**:709-720.

Grundbacher, F. J., and Stanford, E. H. (1962). Genetic factors conditioning resistance in alfalfa to the stem nematode. *Crop Sci.* **2**:211-217.

Haberman, E. (1972). Erforschung der Aktivität und des Isoenzymischen Bestandes der Oxydationsenzyme in der vom Kartoffelnematoden (*Heterodera rostochiensis* Woll.) befallenen Kartoffel. *Eesti NSV Teaduste Akad. Toimetised, Biol.* **21**:348-356.

Hanna, M. R., and Hawn, E. J. (1965). Seedling inoculation studies with the alfalfa stem nematode. *Can. J. Plant Sci.* **45**:357-363.

Hanschke, M., Borris, H., and Kämpfe, L. (1975). Der Einfluss des Befalls durch Kartoffelnematoden (*Heterodera rostochiensis* Woll.) auf den Auxingehalt des Wirtswurzelgewebes. *Biol. Runds.* **13**:235-237.

Hare, W. W. (1965). The inheritance of resistance of plants to nematodes. *Phytopathology* **55**:1162-1167.

Harmuth, P. (1977). Resistenzursachen von Kulturhafer gegenüber dem Getreidezystenälchen (*Heterodera avenae* Wollenweber, 1924). *Mitteil. Biol. Bundesanstalt Land-Forstwirtsch. Berlin-Dahlem* **178**:132.

Hayes, J. D., and Cotten, J. (1971). Breeding for nematode resistance with particular reference to *Heterodera avenae* Woll. *In* "Barley Genetics. II. Proceedings of Second International Barley Genetics Symposium" (R. A. Nilan, ed.), pp. 527-534. Washington State University Press, Seattle.

Haynes, R. L., and Jones, C. M. (1976). Effects of Bi locus in cucumber on reproduction, attraction, and response of the plant to infection by the southern root-knot nematode. *J. Am. Soc. Hort. Sci.* **101**:422-424.

Heijbroek, W. (1977). Partial resistance of sugarbeet to beet cyst eelworm (*Heterodera schachtii* Schm.). *Euphytica* **26**:257-262.

Hesling, J. J., and Wallace, H. R. (1960). Susceptibility of varieties of chrysanthemum to infestation by *Aphelenchoides ritzema-bosi* (Schwartz). *Nematologica* **5**:297-302.

Holm Nielsen, C. (1966). Untersuchungen über die Vererbung der Resistenz gegen den Getreidenematoden (*Heterodera avenae*) beim Weizen. *Nematologica* **12**:575-578.

Holm Nielsen, C. (1972). The test assessment for cereal cyst nematode (*Heterodera avenae*). *XIth Int. Nematol. Symp. Reading, England, Sept. 1972,* pp. 50-51.

Hoopes, R. W., Andersen, R. E., and Mai, W. F. (1978). Internal response of resistant and susceptible potato clones to invasion by potato cyst nematode *Heterodera rostochiensis. Nematropica* 8:13-20.

Howard, H. W., and Cotten, J. (1978). Nematode-resistant crop plants. *In* "Plant Nematology" (J. F. Southey, ed.), pp. 313-332. Ministry of Agriculture, Fisheries and Food GD1, Her Majesty's Stationery Office, London.

Huang, C. S., and Maggenti, A. R. (1969). Mitotic aberrations and nuclear changes in developing giant cells of *Vicia faba* caused by root knot nematode, *Meloidogyne javanica. Phytopathology* **59**:447-455.

Huijsman, C. A. (1974). Host-plants of *Heterodera rostochiensis* Woll. and the breeding for resistance. *Eppo Bull.* **4**:501-509.

Huijsman, C. A., Klinkenberg, C. H., and den Ouden, H. (1969). Tolerance to *Heterodera rostochiensis* Woll. among potato varieties and its relation to certain characteristics of root anatomy. *Eur. Potato J.* **12**:134-147.

Hung, C. L., and Rohde, R. A. (1973). Phenol accumulation related to resistance in tomato to infection by root-knot and lesion nematodes. *J. Nematol.* **5**:253-258.

Hunt, O. J., Peaden, R. N., Faulkner, L. R., Griffin, G. D., and Jensen, H. J. (1969). Development of resistance to root-knot nematode (*Meloidogyne hapla* Chitwood) in alfalfa (*Medicago sativa* L.). *Crop Sci.* **9**:624-627.

Jatala, P., and Russell, C. C. (1972). Nature of sweet potato resistance to *Meloidogyne incognita* and the effects of temperature on parasitism. *J. Nematol.* **4**:1-7.

Jatala, P., Arens, M. L., and Christiansen, J. (1977). Effects of bitter potato root exudates on hatching of *Globodera pallida* and *G. rostochiensis. J. Nematol.* **9**:273.

Jena, R. N., and Rao, Y. S. (1977). Nature of resistance in rice (*Oryza sativa* L) to the root-knot nematode (*Meloidogyne graminicola* Golden and Birchheld). II. Mechanisms of resistance. *Proc. Indian Acad. Sci.* **86B**:31-38.

Jones, F. G. W. (1974a). Host parasite relationships of potato cyst-nematodes: a speculation arising from the gene for gene hypotheseis. *Nematologica* **20**:437-443.

Jones, F. G. W. (1974b). Nematology department. *Rep. Rothamsted Exp. Stn. for 1973, Part 1,* pp. 149-166.

Jones, F. G. W. (1979). Nematology department. *Rep. Rothamsted Exp. Stn. for 1978, Part 1,* pp. 167-189.

Jones, F. G. W., and Parrott, D. M. (1965). The genetic relationships of pathotypes of *Heterodera rostochiensis* Woll. which reproduce on hybrid potatoes with genes for resistance. *Ann. Appl. Biol.* **56**:27-36.

Jones, M. G. K. (1976). Movement of solutes from host to parasite in nematode infected roots. *In* "Transport and Transfer Processes in Plants" (I. F. Wardlaw and J. B. Passioura, eds.). Academic Press, New York.

Jones, M. G. K., and Dropkin, V. H. (1976). Scanning electron microscopy of nematode-induced giant transfer cells. *Cytobios* **15**:149-161.

Jones, M. G. K., and Northcote, D. H. (1972a). Nematode-induced syncytium–a multinucleate transfer cell. *J. Cell. Sci.* **10**:789-809.

Jones, M. G. K., and Northcote, D. H. (1972b). Multinucleate transfer cells induced in coleus roots by the root-knot nematode, *Meloidogyne arenaria. Protoplasma* **75**:381-395.

Kaplan, D. T., Keen, N. T., and Thomason, I. J. (1978). Glyceollin accumulation related to expression of resistance in soybeans to the root-knot nematode. *J. Nematol.* **10**:291-292.

Király, Z., Barna, B., and Érsek, T. (1972). Hypersensitivity as a consequence, not the cause, of plant resistance to infection. *Nature* **239**:456-457.

Kochba, J., and Samish, R. M. (1971). Effect of kinetin and 1-naphthylacetic acid on root-knot nematodes in resistant and susceptible peach rootstocks. *J. Am. Soc. Hort. Sci.* **96**: 458-461.

Kochba, J., and Samish, R. M. (1972). Level of endogenous cytokinins and auxin in roots of nematode resistant and susceptible peach rootstocks. *J. Am. Soc. Hort. Sci.* **97**:111-115.

Koeslag, J. D. (1931). Een onderzoek over de resistentie van verschillende roggerassen tegen het stengelaaltje. *Tijdschr. Plantenziekten* **37**:96-104.

Kort, J., Ross, H., Rumpenhorst, H. J., and Stone, A. R. (1977). An international scheme for identifying and classifying pathotypes of potato cyst-nematodes *Globodera rostochiensis* and *G. pallida. Nematologica* **23**:333-339.

Laterrot, H. (1973). Sélection de variétés de tomate résistantes aux *Meloidogyne. OEPP/EPPO Bull.* **3**:89-92.

Levin, D. A. (1976). The chemical defenses of plants to pathogens and herbivores. *Annu. Rev. Ecol. Syst.* **7**:121-159.

Lewis, S. A. (1973). Identification and metabolism of indoles in *Meloidogyne incognita* and in cotton resistant and susceptible to *Meloidogyne incognita. Dissertation Abstr.* **34B**: 484.

Lewis, S. A., and McClure, M. A. (1975). Free amino acids in roots of infected cotton seedlings resistant and susceptible to *Meloidogyne incognita. J. Nematol.* **7**:10-15.

Lundin, P. (1969). Breeding of lucerne for resistance to stem nematode and *Verticillium* wilt. *Sveriges Utsädesförenings Tidskrift* **79(Suppl).**:133-139.

Malo, S. E. (1964). A review of plant breeding for nematode resistance. *Proc. Soil Crop Sci. Soc. Florida* **24**:354-365.

Malo, S. E. (1967). Nature of resistance of "Okinawa" and "Nemaguard" peach to the root-knot nematode *Meloidogyne javanica. Proc. Am. Soc. Hort. Sci.* **90**:39-46.

Martin, W. J., Hasling, V. C., and Catalano, E. A. (1976). Ipomeamarone content in diseased and nondiseased tissues of sweet potatoes infected with different pathogens. *Phytopathology* **66**:678-679.

Matson, A. L., and Williams, L. F. (1965). Evidence of a fourth gene for resistance to the soybean cyst nematode. *Crop Sci.* **5**:477.

Mayer de Scurrah, M. (1978). Breeding for resistance to Andean pathotypes of *Globodera pallida* and *G. rostochiensis. Abstr. Conf. Papers, 7th Triennial Conf. Euro. Assoc. Potato Res., Warsaw, 25 June–1 July, 1978. Breeding and Varietal Assessment II*, pp. 181-182.

McClure, M. A., Ellis, K. C., and Nigh, E. L. (1974a). Resistance of cotton to the root-knot nematode, *Meloidogyne incognita. J. Nematol.* **6**:17-20.

McClure, M. A., Ellis, K. C., and Nigh, E. L. (1974b). Post-infection development and histopathology of *Meloidogyne incognita* in resistant cotton. *J. Nematol.* **6**:21-26.

McClure, M. A., Misaghi, I., and Nigh Jr., E. L. (1973). Shared antigens of parasitic nematodes and host plants. *Nature* **244**:306.

Milne, D. L., Boshoff, D. N., and Buchan, P. W. W. (1965). The nature of resistance of *Nicotiana repanda* to the root-knot nematode, *Meloidogyne javanica. South Afr. J. Agr. Sci.* **8**:557-567.

Misaghi, I., McClure, M. A., and Kruk, T. H. (1975). Concentration of adenylates and energy charge values in cotton roots infected with *Meloidogyne incognita. Physiol. Plant Pathol.* **6**:85-91.

Nakamura, C., and Tsuchiya, T. (1978). Cytogenetics of nematode-resistant diploid sugar beets. *Agron. Abstr.* **58**.

Netscher, C. (1976). Observations and preliminary studies on the occurrence of resistance breaking biotypes of *Meloidogyne* spp. on tomato. *Cah. ORSTOM Ser. Biol.* **11**:173-178.

Nilsson-Ehle, H. (1908). Om olika angrepp av hafreålen (*Heterodera Schachtii*) på olika kornsorter. *Sveriges Utsädesförenings Tidskrift* **18**:171-173.

Nilsson-Ehle, H. (1920). Über Resistenz gegen *Heterodera schachtii* bei gewissen Gerstensorten, ihre Vererbungsweise und Bedeutung für die Praxis. *Hereditas* **1**:1-34.

Noel, G. R., and McClure, M. A. (1978). Peroxidase and 6-phosphogluconate dehydrogenase in resistant and susceptible cotton infected by *Meloidogyne incognita. J. Nematol.* **10**: 34-39.

Nordenskiöld, H. (1971). The genetic background of the resistance to nematodes (*Ditylench dipsaci*) in red clover (*Trifolium pratense*). *Hereditas* **69**:301-302.

O'Bannon, J. H., Chew, V., and Tomerlin, A. T. (1977). Comparison of five populations of *Tylenchulus semipenetrans* to *Citrus, Poncirus,* and their hybrids. *J. Nematol.* **9**:162-165.

O'Brien, P. C., and Fisher, J. M. (1977). Development of *Heterodera avenae* on resistant wheat and barley cultivars. *Nematologica* **23**:390-397.

O'Brien, P. C., and Fisher, J. M. (1978). Studies on the mechanism of resistance of wheat to *Heterodera avenae. Nematologica* **24**:463-471.

O'Brien, P. C., Sparrow, D. H. B., and Fisher, J. M. (1979). Inheritance of resistance to *Heterodera avenae* in barley. *Nematologica* **25**:348-352.

Owens, R. G. and Novotny Specht, H. (1966). Biochemical alterations induced in host tissues by root-knot nematodes. *Contrib. Boyce Thompson Inst.* **23**:181-198.

Padu, E. (1977). Die Aktivität der Phenylalaninammoniumlyase (EC 4.3.1.5.) in den von der Kartoffelnematode befallenen Kartoffelpflanzen. *Eesti NSV Teaduste Akad. Toimetised, Biol.* **26**:102-107.

Paulson, R. E. (1976). Changes in cell structure induced by *Meloidogyne incognita* and *Meloidogyne hapla* (Nematoda) in susceptible and resistant tomato (*Lycopersicon esculentum*) roots. *Dissertation Abstr. Intern.* **36B**:4275.

Paulson, R. E., and Webster, J. M. (1972). Ultrastructure of the hypersensitive reaction in roots of tomato, *Lycopersicon esculentum* L., to infection by the root-knot nematode, *Meloidogyne incognita. Physiol. Plant Pathol.* **2**:227-234.

Peacock, F. C. (1959). The development of a technique for studying the host/parasite relationship of the root-knot nematode *Meloidogyne incognita* under controlled conditions. *Nematologica* **4**:43-55.

Price, M., Caviness, C. E., and Riggs, R. D. (1978). Hybridization of races of *Heterodera glycines. J. Nematol.* **10**:114-118.

Rebois, R. V. and Webb, R. E. (1979). Reniform nematode resistance in potato clones. *Am. Potato J.* **56**:313-319.

Rebois, R. V., Epps, J. M., and Hartwig, E. E. (1970). Correlation of resistance in soybeans to *Heterodera glycines* and *Rotylenchulus reniformis. Phytopathology* **60**:695-700.

Rebois, R. V., Madden, P. A., and Eldridge, B. J. (1975). Some ultrastructural changes induced in resistant and susceptible soybean roots following infection by *Rotylenchulus reniformis. J. Nematol.* **7**:122-139.

Rebois, R. V., Steele, A. E., Stoner, A. K., and Eldridge, B. J. (1977). A gene for resistance to *Rotylenchulus reniformis* in tomato, and a possible correlation with resistance to *Heterodera schachtii. J. Nematol.* **9**:280-281.

Reed, B. M., Richardson, P. E., and Russell, C. C. (1979). Stem nematode infection of resistant and susceptible cultivars of alfalfa. *Phytopathology* **69**:993-996.

Reynolds, H. W., Carter, W. W., and O'Bannon, J. H. (1970). Symptomless resistance of alfalfa to *Meloidogyne incognita acrita. J. Nematol.* **2**:131-134.

Rich, J. R., Keen, N. T., and Thomason, I. J. (1977). Association of coumestans with the hypersensitivity of Lima bean roots to *Pratylenchus scribneri. Physiol. Plant Pathol.* **10**:105-116.

Riggs, R. D. (1977). Worldwide distribution of soybean-cyst nematode and its economic importance. *J. Nematol.* **9**:34-39.

Riggs, R. D., and Winstead, N. N. (1959). Studies on resistance in tomato to root-knot nematodes and on the occurrence of pathogenic biotypes. *Phytopathology* **49**:716-724.

Riggs, R. D., Kim, K. S., and Gipson, I. (1973). Ultrastructural changes in Peking soybeans infected with *Heterodera glycines. Phytopathology* **63**:76-84.

Robinson, R. A. (1979). Permanent and impermanent resistance to crop parasites; a reexamination of the pathosystem concept with special reference to rice blast. *Z. Pflanzenzücht.* **83**:1-39.

Rohde, R. A. (1965). The nature of resistance in plants to nematodes. *Phytopathology* **55**:1159-1162.

Rohde, R. A. (1972). Expression of resistance in plants to nematodes. *Annu. Rev. Phytopathol.* **10**:233-252.

Ross, J. P. (1958). Host-parasite relationships of the soybean cyst nematode in resistant soybean roots. *Phytopathology* **48**:578-579.

Roy, A. K. (1975). Studies on resistance of rice to the attack of *Meloidogyne graminicola. Z. Pflanzenkrankh. Pflanzensch.* **82**:384-387.

Savitsky, H. (1975). Hybridization between *Beta vulgaris* and *B. procumbens* and transmission of nematode (*Heterodera schachtii*) resistance to sugarbeet. *Can. J. Gen. Cytol.* **17**:197-209.

Sawhney, R., and Webster, J. M. (1975). The role of plant growth hormones in determining the resistance of tomato plants to the root-knot nematode, *Meloidogyne incognita. Nematologica* **21**:95-103,

Sawhney, R., and Webster, J. M. (1979). The influence of some metabolic inhibitors on the response of susceptible/resistant cultivars of tomato to *Meloidogyne incognita. Nematologica* **25**:86-93.

Schindler, A. F., Abadie, A. N., and Webb, R. E. (1977). Effect of plant endoparasitic nematodes, *Meloidogyne hapla* and *M. incognita-acrita*, on the production of toxic stress metabolites in white potatoes. *Proc. Am. Phytopathol. Soc.* **4**:162.

Seinhorst, J. W. (1956). Biologische rassen van het stengelaaltje *Ditylenchus dipsaci* (Kühn) Filipjev en hun waardplanten. I. Reacties van vatbare en resistente planten op aantasting en verschillende vormen van resistentie. *T. Pl. ziekten* **62**:179-188.

Seinhorst, J. W. (1961). Plant-nematode inter-relationships. *Annu. Rev. Microbiol.* **15**:177-196.

Sethi, C. L., and Sharma, N. K. (1976). Nature of resistance of cowpea to the root-knot nematode, *Meloidogyne incognita* and the pigeon pea cyst nematode, *Heterodera cajani. Indian J. Nematol.* **6**:81-85.

Shepherd, R. L. (1974). Transgressive segregation for root-knot nematode resistance in cotton. *Crop Sci.* **14**:872-875.

Sidhu, G., and Webster, J. M. (1973). Genetic control of resistance in tomato. I. Identification of genes for host resistance to *Meloidogyne incognita. Nematologica* **19**:546-550.

Sidhu, G. S., and Webster, J. M. (1975). Linkage and allelic relationships among genes for resistance in tomato (*Lycopersicon esculentum*) against *Meloidogyne incognita. Can. J. Gen. Cytol.* **17**:323-328.

Singh, B., and Choudhury, B. (1973). The chemical characteristics of tomato cultivars resistant to root-knot nematodes (*Meloidogyne* spp.). *Nematologica* **19**:443-448.

Sitaramaiah, K., and Pathak, K. N. (1979). Effect of phenolics and an aromatic acid on *Meloidogyne javanica* infecting tomato. *Nematologica* **25**:281-287.

Slana, L. J., and Stavely, J. R. (1978a). Location of the factor for resistance to *Meloidogyne incognita* in NC 95 tobacco. *J. Nematol.* **10**:300.

Slana, L. J., and Stavely, J. R. (1978b). High soil temperatures break resistance of *Nicotiana tomentosa* and resistant NC 95 tobacco to *Meloidogyne incognita. Phytopathol. News* **12**:133.

Slootmaker, L. A. J., Lange, W., Jochemsen, G., and Schepers, J. (1974). Monosomic analysis in bread wheat of resistance to cereal root eelworm. *Euphytica* **23**:497-503.

Smith, O. D., Boswell, T. E., and Thames, W. H. (1978). Lesion nematode resistance in peanuts. *Crop Sci.* **18**:1008-1011.

Steele, A. E. (1977). Inheritance of resistance to *Heterodera schachtii* in *Lycopersicon* spp. *J. Nematol.* **9**:285.

Sturhan, D. (1966). Rassen bei phytoparasitären Nematoden. *Mitteil. Biol. Bundesanstalt Land Forstwirtschaft Berlin-Dahlem* **118**:40-53.

Sturhan, D. (1969). Das Rassenproblem bei *Ditylenchus dipsaci. Mitteil. Biol. Bundesanstalt Land-Forstwirtschaft Berlin-Dahlem* **136**:87-98.

Sullivan, J. A., and Christie, B. R. (1979). The inheritance of resistance to the northern root-knot nematode (*Meloidogyne hapla* Chitwood) in alfalfa (*Medicago sativa* L.). *Forage Notes* **24**:20-23.

Sylvén, N. (1937). Svalöfs Merkurrödklöver. *Sveriges Utsädesförenings Tidskrift* **47**:15-19.

Szczygiel, A., and Danek, J. (1975). Susceptibility of strawberry cultivars to leaf and bud nematodes (*Aphelenchoides* spp.). *Fruit Sci. Rep. Skierniewice Poland* **11**:47-57.

Taylor, D. P. (1975). Observations on a resistant and a susceptible variety of tomato in a field heavily infested with *Meloidogyne* in Senegal. *Cah. ORSTOM Ser. Biol.* **10**:239-245.

Thomas, J. D., Caviness, C. E., Riggs, R. D., and Hartwig, E. E. (1975). Inheritance of reaction to race 4 of soybean-cyst nematode. *Crop Sci.* **15**:208-210.

Torp, J., and Hedeby Jensen, K. (1978). Glycosidase i rødder af nematoderesistente/modtagelige byg- og havresorter. *Nordisk Jordbrugsforskning* **60**:441-442.

Townshend, J. L., and Baenziger, H. (1976). Evidence of resistance to root-knot and root-lesion nematodes in alfalfa clones. *Can. J. Plant Sci.* **56**:977-979.

Toxopeus, H. J., and Huijsman, C. A. (1952). Genotypical background of resistance to *Heterodera rostochiensis* in *Solanum tuberosum*, var. *andigenum*. *Nature* **170**:1016.

Triantaphyllou, A. C. (1971). Genetics and cytology. *In* "Plant Parasitic Nematodes" (B. M. Zuckerman, W. F. Mai, and R. A. Rohde, eds.), Vol. 2, pp. 1-34. Academic Press, New York.

Triantaphyllou, A. C. (1973). Environmental sex differentiation of nematodes in relation to pest management. *Annu. Rev. Phytopathol.* **11**:441-462.

Triantaphyllou, A. C. (1975). Genetic structure of races of *Heterodera glycines* and inheritance of ability to reproduce on resistant soybeans. *J. Nematol.* **7**:356-364.

Vanderplank, J. E. (1978). "Genetic and Molecular Basis of Plant Pathogenesis." Springer-Verlag, Berlin, Heidelberg, and New York.

Van Staden, J., and Dimalla, G. G. (1977). A comparison of the endogenous cytokinins in the roots and xylem exudate of nematode-resistant and susceptible tomato cultivars. *J. Exp. Bot.* **28**:1351-1356.

Veech, J. A. (1978). An apparent relationship between methoxy-substituted terpenoid aldehydes and the resistance of cotton to *Meloidogyne incognita*. *Nematologica* **24**:81-87.

Veech, J. A. (1979). Histochemical localization and nematoxicity of terpenoid aldehydes in cotton. *J. Nematol.* **11**:240-246.

Veech, J. A. and Endo, B. Y. (1970). Comparative morphology and enzyme histochemistry in root-knot resistant and susceptible soybeans. *Phytopathology* **60**:896-902.

Veech, J. A., and McClure, M. A. (1977). Terpenoid aldehydes in cotton roots susceptible and resistant to the root-knot nematode, *Meloidogyne incognita*. *J. Nematol.* **9**:225-229.

Viglierchio, D. R. (1971). Nematodes and other pathogens in auxin-related plant-growth disorders. *Bot. Rev.* **37**:1-21.

Viglierchio, D. R. (1978). Resistant host response to ten California populations of *Meloidogyne incognita*. *J. Nematol.* **10**:224-227.

Viglierchio, D. R., and Mjuge, S. G. (1975). Auxin inactivation systems of nemic origin. *Nematologica* **21**:471-475.

Viglierchio, D. R., and Yu, P. (1965). Plant parasitic nematodes: a new mechanism for injury of hosts. *Science* **147**:1301-1303.

Wålstedt, I. (1936). Verschiedene Rassen bei *Heterodera Schachtii* Schmidt. *Der Züchter* **8**:201-208.

Webber, H. J., and Orton, W. A. (1902). A cowpea resistant to root-knot (*Heterodera radicicola*). *USDA BPI Bull.* **17**:23-36.

Webster, J. M. (1969). The host-parasite relationships of plant-parasitic nematodes. *Adv. Parasitol.* **7**:1-40.

Webster, J. M. (1975). Aspects of the host-parasite relationship of plant-parasitic nematodes. *Adv. Parasitol.* **13**:225-250.

Webster, J. M., and Paulson, R. E. (1972). An interpretation of the ultrastructural response of tomato roots susceptible and resistant to *Meloidogyne incognita* (Kofoid et White) Chitwood. *OEPP/EPPO Bull.* **6**:33-39.

Wheeler, H. (1975). "Plant Pathogenesis." Springer-Verlag, Berlin, Heidelberg, and New York.

Williams, T. D. (1978). Cyst nematodes: biology of *Heterodera* and *Globodera*. *In* "Plant Nematology" (J. F. Southey, ed.), pp. 156-171. Ministry of Agriculture, Fisheries and Food GD 1, Her Majesty's Stationery Office. London.

Wilski, A., and Giebel, J. (1966). β-Glucosidase in *Heterodera rostochiensis* and its significance in resistance of potato to this nematode. *Nematologica* **12**:219-224.

Yu, M. H. (1978). Meiotic behavior of a disomic nematode-resistant sugarbeet. *Crop. Sci.* **18**:615-618.

Chapter 10

APHID PROBING AND FEEDING, ELECTRONIC MONITORING, AND PLANT BREEDING

T. Richard Tarn
Jean B. Adams

10.1 INTRODUCTION

Plant resistance to aphids depends upon the ability of plant characteristics to prevent successful aphid probing and feeding. Progress in breeding aphid-resistant cultivars depends upon our knowledge of aphid probing and feeding behavior and our success in incorporating the necessary plant characteristics into new crop cultivars. We review here a monitoring technique that advances knowledge of this behavior and provides a new means of identifying plant resistance to aphids.

Conventional means of evaluating plant acceptability to aphids have depended on various methods of observing patterns of settling and feeding and of measuring the rate of aphid reproduction. These studies have been carried out in the field, in the greenhouse, and in the laboratory, and significant progress has been made in our knowledge of probing and feeding behavior (e.g., Pollard, 1973, 1977; Garrett, 1973; Harris and Bath, 1973; Harris, 1977). Such studies have been too dependent upon subjective judgments: generally they have been indi-

PATHOGENS, VECTORS, AND PLANT DISEASES

ISBN 0-12-326440-5

rect and have not permitted the precise measurement of the initial aphid contact with the plant and subsequent probing and feeding. Observations on living aphids used the extension of the labium as an indication of preliminary probing, while the results of penetration have been assessed indirectly on the basis of salivary sheaths observed in fixed plant tissue.

The direct monitoring of aphid probing and feeding activities became possible 16 years ago with the development of McLean and Kinsey (1964, 1965) of an electronic recording technique and the establishment of a correlation between the electrophysiological readout patterns and the penetration behavior of some crop-infesting aphid species. One of the uses of this technique of measuring plant-aphid interactions is to evaluate plants for resistance to aphids. The technique has already been applied for this purpose by Tarn and Adams (1973), Haniotakis and Lange (1974), Nielson and Don (1974), Adams and Wade (1976), Shanks and Chase (1976), Kennedy *et al.* (1978) and Holbrook (1980). Today, plant resistance to insects is considered one of a number of components that together form an integrated pest control system (Luckmann and Metcalf, 1975). A level of aphid resistance may delay the occurrence of economic damage (van Emden and Wearing, 1965), provide a means of reducing the undesirable pesticide load in the agro-ecosystem (Maxwell *et al.*, 1972), and reduce virus spread by aphid vectors (Kennedy, 1976). The latter result depends upon the interactions of plant, insect, and virus. Each case needs individual consideration since insect resistance may also increase the spread of viruses, especially of nonpersistent viruses, as Kennedy (1976) recognized. However, crop losses to aphid-transmitted viruses are considerable (Eastop, 1977) and virus control through insect control must be considered. Historically, most examples of resistance have been of antibiosis (Painter, 1951), that is, those situations in which the host plant adversely affected the development and reproduction of the aphid. Nonpreference, in which the aphid quickly rejects the host as a source of food, is a little-used but powerful source of resistance, and one that is well suited to investigation by the electronic monitoring technique of McLean and Kinsey.

In reviewing aphid probing and feeding behavior, electronic monitoring, and plant breeding, we look carefully at the technique, its potential and its limitations. As a basis for the interpretation of monitoring results, we briefly review the present understanding of probing and feeding, and relate printout characteristics to different types of plant resistance. We also discuss the use of the technique in variety development programs.

10.2 THE HOST-PARASITE RELATIONSHIP

Plants and aphids have co-evolved for over 200 million years in a reciprocally beneficial relationship (Kennedy and Fosbrooke, 1973). The development and maintenance of this complex and subtle relationship are reflected in many aspects of the structure, functioning, and life cycles of both the plant host and the aphid parasite.

The fallout of aerial aphid plankton is haphazard and the nature of a host plant probably has no impact until an aphid is only a few meters away or is already on the plant surface. Gibson and Plumb (1977) review plant characteristics, such as color, pubescence, and nature of the cuticle, which may be related to the failure of aphids to settle. Once on a leaf, an aphid may respond to unacceptable tactile stimuli or may reject a plant at the point of proboscal contact, stylet penetration, or ingestion. At each link in this chain of events, a plant may present a lesser or greater barrier to aphid feeding (Kennedy, 1965). Most plants generally support few or no aphids, demonstrating the success of plant barriers in discouraging or preventing aphid feeding.

The evolutionary responses of an aphid are reflected in its life cycle and the sophistication of its feeding behavior. Aphids enhance their chances of survival by developing large populations produced by live births of parthenogenic females by parthenogenic females. A single female that arrives on an acceptable host can quickly give rise to a large number of genetically similar females that can all survive.

Active host selection commences once an aphid has alighted on a plant. The insect is able to investigate its host through an abundance of mechanoreceptors on its antennae, legs, and mouthparts, and with chemoreceptors on at least the mouthparts (Forbes and Mullick, 1970). The behavior of an aphid in its orientation to the plant, its examination, probing, selection and location of a food source, its ingestion, and withdrawal of its stylets have been effectively described by Pollard (1973, 1977). In the first of the sequence of events that lead to penetration, the proboscis is protracted, the antennae are extended and vibrate, the proboscis moves about on the surface, and a slight exudate of saliva is usually detectable at the interface between proboscal tip and plant surface. The shape of the gelled saliva at this site depends on the characteristics of the labial apex (Nault and Gyrisco, 1966) and may be regarded as a labial imprint on the leaf surface (Harris and Bath, 1973; Harris *et al.*, 1981). Many authors have suggested that the aphid can sense the suitability of its host tissue at this point, or during the 60 sec that the aphid typically takes to penetrate the epidermis (Pollard, 1973; Harris, 1977). However, until now, there has been no evidence of sense receptors with the potential for detecting host acceptability at the time of proboscal contact.

The labium bears many tactile hairs and the long pointed sensory cones at the labial tip surrounding the opening from which the stylets emerge are also probably mechanoreceptors (Forbes, 1977), (Fig. 1). In addition to the mechanoreceptors, there are short, stubby cones with terminal pores (Fig. 1C, D) characteristic of chemoreceptors (Adams, unpublished; Harris, unpublished). We postulate that the pore openings on these receptors play a role in detecting the end products of the interaction between aphid saliva and plant epidermis. This conclusion is in contrast to that of Tjallingii (1978b) who concludes that "aphid mouthparts are devoid of external chemoreceptors." Following proboscal contact, penetration of plant tissue by the stylets is usually accompanied by the secretion of more saliva, which in turn reacts with the surrounding cell walls. The

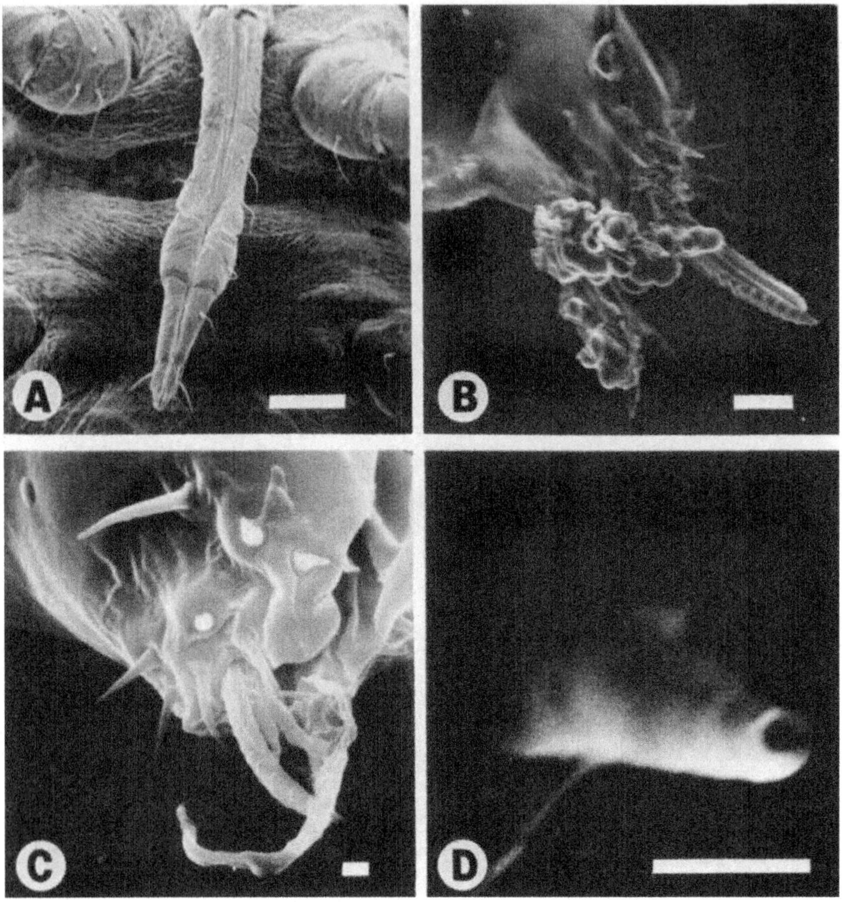

FIG. 1. Scanning electron micrographs of labium and stylets of *Myzus persicae* (Sulzer). (A) Labium showing long mechanoreceptors (bar = 50 μm). (B) Proboscal tip showing extruded serrated stylets and saliva (bar = 5 μm). (C) Proboscal tip showing slightly extruded stylets coated with saliva and two types of sensory setae–long, tapered mechanoreceptors and short, stubby receptors presumed to be chemoreceptors (bar = 1μm). (D) Further enlargement of presumed chemoreceptor showing terminal pore (bar = 1 μm).

patterns of stylet tracks in the tissue have been studied extensively (Pollard, 1973), and the method of penetration, either hydrolytic or mechanical, depends on the presence or absence of pectin-hydrolyzing enzymes in the saliva (McAllan and Adams, 1961). When ingestion takes place, aphids have a further opportunity to evaluate their food source through gustatory receptors in the pharyngeal canal (Wensler and Filshie, 1969).

The stages of feeding, from initial proboscal contact to ingestion, can be followed in the readouts provided by the electronic monitoring technique, as we illustrate in Section 10.4. Furthermore, the readouts make it possible, for the first time, to recognize the stage at which further penetration or ingestion ceases. In Section 10.5, we consider our understanding of resistance mechanisms based on this information.

10.3 ELECTRONIC MONITORING SYSTEMS

The first noted use of the electronic monitoring system was that by Roche *et al.* (1962) to measure the feeding behavior of hookworms. Two years later, McLean's laboratory published the first (McLean and Kinsey, 1964) of a series of papers in which the technique was used to study the feeding behavior of pea aphids, *Acrythosiphon pisum* (Harris), on *Vicia faba* L. This technique is a valuable research tool that can be applied by entomologists with input from electronic specialists to fine-tune the system to particular plant-insect systems. We review the technique as it was developed by McLean and others, and consider some work with other insects that may have application to aphid studies.

10.3.1 The Basic System

The basic system used in most of the aphid work has been described by McLean and Kinsey (1964). More detailed accounts of the system and its modifications were given by McLean and Weigt (1969) and McLean (1977). Electrical measurement systems are discussed in full by Weber and McLean (1975). The system used by McLean and described by McLean and Weigt (1969) used the main 120-V AC power supply fed into a voltage regulator to eliminate line voltage fluctuations. Connected in series was a transformer to reduce the voltage to about 6 V AC and a potentiometer to adjust the voltage supply to the leaf. A clamp connection was made to the leaf. In the earliest work, an aphid was connected to the system by tarsal contact with a wire grid that was insulated from the leaf. This provided a poor electrical contact and was replaced by a fine wire attached, at one end, to the aphid's back with rapid drying conducting cement and, at the other, connected directly to an amplifier (McLean and Kinsey, 1967). An audio amplifier was necessary to increase the voltage sufficiently to operate the recording devices. Published data from the system have been from strip-chart recorders operated with a rectifier circuit. Ouput has also been through a speaker or, to monitor distortion and detect system malfunctions, an oscilloscope. The latter two are not essential to the basic system.

Aphid saliva and plant sap are both electrically conductive; therefore, when the aphid salivates on the plant surface or penetrates the leaf, the circuit is completed. The precise nature of the labial and stylet contact with leaf tissue, the conductivity of fluids in the salivary and food canals, and the rate at which these fluids are moving contribute to the current and voltage fluctuations that are recorded (McLean and Kinsey, 1964, 1965).

10.3.2 Adaptations

Most adaptations attempted to increase the sensitivity of the system by improving grounding and eliminating "noise." Nielson and Don (1974) raised the electrical potential of the plant by inserting a copper lead into the moist soil. This was more convenient than the leaf connection and reduced the significance of changes in leaf resistance as the aphid moved in relation to the contact point. They also included a voltage regulator and frequency oscillator to eliminate day-to-day and plant-to-plant voltage variation.

Elimination of power line interference was achieved by Adams and Wade (1976) working in collaboration with the Bio-Engineering Institute, University of New Brunswick, Fredericton, by using a frequency lower than 60 Hz. This group demonstrated that frequencies below 20 Hz and amplitudes less than 2 V peak to peak, did not appear to interfere with aphid feeding behavior. The system operated on transformed mains power supply with selective amplification of the 18-Hz input frequency, and the nonrectified output of the 18-Hz input frequency was recorded (Fig. 2). These adaptations made the use of a shielded room unnecessary and simplified the system. Brown and Holbrook (1976) used a similar system.

DC systems have the advantages of simplified circuitry and reduced line frequency noise (Schaefers, 1966). They also have potential for the development of portable systems for use in field studies. Schaefers (1966) used 1.5-V mercury batteries for a DC power source. This system required a DC strip-chart recorder but no transformer, potentiometer, amplifier, or rectifier. A similar DC system was used by Shanks and Chase (1976), who took considerable precautions to avoid electrical interference, by using electrocardiogram cream over the conducting silver paint to improve the quality of recordings. A similar system was used by Tjallingii (1978a).

10.3.3 Output Management

Most reports on the electronic monitoring of feeding are based on data from strip-chart recorders. McLean's group apparently began with chart speeds of 18.75

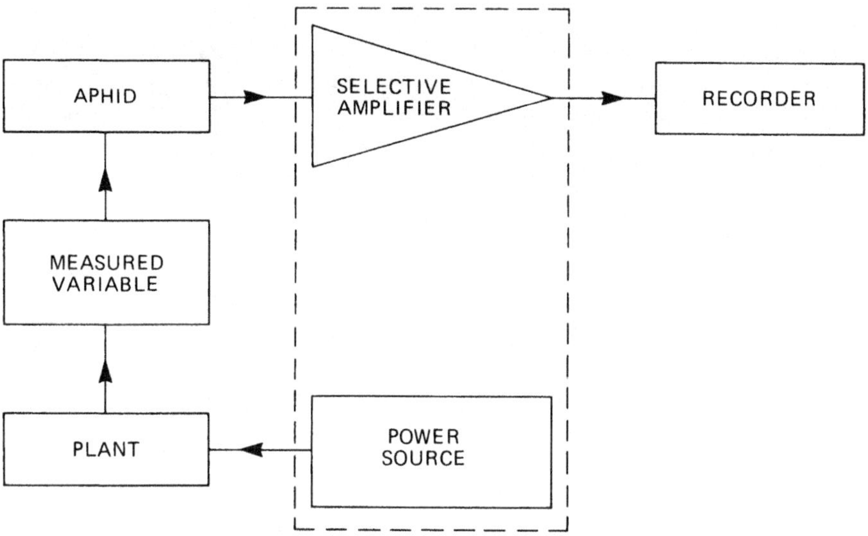

FIG. 2. Block diagram showing the system components used by Adams and Wade (1976). The power source and selective amplifier constitute the "biobox." The measured variable is the varying conductivity of aphid saliva and plant "sap" as the aphid probes and feeds.

cm/min. They reduced them to 4.69 cm/min (McLean and Kinsey, 1967) and, subsequently, to 1.91 cm/min for 24-hr recordings (McLean and Kinsey, 1968a). Most aphid studies have since used similar low chart speeds. Even at these slow speeds, 24-hr recordings produce 27.43 m of tape (McLean and Kinsey, 1968a) and data tabulation is long and tedious. Nevertheless, Nielson and Don (1974) used 7.62 cm/min, and Adams and Wade (1976) used 6 cm/min during standard 4-min recordings when screening for host resistance on the basis of aphid behavior during proboscal contact and penetration. A variable-speed recorder allows the tape speed to be varied, depending on the information required. In cases where plant-host rejection is being evaluated, specific and relatively short exposure times may be acceptable. Short-term events may be resolved with fast chart speeds: an example is the use by Kashin (1966) of speeds up to 600 cm/min to study engorgement of the mosquito (*Aedes aegypti* L.). Not all authors have reported chart speeds. Absence of this information makes comparison of data difficult, and confusion may result when some charts are printed from left to right and others from right to left. We suggest that all charts be printed from left to right and that the speed always be given.

Flexibility was introduced by the use of a tape recorder to store the information that can be played back later to a strip-chart recorder (McLean and Weigt, 1969). Automated recording has been used for evaluating mosquito repellent compounds. Kashin and Kardatzke (1969) developed a computer program that used the percentage of engorged mosquitoes, plus the percentage of biting time, to calculate repellency indices that discriminated between effective and ineffective repellents. Kashin and Arneson (1969) used an electronic digital timer to register mosquito bites (bite time) and a second timer to record total test time. This system eliminated the need for laborious chart recording measurements, leaving only the number of engorged mosquitoes and the total number of live mosquitoes to be counted. Such an approach may be applicable to monitoring aphid feeding behavior for periods that extend beyond initial probing and include interrupted feeding.

In a different development, the feeding activities of *Rhodnius prolixus* Stal. were studied by Smith and Friend (1971), who used split-screen video recording. Interpretation of behavior was facilitated by displaying side by side the chart record and the film of feeding. These workers found the technique useful for the analysis of rapid and transitory events, and noted its application to any species of insect that can be induced to feed through a membrane on a liquid diet contained in an optically transparent chamber.

Finally, a modification to monitor tick (*Hyalomma dromedarii* Koch) feeding in natural micrometeorological conditions was described by Sweatman *et al.* (1976). The electronic feeding data from ticks on the backs of free-ranging cattle were transmitted to a receiver located in a control room. Perhaps aphid feeding in the field could be monitored this way.

10.3.4 System Effects on Plant and Aphid

This technique inevitably subjects plant and aphid to unnatural environments and may cause trauma to the aphid. There appears to be little system effect on the plant, although the use of a DC system could cause movement of proteins and other electrolytes and subsequent polarization and physiological changes in both plants and animals (McLean and Weigt, 1969). However, in work with *Rhodnius*, Smith and Friend (1971) did not find such effects in their system, where small currents, relatively large electrodes, and short-term experiments were the norm.

Aphids did not appear to exhibit abnormal behavior either when McLean and Kinsey (1964) and Brown and Holbrook (1976) used conducting paint to attach the wire to the abdominal dorsum, or when Schaeffers (1966) used acetone-thinned paint. The flexible, light-weight wire allows the aphid free movement within the range of the tether (McLean and Kinsey, 1967), and Tjallingii (1978a) noted that aphids connected to 25-μm diam gold wire penetrated sooner than those attached to a more rigid 0.2-mm copper wire. Tjallingii also observed that specimens of *A. pisum* wired for 10 days produced more than 20 nymphs. In our studies the interference did not materially lengthen the instar of immature *Myzus persicae* (Sulzer), and moulting (along with the adherent wire) appeared normal. The size of the glue spot, which varies with the will of the applicator and the size of the aphid and its dorsum, had a direct affect on the amplitude of the trace. A relatively large spot produced a greater amplitude and a smaller spot a lower one. Tjallingii (1978a) reported similar variations in amplitude.

It is important that a test aphid's behavior not be affected by the insect's place in an electrical circuit. McLean and Weight (1969) calculated the current through the aphid to be 3.72×10^{-9} A with 0.495 V across the aphid. In a similar system Brown and Holbrook (1976) reported a maximum current through the aphid of 2×10^{-9} A during initial salivary surges, and noted that voltage could be increased to from 10 to 100 times its usual recording value before an aphid withdrew its mouthparts. Our tests used 1.5 V because of previously observed aphid restlessness at higher voltages.

In the system used by Kashin and Wakely (1965) mosquitoes were subjected to a current of 6.0×10^{-9} A with an applied voltage of 1.35 V DC. These authors detected no changes in mosquito behavior and stated that this amount of electricity is "probably no more than the static electrical charge that the insect attains during free flight." Smith and Friend (1971) observed no effect on the feeding behavior of *Rhodnius* when they used 0.13 V and a maximum current of 5.0×10^{-8} A.

10.3.5 Nonsystem Effects on Plant and Aphid

The aphid is so much "a creature of its substrate" (Adams and Drew, 1963) that studies of aphid feeding behavior on test plants must consider both prior nutritional history of the aphid and the state of the test plant. Many studies, including those of McLean (1971) using the electronic monitoring technique, have shown that aphid feeding behavior varies with the host or diet on which the aphid was reared. Aphids used in screening for resistant crop cultivars should be raised on host plants of the same or a closely related species, and at the same stage of growth and nutrition as the test plants.

Many aphid studies use excised leaves or leaf disks. Adams and Wade (1976) have shown that this treatment enhances host acceptance by the aphid. Chart records have shown that aphids that would not settle and feed on an unacceptable leaf attached to the plant, commenced probing and feeding within seconds of the leaf petiole being severed. Leaf resistance to aphids was apparently associated with plant metabolism and was not dominantly a morphological feature. Excision is known to produce changes in leaf physiology (Farkas *et al.*, 1964) but these have not been recorded to be as rapid as the aphid response.

Individual aphids may also vary, and so it is important to use aphids of the same species, the same biotype, the same morph, and the same age, unless these are specific variables of an experiment. In our tests it was often difficult to get consistent probing behavior from individual aphids even from the same clone reared on the same host plant. Prior feeding status was found to be a variable. Some aphids appeared to be satiated and hence restless; others, which had been deprived of food, fed readily and consistently. In one test, a specimen of *M. per-*

sicae ingested without ceasing for 4.5 hr. On this tape there was no indication that the stylet canals were ever emptied, thus breaking the electrical connection, until the aphid withdrew its stylets and moved away from the probe site (Adams and Wade, 1976).

Aphids have a wide range of temperature tolerance, but their activity changes at both ends of the range. This has been confirmed in electronic monitoring readouts (McLean and Kinsey, 1968b). In our use of the monitoring technique, we found it necessary to avoid temperatures above 30°C if "consistent" behavior was desired; restlessness and refusal to settle were customary above this temperature.

Work by Helson and Penman (1970) suggests that barometric changes can influence insect behavior. Our results appear to support this. Aphids interrupted their normal settling and probing on susceptible hosts more frequently during periods of barometric change. This observation is made in hindsight since our pressure instrumentation was primitive and did not show the difference between negative and positive ionic fields in such pressure changes. Helson and Penman had associated insect "misbehavior" with negative fields. Other factors that have seemed to influence test aphid behavior (although we have not conducted properly controlled experiments to demonstrate the points!) have been "tobacco residue" on the operator's hands and groups of visitors viewing the demonstration.

10.4 INTERPRETATION OF THE READOUT

To recognize important variations in readout patterns, such as those characteristic of aphid feeding on unacceptable "resistant" hosts, it is necessary to understand feeding behavior and the corresponding readout patterns on acceptable "susceptible" hosts. We consider first the interpretation of readouts on acceptable hosts and then the differences in this behavior on unacceptable hosts.

10.4.1 Aphid Behavior on Acceptable Hosts

When the aphid secretes its first saliva on the leaf surface, the saliva appears to dissolve some material from the cuticle. We postulate that chemoreceptors at the tip of the labium (see Section 10.2), which are immersed in this saliva, can sense such materials and provide the aphid with information relating to the acceptability of the substrate. This is also the first point at which host suitability can be recognized. If this first sensory response is acceptable to the aphid, the stylets penetrate the epidermis and continue into the mesophyll.

In the literature many terms are used to describe the initial exploration of the plant surface by the aphid. These terms, which include test probes, superficial probes, access probes, and proboscal contact, are often not well defined and, with the exception of the last one, imply some penetration of the plant by the stylets. However, they all recognize the increasing importance recently given to "test" probes before penetration is completed and ingestion begins. We favor the concept of "proboscis contact," first used by Tjallingii (1976), to describe this initial investigation of the substrate. Proboscal contact involves little or no stylet penetration and recognizes the ability for sensory discrimination at this stage. The same distinction was made by Lehmann *et al.* (1975), who used "tasten" to describe repeated short placement of the labium on the leaf surface without penetration. We restrict the use of "penetration" for movement of the stylet bundle through the plant epidermis and tissue to the feeding site. The term "probe" is retained to include proboscal contact, penetration, and ingestion.

The first interpretations of electronic readout patterns of aphid feeding were by McLean and colleagues (McLean and Kinsey, 1965, 1967), and we draw heavily on their work in reviewing aphid behavior on acceptable hosts. The initial proboscal contact with the epidermis is followed almost immediately by salivation (McLean and Kinsey, 1965). Time lags of up to 12.5 sec were noted by Hodges and McLean (1969) and Nault and Styer (1972), before the high voltage waveform corresponding to salivation was recorded. Nault and Styer reported test probes in which salivary activity continued a further 10-30 sec before there was withdrawal without ingestion.

The first electronic monitoring of penetration involved liquid substrates. McLean and Kinsey (1965) recorded the secretion of sheath saliva and its expansion by watery saliva secreted within the sheath. Later, McLean and Kinsey (1968a) reported that the stylets were pushed through the sheath before the secretion of fresh saliva and the repetition of the sequence. Electronic readouts show high voltages corresponding to salivation, low voltages during brief periods when the sheath is presumed closed and conductivity lowered, and low voltages during ingestion. Aphid penetration of plant tissue produced some differences in the waveforms, though Tjallingii (1978a) found no clear differences in patterns between aphids feeding on plants and on feeding solutions. The same high-amplitude peaks corresponding to salivation were separated by lower voltage traces lasting about 10 sec. Low-voltage peaks had a frequency of about 180/min, corresponding to the peak frequency during ingestion from liquid substrates, and were believed to reflect cibarial pump activity. This pattern of alternate salivation peaks and brief ingestion periods was designated the "X waveform" (McLean and Kinsey, 1965).

A relatively low voltage record is produced by steady ingestion (McLean and Kinsey, 1965) when the aphid is feeding through the open end of its stylet sheath (Miles *et al.*, 1964; Kinsey and McLean, 1967). We have found the amplitude of the ingestion trace to vary with aphid species. For example, it is very low for *Aphis nasturtii* Kaltenbach compared with *M. persicae* on *Solanum tuberosum* L. cv. Katahdin (Adams and Wade, 1976). Variation in the amplitude of the peaks during ingestion appears dependent upon cibarial activity. Peak amplitudes for aphids feeding in nutrient solution is greater than that of aphids feeding in phloem (McLean and Kinsey, 1965). During ingestion from the phloem, cibarial pump activity may be very low because phloem cell pressure is maintained by the plant (Mittler, 1957) during ingestion. From mesophyll cells, peak amplitudes are large since cell pressure is low (McLean and Kinsey, 1968a). Nault and Styer (1972) found that when aphids ingesting from sieve elements were anesthetized and their stylets amputated, ingestion waveforms continued and sap was exuded from the amputated stylets. The same procedure applied to aphids ingesting from mesophyll parenchyma resulted in the waveform dropping close to the baseline and no sap being exuded from the cut stylets.

McLean and Kinsey (1967), using readouts and sectioned plant material, related waveforms to the location of the stylets and to ingestion sites. Histological studies have been carried out by Nault and Styer (1972) and Kennedy *et al.* (1978) to confirm the ingestion sites. These studies appear most successful in relating X waveform to phloem sieve-cell ingestion, and this correlation has been used by other workers in interpreting their own results. Several attempts have been made to explain the origin of X waveforms in more detail (McLean and Kinsey, 1967, 1969; McLean, 1977) but no single explanation may be possible since penetration is a complex series of events. When stylet movement changes from being intercellular to penetrating the cell wall, a different activity and perhaps a different readout pattern may be expected. Other waveforms have been identified (McLean and Kinsey, 1965, 1967) but not fully explained because of the complexity of penetration. Withdrawal is accompanied by further salivation.

In some interesting experiments with *A. pisum* and *Brevicoryne brassicae* (Linnaeus), Tjallingii (1978a) used myograms of cibarial and saliva pump muscle activity and radioactive tracers to measure ingestion and salivation, together with visual observations, to interpret his "detection-signal" patterns. Large B patterns at the beginning of each trace, and elsewhere, were associated with the secretion of sheath saliva. Results from the tracer studies confirmed the correlation of saliva production with the total time of A, B, and C patterns, but showed no correlation with D, E, and F patterns. Cibarial pump activity was and saliva pump activity was not correlated with E pulses. Pattern D, usually occurring with pattern E, showed a strong correlation with ingestion in the tracer studies. There was only low ingestion during A, B, and C, the periods of high saliva production.

10.4.2 Aphid Behavior on Unacceptable Hosts

This electronic monitoring technique has been used by a number of authors to identify qualitative and quantitative differences in aphid behavior at proboscal contact, during penetration of the stylet bundle, and when the stylets reach the sieve elements, and to correlate these with different levels of host acceptability.

The highest level of resistance identified is rejection at the time of proboscal contact. Adams and co-workers found that on selections of some potato species *M. persicae* did little more than salivate and, even during repeated testing, never settled into ingestion (Adams and Wade, 1976) (Fig. 3B). Figure 3 shows the performance of *M. persicae* on resistant potato hybrid FT129-10 in which the 125-sec salivation illustrated was the only attempt to feed in 4 min. It is unfor-

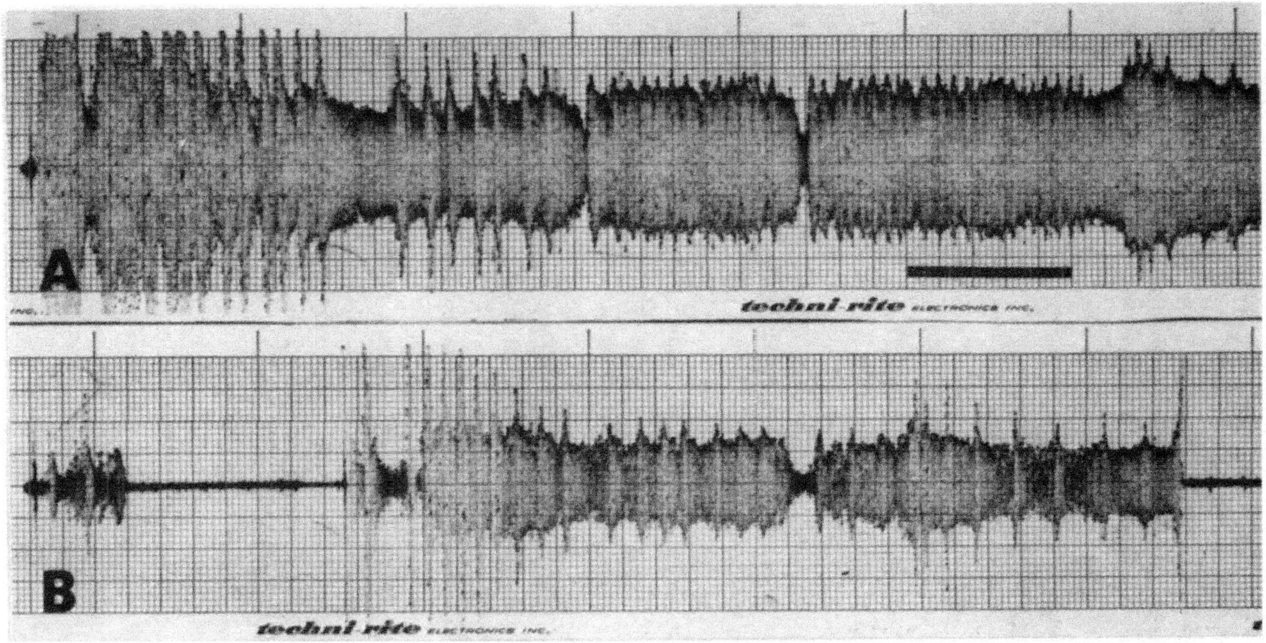

FIG. 3. Readouts of an apterous *M. persicae* on susceptible and resistant potatoes. Chart speed 6 cm/min (bar = 25 sec) reading from left to right. Barometric pressure 100.2 kpA. (A) *M. persicae* on susceptible *S. tuberosum* cv. "Katahdin." Approximately 50 sec of intense salivation preceeds penetration; the ingestion pattern continued for the duration of the 4 min test. (B) *M. persicae* on resistant hybrid FT129-10 (*S. canasense* Hawkes × *S. multidissectum* Hawkes). The initial contact and slight salivation is aborted. Following a 33 sec interval, there is a second contact, abundant salivation (~125 sec), and another abortion. There were no further feeding attempts during the 4 min test.

tunate that Nault and Styer (1972), in using this technique, and many authors who have not used it, have not reported proboscal contact (test probe) data.

Lower levels of resistance have been associated with several characteristics, including short probe times. During 24-hr access periods, *A. pisum*, while feeding on the nonhosts *Vigna sinensis* Endl. and *Lactuca sativa* L., had short total probe times, a large number of probes, little or no sieve-element ingestion, and little other ingestion when compared with their behavior on healthy and diseased *Vicia faba* L. hosts (McLean and Kinsey, 1968a). During 12-hr access periods, *M. persicae* had a higher frequency of access probes, total ingestion probes, and all other probes, but less sieve-element ingestion, when on resistant sugarbeets compared with susceptible beets (Haniotakis and Lange, 1974). No such differences were found in the behavior of *Aphis gossypii* Glover feeding on susceptible and resistant muskmelons (Kennedy *et al.*, 1978).

Aphids that feed more readily on resistant hosts may still show behavioral differences from similar aphids feeding on susceptible hosts. Nielson and Don (1974) reported frequent withdrawals of *Therioaphis maculata* (Buckton) [= *T. trifolii* forma *maculata* (Buckton)] feeding on resistant alfalfa clones. A similar observation was made by us for *M. persicae* feeding on some more resistant potato clones. Nielson and Don (1974) described some salivation and X patterns as differing from the normal sequences leading to ingestion by *T. maculata* feeding on alfalfa. When the aphid was feeding on resistant alfalfa clones, such "abnormal" sequences were of long duration while normal patterns were short. Nault and Styer (1972), using detached leaves and 4-hr access periods, reported that salivation time per probe was shorter when *Hyadaphis erysimi* Kaltenbach (= *Lipaphis erysimi* Kaltenbach) was feeding on nonhost *V. faba* than on acceptable hosts. When this aphid fed on nonhosts, X waveform time and sieve-element ingestion were both zero, whereas other ingestion exceeded that on susceptible hosts.

Some authors analyzed tapes quantitatively and used X waveform patterns to distinguish between ingestion from sieve elements and other ingestion from such cells as mesophyll and xylem. When *A. pisum* was fed on nonhosts, both sieve element ingestion and total ingestion were shorter than when feeding was on hosts (McLean and Kinsey, 1968a). Sieve-element ingestion was lacking when *H. erysimi* was feeding on nonhost *V. faba*, although ingestion time from other tissues was long (Nault and Styer, 1972). Relatively short total ingestion and sieve-element ingestion times and a longer time ingesting from other tissues were characteristic of *M. persicae* feeding on resistant sugarbeet (Haniotakis and Lange, 1974). Little or no total ingestion and sieve-element ingestion were characteristic of *T. maculata* feeding on resistant alfalfa (Nielson and Don, 1974). These same authors noted broken sequences 1-2 min after stylet penetration of phloem, suggesting a plant defense reaction. Kennedy *et al.* (1978) studied the behavior of *A. gossypii* on resistant and susceptible muskmelon (*Cucumis melo* L.). In histological studies they found more stylet sheaths terminated in the mesophyll than in the phloem in resistant plants. Also, in resistant plants, there were more stylet sheaths with branches, and more branches per sheath. In parallel electronic monitoring studies, contact with sieve-element cells occurred more often in resistant than in susceptible plants. If these contacts were correctly identified, there appears to be a conflict between the two sets of results. However, electronic monitoring showed the duration of sieve-element ingestion to be three times longer on susceptible than on resistant plants.

An interesting set of results that parallel resistant behavior was obtained by McLean and Kinsey (1968b) for *A. pisum* feeding on acceptable *V. faba* under low and high temperatures. Relatively long nonprobing times, greater-than-normal salivation, and reduced ingestion by the aphids at extreme temperatures were attributed to the environmental stresses.

Generalizations about aphid behavior on unacceptable hosts must be made with caution because of the diversity of aphid species and host plants, the variations in expression of plant resistance, and our often incomplete knowledge of what is "normal" behavior. However, to date, electronic monitoring results indicate that plant resistance may be expressed by some or all of the following characteristics: short probes, long periods with no probing, an increased frequency of unsuccessful probes, an increased amount of salivation, reduced sieve-element ingestion, increased ingestion from outside the phloem, reduced total ingestion, and increased difficulty in reaching acceptable tissue as shown by an increase in the amount of branching in stylet sheaths.

10.5 MECHANISMS OF RESISTANCE

In the chain of events that leads to successful aphid feeding, plant resistance can provide an effective barrier at the time of proboscal contact with the epidermis, during penetration of the epidermis and mesophyll, and when the stylet bundle reaches the phloem sieve elements. This was shown in our earlier consideration of the host-parasite relationship (Section 10.2). In this section we add behavioral observations, including the results of electronic monitoring, and propose a new site for the expression of plant resistance to aphids.

The highest resistance for which there is direct evidence is the immediate rejection by *M. persicae* of selections of *Solanum* species hybrids (Adams and Wade, 1976). This resistance is probably detected by the labial chemoreceptors responding to the end-products of the saliva-plant epidermis interaction. Nault and Styer (1972) report similar very brief probes and speculate that *H. erysimi* can determine the level of acceptability of its host without ingestion. Relying on direct observation of aphid behavior, von Klingauf (1972) showed that the chemical nature of the plant epidermis was important in host acceptance by *A. pisum*. Similarly, Tjallingii (1976) showed that in *B. brassicae* proboscal contact occurred on hosts and nonhosts, but on nonhosts contacts were much shorter though more frequent. Müller (1958) studied the settling behavior of *Aphis fabae* Scopoli on different cultivars of *V. faba* and found that during the search time aphids probed for only 20-40 sec and very rarely penetrated beyond the epidermis. A similar resistance to settling of aphids on certain sugarbeet selections was reported by Russell (1966).

Work with chemicals and plant extracts also indicates that the aphid can evaluate the acceptability of its host very quickly. Crucifer feeding aphids probe and feed on nonhosts treated with the mustard oil glucoside sinigrin, which is characteristic of the Cruciferae. The normally unacceptable host *V. faba* becomes acceptable to both *B. brassicae* (Wensler, 1962) and *H. erysimi* (Nault and Styer, 1972) following such treatment. Another glucoside characteristic of apples, phlorizin, was shown by von Klingauf (1971) to be important in the first phase of host testing. Montgomery and Arn (1974) concluded that the apple-infesting *Aphis pomi* de Geer tolerated phlorizin, while non-apple-feeding aphids were deterred by it. The concentration of an alkaloid, sparteine, in *Sarothamnus scoparius* L. was found to be positively correlated with the distribution of *Acrythosiphon spartii* (Koch) on the plant; *A. spartii* stayed longer on nonhost peas and beans when the leaves were treated with sparteine (Smith, 1966). In another example, the strawberry aphid, Chaetosiphon fragaefolii (Cockerell), was shown by Shanks and Finnigan (1970) to respond to extracts of *Fragaria vesca* L. cv. Alpine. This aphid probed as much on pea leaves vacuum infiltrated with strawberry leaf extract as it did on strawberry leaves. However, for longer survival, the treated pea leaves were no more attractive than untreated ones. The authors concluded that final nonacceptability was determined by arrestants encountered during deeper probing. Branson (1972), who used leaf disks and leaf extracts in sachets, found Florida gamagrass (*Tripsacum floridanum* Porter ex Vasey) to be nearly immune to the corn leaf aphid, *Rhopalosiphum maidis* (Fitch). Branson described the resistance as antibiosis or extreme nonpreference, and suggested that a water soluable substance in *T. floridanum* acted as a toxin or interfered with the feeding behavior of the aphids. In leaves of *Citrus sinensis* (L.) Osbeck cv. Pineapple Sweet Orange, a factor that is apparently acquired upon maturity and present in the cuticle or outer cell wall of the epidermis, or both, was found responsible for repelling four species of aphids (Zettler *et al.*, 1969). Over half the probes made by these aphids on mature citrus leaves lasted less than 15 sec. In a final example of this kind, Juneja *et al.* (1972) found that benzyl alcohol extracted from aphid-resistant barleys was a possible resistance factor. They found that susceptible barleys grown in a medium containing this extract were as resistant to the green bug *Schizaphis graminum* (Rondani) as were resistant barley cultivars.

The examples cited so far are of early termination of penetration. Other studies indicate that resistance involves interference with aphids' stylets reaching the phloem and with continuing ingestion. McMurtry and Stanford (1960) found that *T. maculata* was less successful in reaching phloem when feeding on the resistant alfalfa clone C-84 than when feeding on the susceptible clone Caliverde. When aphids on C-84 did locate the phloem, no honeydew was produced, indicating very limited ingestion. All aphids placed on leaf disks of C-84 died within 24-42 hr. Working with the same aphid species on other alfalfa cultivars, Nielson and Don (1974) found that normal salivation and sieve-element ingestion were reduced on nonhosts; the aphids ingested from other tissues. The aphids made repeated attempts to reach the phloem of resistant cultivars, but only a small percentage succeeded, and success was always followed by sudden withdrawal from the phloem after 1 to 2 min of ingestion. One aphid biotype did not ingest from any tissue of the resistant alfalfa clone C-908. Several biotypes of *S. graminum* have been identified on the basis of preferential feeding in barley cultivars (Saxena and Chada, 1971). Penetration by biotype A was intercellular; phloem cells generally collapsed when penetrated, and some mesophyll cells around the stylet path were also affected. Penetration by biotype B was mostly intracellular; the penetration paths were mostly straight, and in only 14 of 182 observations did the stylets reach the vascular bundles. This biotype fed mostly from the mesophyll tissue, and mesophyll cells collapsed following stylet penetration.

A hypersensitive plant reaction to feeding by the rosy apple aphid, *Dysaphis plantaginea* (Pass.), was reported by Alston and Briggs (1970), who found that in six species of applie the aphids left their host plants as necrosis developed. A particularly interesting plant-aphid interaction is that of the *Tropaeolum majus* L., infesting aphid of the *A. fabae* complex, which persists in settling on *V. faba* on which neither it nor its young can survive. Although this aphid settles and probes on *V. faba*, it is apparently unable to ingest, as evidenced by the lack of honeydew production (Taylor, 1959). Finally, Kennedy *et al.* (1978) used histological and electronic recording techniques to study muskmelon inbred 91213 resistance to *A. gossypii*. They concluded that the resistance factor(s) interfered with ingestion from the sieve cells and suggested that resistance was associated either with components of the phloem such as the companion cells or the cell walls, or with a response(s) of the phloem cells themselves to the aphid's stylets or saliva.

Some authors [e.g., Gibson (1972) and Lowe and Russell (1974)] have noted situations in which young aphids have been unable to reach the phloem because the distance between the epidermis and the phloem was greater than the length of their stylets. This may effectively reduce the rate of aphid population increase, but we do not consider it to be resistance in the same sense as the examples we have just considered.

Histological studies and investigations using electronic monitoring of aphid feeding have provided many examples of high levels of plant resistance to aphids, but little progress has been made in explaining the nature of this resistance at the cellular or subcellular levels. Nault and Gyrisco (1966) noted that probing aphids undoubtedly tear plasmodesmata connecting adjacent cells. The stylet contact with the plasmodesmata seems to us to present an opportunity for the expression of plant resistance to the aphid. Lowe (1967) recognized that small aphids probing extensively in the mesophyll may suck up fluid from the free surface of the mesophyll cells, but he made no comment on the nature of the substances that may be in this fluid. Kennedy *et al.* (1978) considered the resistance factor in muskmelon to be associated with companion cells or phloem cell walls, or to be produced by the phloem cells in response to aphid probing. Plant resistance to fungal diseases has been associated with the rapid production of phytoalexins, and Nielson and Don (1974) adopted this hypothesis to suggest that phytoalexings may be involved in the resistance of an alfalfa clone to *T. maculata*. However, the role of phytoalexins may be different from that first put forward. A recent review of compounds associated with disease resistance (Kuć and Shain, 1977) suggests that phytoalexins may be a part of a general tissue repair mechanism, which may also have a role in resistance. Phytoalexins may occur in tract quantities in apparently healthy plants and may accumulate in response to a wide variety of stimulants. Interestingly, Kuć and Shain note that the rapidity of many incompatible disease interactions suggests a surface phenomenon based on components on the surfaces of cell walls or membranes.

We suggest that the site of action of many of the examples of plant resistance to aphids cited above is the apoplast. The apoplast forms a continuous system throughout the plant that is made up of the interconnected cell wall system, associated structures such as the cuticle, and the conducting elements of the xylem. The apoplast is considered to be the major route for water flow and for translocation of water-soluable substances such as herbicides, fungicides, and pesticides throughout the plant (Edgington and Peterson, 1977). Enzyme-containing lysosomes within plant cell walls are also a part of the apoplast and may be involved in plant-aphid interactions. The symplast, the continuum of interconnected plant cell protoplasts, may also have a role in aphid resistance. The plasmodesmata mentioned by Nault and Gyrisco (1966), and above, are part of the symplast. However, during penetration of plant tissues the aphid stylet is in more frequent contact with the apoplast and it seems more reasonable to us to expect the primary plant-aphid interactions to be expressed through this system.

10.6 BREEDING FOR PLANT RESISTANCE TO APHIDS

The main requirements for developing aphid resistant crop cultivars have been identified in a number of recent reviews (Dahms, 1972; Gallun *et al.* 1975; Kogan, 1975; Gibson and Plumb, 1977): (1) The first requirement is to have reliable evaluation techniques based on sound knowledge of host-parasite relationships. Laboratory evaluation should correlate well with field performance. (2) Within the crop germplasm pool, it is necessary to have variation in plant response to the parasite, so that there is potential for improving resistance through selection. (3) It is desirable to know that the plant resistance is durable and is not easily overcome by new parasite biotypes. This can be judged with knowledge of the resistance mechanism and its mode of inheritance. (4) Finally, the resistance must be incorporated into new cultivars. This is a time-consuming and often difficult step, depending upon the genetics of the resistance, the breeding system of the crop in question, and the usually large array of other traits that are of importance in the breeder's considerations. To these requirements may be added the necessity of teamwork that includes entomologists, plant breeders, and other specialists whose input is needed in a successful cultivar development program (see International Maize and Wheat Improvement Center, 1974).

The electronic monitoring of aphid probing and feeding behavior facilitates selection for very high levels of host resistance and greatly enhances our knowledge of the type of resistance being selected. In extreme cases of nonpreference, the aphid may reject the plant during proboscal contact or superficial penetration, without any substantial ingestion taking place. In cases where aphids ingest, the duration of penetration will be related to the energy expended in searching for food. The greater the energy cost of feeding, the less well the aphid will survive on that host. De Reggi (1971) similarly concluded that increases in energy costs of feeding reduced survival rates. Aphids arriving on a nonpreferred crop may move on and cause problems elsewhere (Gallun, 1972), but during further dispersal the aphid population will be subject to more losses. Given that aphid pests of crop species are mainly host-alternating or polyphagous, nonhost-alternating species (Dixon, 1977), the increasing use of aphid resistant cultivars may be expected to reduce the aphid population.

Since most aphid species are parthenogenic, a single individual that can feed and reproduce more successfully than others on a resistant plant may initiate a new biotype. The need for durable resistance that reduces the possibility of biotype evolution is emphasized by the seven crop-infesting aphid species in which biotypes are known (Gibson and Plumb, 1977). Such resistance is provided by multiple gene control (Pimentel and Bellotti, 1976) and may account for the big advantage of nonpreference, where selection for aphid biotypes is less likely (Gallun, 1972). The several distinct steps in successful proboscal contact and penetration offer the possibility of multiple gene control and provide the durability expected for most types of nonpreference.

The availability of very high levels of crop resistance to aphids increases the potential for control of certain types of cirus diseases (Kennedy, 1976). Any factor consistently reducing the vector population may be expected to have a similar or greater impact on virus spread (Swenson, 1968).

10.6.1 Electronic Monitoring as a Screening Technique

Electronic monitoring is significantly different from other screening techniques presently available. The advantages have already been made clear, but there are also some limitations.

First, it is necessary to establish, for the germplasm pool of each crop and its aphid parasite, a correlation between field performance and resistance parameters that can be identified on the readout. This has been accomplished to varying degrees for potatoes (Tarn and Adams, 1973; Adams and Wade, 1976; Holbrook, 1980), sugarbeet (Haniotakis and Lange, 1974), alfalfa (Nielson and Don, 1974), and muskmelon (Kennedy, *et al.*, 1978). Considerable initial searching may be necessary to identify resistance parameters on electronic readouts;

Shanks and Chase (1976) could not distinguish susceptible and resistant strawberry cultivars by this method and attributed resistance to nutritional factors that were effective following ingestion. Comparisons between species and genera should be made with care because of the possible existence of different resistance mechanisms.

Second, electronic monitoring is not adapted to handle thousands of selections at the early stages of a breeding program. A prescreening technique, such as infesting seedling flats with aphids, is a necessary preliminary evaluation. When plant resistance can be identified on the basis of proboscal contact, or penetration with little or no ingestion, screening time need only be related to the time taken to reach sieve elements on an acceptable host, which is normally only a few minutes (Pollard, 1973). Short recording times, plus an appropriate recording speed, reduce the amount of readout to be stored and analyzed. How data are to be used will determine whether or not computerized handling is desirable. Susceptible breeding selections can be discarded immediately, whereas those showing some resistance should be retested. It is also necessary to watch for any differences that may result from different aphid biotypes, or alate versus apterous forms.

Third, when large numbers of seedlings are being evaluated, a simple means of assessment is all that is necessary. Haniotakis and Lange (1974) used sieve-element ingestion as a percentage of recording time to give a "relative resistance index." They used this index to test the uniformity of individual plants within two sugarbeet lines, and showed differences between lines. Five categories of host plant acceptability to *M. persicae* were used by Adams and Wade (1976) to evaluate potato clones on the basis of (1) repeated and prompt rejection of the host, (2) frequent interruptions of penetration, (3) considerable salivation with some ingestion, (4) less salivation and considerable ingestion, and (5) brief salivation and long periods of uninterrupted ingestion. Holbrook (1980) compared resistances of potato cultivars to aphids using the mean number of probes and total time of probing of aphids on test and control plants. He found a 15 min test satisfactory for ranking cultivars.

Finally, because of the sensitivity of this monitoring method, the nonsystem effects must be considered carefully (see Section 10.3.5). It is particularly important that the aphids be maintained on a susceptible cultivar of the crop in which resistance is being sought. When the duration of each test is short, one aphid may be used to evaluate more than one selection. In our work with potatoes, aphid behavior was first checked on the susceptible cultivar Katahdin before being used to evaluate three subsequent selections.

10.6.2 Achievements and Opportunities

Russell (1978) considers a variety of examples of breeding for aphid resistance, and her book reverses a trend, noted by Dahms (1972), for little attention to be given to insect resistance in plant-breeding textbooks. Reviews by Maxwell *et al.* (1972) and Knight and Alston (1974) also describe many achievements in aphid resistance breeding. For example, the need to reduce both yield losses and pesticide use on alfalfa led to the introduction of cultivars with high levels of resistance to one or more aphids. Among perennial fruits, immunity and resistance to aphids have ben used to reduce the spread of aphid-borne viruses. These accomplishments have been made using long-established methods of evaluating aphid resistance.

Use of the electronic-monitoring technique has brought significant advances in the knowledge of aphid feeding behavior and has been shown to be a practical means of identifying plants with high levels of aphid resistance. It has also been used to study the effects of systemic insecticides on feeding aphids (Holbrook, 1977; Weidemann, 1979), the effects of mineral oil on aphid probing and virus transmission (Simons *et al.*, 1977), and the acquisition and transmission of viruses (Hodges and McLean, 1969; Holbrook, 1978; Leonard and Holbrook, 1978).

The achievements of the past 15 years point to the tremendous opportunities ahead. The technique still has much to add to our knowledge of aphid behavior: in Section 10.5, we cited many examples in which electronic monitoring has not yet been applied, e.g., aphid immunity in raspberries. In plant-breeding programs the technique should be valuable in those crops where aphid resistance is already a breeding objective. In crops where resistance is desirable but is not currently being incorporated, the technique facilitates the identification of high levels of resistance and provides a valuable screening procedure.

Electronic monitoring of aphid probing and feeding has become available at a time when the concept of pest management and the role of plant resistance as a pest management tool has been firmly reestablished in the context of present-day agriculture. Studies of aphid behavior and virus transmission by aphids continue to receive attention. Plant resistance to aphids must become an objective of more breeding programs. We have no doubt that this technique represents an idea whose time has come and that it will have a key role as breeders and entomologists manipulate the evolution of resistance to insects on the side of the plant by selective breeding as has been done with other desirable plant characteristics.

10.7 SUMMARY

When an aphid probes into plant tissue, the presence of aphid saliva and plant sap, both electrically conductive, makes it possible to include the feeding aphid in an electrical circuit. When the circuit is designed to avoid adverse effects on the aphid and to minimize electrical interference, the recorded variations in voltage and current are found to be related to the aphid's probing and feeding behavior. This technique has many applications including the evaluation of host resistance to aphids. High levels of plant resistance are characterized by readouts indicating much salivation, little ingestion, and many interruptions. Low levels of plant resistance are characterized by salivation and rapid penetration, and long periods of steady ingestion. This technique is of greatest value in a breeding program in selecting from among relatively few seedlings that have survived an earlier mass-screening procedure. The technique is also of value for investigating the nature of plant resistance.

10.8 ACKNOWLEDGMENTS

We are pleased to express our appreciation to the many who helped by their original studies and by contributing to the preparation of this chapter. For engineering assistance, we thank D. J. Buckley, Engineering Research Institute, Agriculture Canada, Ottawa, and Professors Robert Scott and Vaughn Dunfield, Bio-Engineering Institute, University of New Brunswick, Fredericton. For guidance and use of the stereoscanning microscope equipment in their Departments, we thank Dr. F. W. Fyfe, Faculty of Medicine, Dalhousie University, Halifax, N. S., and Dr. William Seabrook, Biology Department, University of New Brunswick, Fredericton. For preparation of the plates, we thank Mr. Roger Smith of the Biology Department, University of New Brunswick, Fredericton. At the Fredericton Research Station we acknowledge the help of Dr. M. E. MacGillivray, who checked the accuracy of aphid names and reviewed the manuscript; Mr. Donald Gammon and Ms. Norma Taniguchi, who did the library searches; Dr. M. C. Clark, who contributed through discussions of resistance mechanisms; Dr. H. De Jong, who translated several German language papers, and Mrs. Donna Pollock, who gave unstinted clerical help. Two other associates deserve special mention: Dr. D. L. McLean, University of California, Davis, the "father of the system," for his always generous encouragement and Mr. C. V. Wade, until recently associated with the junior author at this station, who carried out the demanding work of preparing the readouts.

10.9 REFERENCES

Adams, J. B., and Drew, M. E. (1963). A cellulose-hydrolyzing factor in aphids. *Can. J. Zool.* **41**:1205-1212.

Adams, J. B., and Wade, C. V. (1976). Aphid behavior and host-plant preference demonstrated by electronic patterns of probing and feeding. *Am. Potato J.* **53**:261-267.

Alston, F. H., and Briggs, J. B. (1970). Inheritance of hypersensitivity to rosy apple aphid *Dysaphis plantaginea* in apple. *Can. J. Gen. Cytol.* **12**:257-258.

Branson, T. F. (1972). Resistance to the corn leaf aphid in the grass tribe Maydeae. *J. Econ. Entomol.* **65**:195-196.

Brown, C. M., and Holbrook, F. R. (1976). An improved electronic system for monitoring feeding of aphids. *Am. Potato J.* **53**:457-462.

Dahms, R. G. (1972). Techniques in the evaluation and development of host-plant resistance. *J. Environ. Quality* **1**:254-259.

De Reggi, L. M. (1971). Interactions entre le comportement alimentaire et le comportement de reproduction du puceron *Myzus persicae*. *Ann. Zool. Ecol. Animale* **3**:43-48.

Dixon, A. F. G. (1977). Aphid ecology: life cycles, polymorphism, and population regulation. *Annu. Rev. Ecol. Syst.* **8**:329-353.

Eastop, V. F. (1977). Worldwide importance of aphids as virus vectors. *In* "Aphids as Virus Vectors" (K. F. Harris and K. Maramorosch, eds.), pp. 3-62. Academic Press, New York.

Edgington, L. V., and Peterson, C. A. (1977). Systemic fungicides: theory, uptake, and transportation. *In* "Antifungal Compounds," Vol. 2, Interactions in Biological and Ecological Systems (M. R. Siegel and H. D. Sisler, eds.), pp. 51-89. Marcel Dekker, New York.

Farkas, G. L., Dézsi, L., Horváth, M., Kisbán, K., and Udvardy, J. (1964). Common pattern of enzymatic changes in detached leaves and tissues attacked by parasites. *Phytopathol. Z.* **49**:343-354.

Forbes, A. R. (1977). The mouthparts and feeding mechanism of aphids. *In* "Aphids as Virus Vectors" (K. F. Harris and K. Maramorosch, eds.), pp. 83-103. Academic Press, New York.

Forbes, A. R., and Mullick, D. B. (1970). The stylets of the balsam woolly aphid, *Adelges piceae* (Homoptera: Adelgidae). *Can. Entomol.* **102**:1074-1082.

Gallun, R. L. (1972). Genetic interrelationships between host plants and insects. *J. Environ. Quality* **1**:259-265.

Gallun, R. L., Starks, K. J., and Guthrie, W. D. (1975). Plant resistance to insects attacking cereals. *Annu. Rev. Entomol.* **20**:337-357.

Garrett, R. G. (1973). Non-persistent aphid-borne viruses. *In* "Viruses and Invertebrates" (A. J. Gibbs, ed.), pp. 476-492. North-Holland, Amsterdam and London.

Gibson, R. W. (1972). The distribution of aphids on potato leaves in relation to vein size. *Entomol. Exp. Appl.* **15**:213-223.

Gibson, R. W., and Plumb, R. T. (1977). Breeding plants for resistance to aphid infestation. *In* "Aphids as Virus Vectors" (K. F. Harris and K. Maramorosch, eds.), pp. 473-500. Academic Press, New York.

Haniotakis, G. E., and Lange, W. H. (1974). Beet yellow virus resistance in sugar beets: mechanism of resistance. *J. Econ. Entomol.* **67**:25-28.

Harris, K. F. (1977). An ingestion-egestion hypothesis of noncirculative virus transmission. *In* "Aphids as Virus Vectors" (K. F. Harris and K. Maramorosch, eds.), pp. 185-220. Academic Press, New York.

Harris, K. F., and Bath, J. E. (1973). Regurgitation by *Myzus persicae* during membrane feeding: its likely function in transmission of nonpersistent plant viruses. *Ann Entomol. Soc. Am.* **66**:793-796.

Harris, K. F., Treur, B., Tsai, J. and Toler, R. (1981). Observations on leafhopper ingestion-egestion behavior: its likely role in the transmission of semipersistent viruses and other plant pathogens. In press.

Helson, G. A. H., and Penman, J. E. R. (1970). Biological transformations associated with weather changes. An hypothesis on the flight activity of *Wiseana* spp. *Intern. J. Biometeorol.* **14**:227-246.

Hodges, L. R., and McLean, D. L. (1969). Correlation of transmission of bean yellow mosaic virus with salivation activity of *Acyrthosiphon pisum* (Homoptera: Aphididae). *Ann. Entomol. Soc. Am.* **62**:1398-1401.

Holbrook, F. R. (1977). Aldicarb and thiofanox: effect on the feeding activity of green peach aphids. *J. Econ. Entomol.* **70**:742-744.

Holbrook, F. R. (1978). Transmission of potato leaf roll virus by the green peach aphid. *Ann. Entomol. Soc. Am.* **71**:830-831.

Holbrook, F. R. (1980). An index of acceptability to green peach aphids for *Solanum* germplasm and for a suspected non-host plant. *Am. Potato J.* **57**:1-6.

International Maize and Wheat Improvement Center (1974). CIMMYT Report on Maize Improvement 1973. Mexico, D. F.

Juneja, P. S., Gholson, R. K., Burton, R. L., and Starks, K. J. (1972). The chemical basis for greenbug resistance in small grains. 1. Benzyl alcohol as a possible resistance factor. *Ann. Entomol. Soc. Am.* **65**:961-964.

Kashin, P. (1966). Electronic recording of the mosquito bite. *J. Insect Physiol.* **12**:281-286.

Kashin, P., and Arneson, B. F. (1969). An automated repellency assay system. II. A new electronic "bitometer-timer." *J. Econ. Entomol.* **62**:200-205.

Kashin, P., and Kardatzke, M. L. (1969). An automated repellency assay system. I. Statistical methods. *J. Econ. Entomol.* **62**:10-13.

Kashin, P., and Wakeley, H. G. (1965). An insect "bitometer." *Nature* **208**:462-464.

Kennedy, G. G. (1976). Host plant resistance and the spread of plant viruses. *Environ. Entomol.* **5**:827-832.

Kennedy, G. G., McLean, D. L., and Kinsey, M. G. (1978). Probing behaviour of *Aphis gossyppi* on resistant and susceptible muskmelon. *J. Econ. Entomol.* **71**:13-16.

Kennedy, J. S. (1965). Mechanisms of host plant selection. *Ann. Appl. Biol.* **56**:317-322.

Kennedy, J. S., and Fosbrooke, I. H. M. (1973). The plant in the life of an aphid. *In* "Insect/Plant Relationships" (H. F. van Emden, ed.). *Roy. Entomol. Soc. London Symp.* **6**: 129-140. Blackwell Scientific Publications, Oxford.

Kinsey, M. G., and McLean, D. L. (1967). Additional evidence that aphids ingest through an open stylet sheath. *Ann. Entomol. Soc. Am.* **60**:1263-1265.

Knight, R. L., and Alston, F. H. (1974). Pest resistance in fruit breeding. *In* "Biology in Pest and Disease Control" (D. P. Jones and M. E. Solomon, eds.). *13th Symp. Br. Ecol. Soc., 1972,* pp. 73-86. Blackwell Scientific Publications, Oxford.

Kogan, M. (1975). Pest resistance in pest management. *In* "Introduction to Insect Pest Management" (R. L. Metcalf and W. H. Luckman, eds.), pp. 103-146. Wiley, New York.

Kuć, J., and Shain, L. (1977). Antifungal compounds associated with disease resistance in plants. *In* "Antifungal Compounds," Vol. 2, Interactions in Biological and Ecological Systems (M. R. Siegel and H. D. Sisler, eds.), pp. 497-535. Marcel Dekker, New York.

Lehmann, W., Claus, S., and Karl, E. (1975). Das Verhalten von Aphiden unter dem Einfluss verschiedener Pflanzenarten und ihrer chemischen Behandlung, bezogen auf den Anteil einstechender Tiere und die lcurzfristigen Probesaugstiche. *Arch. Phytopathol. Pflanzenschutz* **11**:273-282.

Leonard, S. H., and Holbrook, F. R. (1978). Minimum acquisition and transmission times for potato leaf roll virus by the green peach aphid. *Ann. Entomol. Soc. Am.* **71**:493-495.

Lowe, H. J. B. (1967). Interspecific differences in the biology of aphids (Homoptera: Aphididae) on leaves of *Vicia faba*. I. Feeding behaviour. *Entomol. Exp. Appl.* **10**:347-357.

Lowe, H. J. B., and Russell, G. (1974). Probing by aphids in the leaf tissues of resistant and susceptible sugar beet. *Entomol. Expt. Appl.* **17**:468-476.

Luckman, W. H., and Metcalf, R. L. (1975). The pest-management concept. *In* "Introduction to Insect Pest Management" (R. L. Metcalf and W. H. Luckman, eds.), pp. 3-35. Wiley, New York.

McAllan, J. W., and Adams, J. B. (1961). The significance of pectinase in plant penetration by aphids. *Can. J. Zool.* **39**:305-310.

McLean, D. L. (1971). Probing behaviour of the pea aphid, *Acyrthosiphon pisum*. V. Comparison of *Vicia faba, Pisum sativum,* and a chemically defined diet as food sources. *Ann. Entomol. Soc. Am.* **64**:499-503.

McLean, D. L. (1977). An electrical measurement system for studying aphid probing bebehavior. *In* "Aphids as Virus Vectors" (K. F. Harris and K. Maramorosch, eds.), pp. 277-290. Academic Press, New York.

McLean, D. L., and Kinsey, M. G. (1964). A technique for electronically recording aphid feeding and salivation. *Nature* **202**:1358-1359.

McLean, D. L., and Kinsey, M. G. (1965). Identification of electronically recorded curve patterns associated with aphid salivation and ingestion. *Nature* **205**:1130-1131.

McLean, D. L., and Kinsey, M. G. (1967). Probing behavior of the pea aphid, *Acyrthosiphon pisum*. I. Definitive correlation of electronically recorded waveforms with aphid probing activities. *Ann. Entomol. Soc. Am.* **60**:400-406.

McLean, D. L., and Kinsey, M. G. (1968b). Probing behavior of the pea aphid, *Acyrthosiphon pisum*. II. Comparisons of salivation and ingestion in host and non-host plant leaves. *Ann. Entomol. Soc. Am.* **61**:730-739.

McLean, D. L., and Kinsey, M. G. (1968b). Probing behaviour of the pea aphid, *Acyrthosiphon pisum.* III. Effect of temperature differences on certain probing activities. *Ann. Entomol. Soc. Am.* **61**:927-933.

McLean, D. L., and Kinsey, M. G. (1969). Probing behavior of the pea aphid, *Acyrthosiphon pisum.* IV. Effects of starvation on certain probing activities. *Ann. Entomol. Soc. Am.* **62**:987-994.

McLean, D. L., and Weigt, W. A., Jr. (1969). An electronic measuring system to record aphid salivation and ingestion. *Ann. Entomol. Soc. Am.* **61**:180-185.

McMurtry, J. A., and Stanford, E. H. (1960). Observations of feeding habits of the spotted alfalfa aphid on resistant and susceptible alfalfa plants. *J. Econ. Entomol.* **53**:714-717.

Maxwell, F. G., Jenkins, J. N., and Parrott, W. L. (1972). Resistance of plants to insects. *Adv. Agron.* **24**:187-265.

Miles, P. W., McLean, D. L., and Kinsey, M. G. (1964). Evidence that two species of aphid ingest food through an open stylet sheath. *Experientia* **20**:582.

Mittler, T. E. (1957). Studies on the feeding and nutrition of *Tuberolachnus salignus* (Gmelin) (Homoptera, Aphididae). I. The uptake of phloem sap. *J. Exp. Biol.* **34**:334-341.

Montgomery, M. E., and Arn, H. (1974). Feeding response of *Aphis pomi, Myzus persicae,* and *Amphorophora agathonica* to phlorizin. *J. Insect Physiol.* **20**:413-421.

Müller, H. J. (1958). The behavior of *Aphis fabae* in selecting its host plants, especially different varieties of *Vicia faba. Entomol. Exp. Appl.* **1**:66-72.

Nault, L. R., and Gyrisco, G. G. (1966). Relation of the feeding process of the pea aphid to the inoculation of pea enation mosaic virus. *Ann. Entomol. Soc. Am.* **59**:1185-1197.

Nault, L. R., and Styer, W. E. (1972). Effects of sinigrin on host selection by aphids. *Entomol. Exp. Appl.* **15**:423-437.

Nielson, M. W., and Don, H. (1974). Probing behavior of biotypes of the spotted alfalfa aphid on resistant and susceptible alfalfa clones. *Entomol. Exp. Appl.* **17**:477-486.

Painter, R. H. (1951). "Insect Resistance in Crop Plants." University Press of Kansas, Lawrence.

Pimentel, D., and Bellotti, A. C. (1976). Parasite-host population systems and genetic stability. *Am. Nat.* **110**:877-888.

Pollard, D. G. (1973). Plant penetration by feeding aphids (Hemiptera, Aphidoidea): a review. *Bull. Entomol. Res.* **62**: 631-714.

Pollard, D. G. (1977). Aphid penetration of plant tissues. *In* "Aphids as Virus Vectors" (K. F. Harris and K. Maramorosch, eds.), pp. 105-118. Academic Press, New York.

Roche, M., Martinez-Torres, C., and Macpherson, L. (1962). Electroesophagogram of individual hookworm (*Ancylostoma caninum*). *Science* **136**:148-150.

Russell, G. E. (1966). Observations on the settling behavior of aphids on sugar-beet plants in the glasshouse. *J. Agr. Sci.* **67**:405-410.

Russell, G. E. (1978). "Plant Breeding for Pest and Disease Resistance." Butterworths, London.

Saxena, P. N., and Chada, H. L. (1971). The greenbug, *Schizaphis graminum.* 1. Mouth parts and feeding habits. *Ann. Entomol. Soc. Am.* **64**:897-904.

Schaefers, G. A. (1966). The use of direct current for electronically recording aphid feeding and salivation. *Ann. Entomol. Soc. Am.* **59**:1022-1024.

Shanks, C. H., Jr., and Chase, D. (1976). Electrical measurement of feeding by the strawberry aphid on susceptible and resistant strawberries and nonhost plants. *Ann. Entomol. Soc. Am.* **69**:784-786.

Shanks, C. H., Jr., and Finnigan, B. (1970). Probing behavior of the strawberry aphid. *Ann. Entomol. Soc. Am.* **63**:734-737.

Simons, J. N., McLean, D. L., and Kinsey, M. G. (1977). Effects of mineral oil on probing behavior and transmission of stylet-borne viruses by *Myzus persicae. J. Econ. Entomol.* **70**:309-315.

Smith, B. D. (1966). Effect of the plant alkaloid sparteine on the distribution of the aphid

Acyrthosiphon spartii (Koch). *Nature* **212**:213-214.

Smith, J. J. B., and Friend, W. G. (1971). The application of split-screen television recording and electrical resistance measurement to the study of feeding in a blood-sucking insect *(Rhodnius prolixus). Can. Entomol.* **103**:167-172.

Sweatman, G. K., Tomey, G., and Katul, G. (1976). A technique for the continuous recording of tick feeding electrograms and temperature by telemetry from free-ranging cattle. *Intern. J. Parasitol.* **6**:299-305.

Swenson, K. G. (1968). Role of aphids in the ecology of plant viruses. *Annu. Rev. Phytopathol.* **6**:351-374.

Tarn, T. R., and Adams, J. B. (1973). Resistance to the green peach aphid in some potato species. *Am. Potato J.* **50**:383.

Taylor, L. R. (1959). Abortive feeding behavior in a black aphid of the *Aphis fabae* group. *Entomol. Exp. Appl.* **2**:143-153.

Tjallingii, W. F. (1976). A preliminary study of host selection and acceptance behavior in the cabbage aphid, *Brevicoryne brassicae* L. *Symposia Biol. Hung.* **16**:283-285.

Tjallingii, W. F. (1978a). Electronic recording of penetration behavior by aphids. *Entomol. Exp. Appl.* **24**:521-530.

Tjallingii, W. F. (1978b). Mechanoreceptors of the aphid labium. *Entomol. Exp. Appl.* **24**: 531-537.

van Emden, H. F., and Wearing, C. H. (1965). The role of the aphid host plant in delaying economic damage levels in crops. *Ann. Appl. Biol.* **56**:323-324.

von Klingauf, F. (1971). Die Wirkung des Glucosids Pflorizin auf das Wirtswahlverhalten von *Rhopalosiphum insertum* (Walk.) und *Aphis pomi* de Geer (Homoptera: Aphididae). *Z. Ang. Entomol.* **68**:41-55.

von Klinghauf, F. (1972). Die Bedeutung von peripher vorliegenden Pflanzensubstanzen für die Wirtswahl von phloemsaugenden Blattläusen (Aphididae). *Z. Pflanzenkrankh. Pflanzenschutz* **79**:471-477.

Weber, L. J., and McLean, D. L. (1975). "Electrical Measurement Systems for Biological and Physical Scientists." Addison-Wesley, Reading, Massachusetts.

Weidemann, H. L. (1979). Das Saugverhalten von Blattläusen auf Insektizidbehandelten Kartoffelpflanzen. *Potato Res.* **22**:153-159.

Wensler, R. J. D. (1962). Mode of host selection by an aphid. *Nature* **195**:830-831.

Wensler, R. J., and Filshie, B. K. (1969). Gustatory sense organs in the food canal of aphids. *J. Morphol.* **129**:473-492.

Zettler, F. W., Smyly, M. O., and Evans, I. R. (1969). The repellency of mature citrus leaves to probing aphids. *Ann. Entomol. Soc. Am.* **62**:399-402.

Chapter 11

THE ESS OF AN APHID PATHOSYSTEM

Raoul A. Robinson

Department of Biological Sciences
Simon Frazier University
Vancouver, British Colombia, Canada

11.1 INTRODUCTION

In this chapter, an attempt is made to link several disciplines within biology with a view to obtaining a new synthesis. In particular, some of the newer principles of plant disease epidemiology are applied to entomology, and some concepts of animal ethology are applied to plant behavior. This has resulted in a new conceptual model of aphid behavior strategies. This model is presented for the purpose of stimulating new thinking, regardless of any factual corroboration or refutation that may occur in the future.

11.2 PLANT PATHOSYSTEMS

11.2.1 Definition

A pathosystem is a subsystem of an ecosystem and is defined by the phenomenon of parasitism (Robinson, 1976). As with an ecosystem, the geographical,

PATHOGENS, VECTORS, AND PLANT DISEASES

ISBN 0-12-326440-5

biological, conceptual, and other boundaries of a pathosystem may be specified as convenient. A plant pathosystem is one in which the host species is a plant; the parasite is any organism that spends most of its individual lifespan inhabiting and obtaining nutrients from one host individual. The parasite species may thus be an insect, mite, nematode, angiosperm parasitic weed, or any of the various categories of plant pathogens. However, avian and mammalian herbivores that graze a population of plants are outside the conceptual boundaries of a plant pathosystem.

There are two categories of plant pathosystems. The *wild* pathosystem is an autonomous system (i.e., it is entirely self-regulating). It is also a balanced system; that is, it is a dynamically stable system within the limits of fluctuation due to seasonal and other variations. This is self-evident from the fact of its evolutionary survival, as a system. The *crop* pathosystem differs in both these respects; it has a degree of artificial control imposed by man, and it is usually (but not necessarily) an unbalanced system. Many crop pathosystems are unstable to the point of being self-destructive; typically, this occurs with the failure of the resistance of a cultivar, which is then abandoned.

11.2.2 Host and Parasite Are One System

Perhaps the most important feature of the pathosystem concept is that the behavior of the host population and the behavior of the parasite population are treated as one system. It follows that any subsystem of a pathosystem should be defined in terms of both the host and the parasite. Three components of a pathosystem become meaningless if either the host or the parasite is absent: infection, the differential interaction, and the gene-for-gene relationship.

Infection is the first contact made by an aphid individual with a host individual for the purposes of parasitism. It is *alloinfection* when the aphid individual originates away from the host individual it infects; and it is *autoinfection* when the aphid originates on the same host it infects. The part of the epidemic that involves alloinfection only is called the *exodemic*; and the part that involves autoinfection only is called the *esodemic*.

Subpopulations within the pathosystem must also be defined. A *pathotype* is a subpopulation of the parasite in which all individuals have a specified character of parasitic ability in common; and a *pathodeme* is a subpopulation of the host in which all individuals have a specified character of resistance in common (Robinson, 1969). If a series of different pathodemes is inoculated with a series of different pathotypes, there either is or is not a differential interaction. A differential interaction means that a series of parasite differentials is necessary to identify any one resistance; and a series of host differentials is necessary to identify any one parasitic ability. This is the basis of the old plant pathological concept of "physiologic specialization" in which the resistance operates against some populations of the parasite but not others. Such resistance is normally temporary resistance in agriculture because, given an appropriate change in the parasite population, the resistance fails to operate and, in common usage, is said to have "broken down." When there is no differential interaction, there is a constant ranking, that is, one pathodeme is either more resistant or less resistant than another, regardless of which pathotype it is tested against. There is a similar ranking of pathotypes.

The concept of the gene-for-gene relationship (Flor, 1942) involves pairs of "matching" (complementary) genes; one of each pair occurs in the host and confers resistance, while the other occurs in the parasite and confers parasitic ability. If a host individual (or a pathodeme) has a number of these resistance genes, it cannot be parasitized by any parasite individual (or any pathotype) that lacks one or more of the complementary genes for parasitic ability. But it can be parasitized by any parasite individual that possesses all (or more than all) of them. In this situation, when parasitism occurs, the host and parasite (as well as their genes) are said to *match*; if parasitism does not occur, they are described as *nonmatching*.

Originally, a gene-for-gene relationship could only be demonstrated by undertaking genetic studies in both the host and the parasite and relating these to both the presence and the absence of parasitism. Such a procedure is technically very difficult with aphids and it is impossible with plant pathogens in which the sexual phase does not occur. Person (1959) first showed that a gene-for-gene relationship can be demonstrated without genetic studies if all known interactions can be accommodated to a special category of differential interaction now known as the Person differential interaction. Subsequently, Robinson (1976) applied the Habgood (1970) nomenclature to the Person differential interaction and obtained a considerably increased symmetry and ease of use; this is known as the Person/Habgood differential interaction and has been described in detail elsewhere (Robinson, 1979).

11.2.3 Subsystems of a Pathosystem

The subsystems of a pathosystem can be defined in terms of these three, combined, host-parasite components. The *vertical subsystem* is named after Vanderplant's (1963) concept of vertical resistance; it involves both a gene-for-gene relationship and a Person/Habgood differential interaction. The *horizontal subsystem* is named after Vanderplank's (1963) concept of horizontal resistance; it has no gene-for-gene relationship and exhibits no differential interaction (i.e., there is a constant ranking). In addition, a number of other subsystems have no gene-for-gene relationship but do exhibit a differential interaction, which, however, is not a Person/Habgood differential interaction (Robinson, 1979, 1980a). (Readers unfamiliar with the terms "vertical" and "horizontal" should note that they are abstract terms without any descriptive or literal connotations; Vanderplank might equally have chosen words such as hard and soft, or alpha and beta, for the description of his new concepts. Being abstract terms, they can be used at all systems levels to label subsystems, subpopulations such as pathotypes and pathodemes, individuals, mechanisms, genes, and so on.)

Possibly the most important discovery to emerge from the pathosystem concept is that the vertical subsystem can only control alloinfection (the exodemic); and that autoinfection (the esodemic) can only be controlled by the horizontal subsystem.

11.2.4 Behavior Strategies

All living organisms have the two properties of structure and behavior. A behavior pattern is a response to a stimulus; if the response is controlled by the genetic code, it is an inherited behavior pattern; and if the response is controlled by learning, it is an acquired behavior pattern. It is safe to assume that all behavior patterns in a plant pathosystem are inherited.

A strategy is a coherent system of behavior patterns and, again, it may be assumed that all plant pathosystem strategies are inherited. Gould (1977) coined the terms *r-strategist* and *K-strategist* following MacArthur and Wilson's (1967) concepts of r-selection and K-selection. An r-strategist has a very high reproductive rate in order to exploit favorable growing conditions that are strictly limited in space, time, or both. Aphids are typical r-strategists. K-strategists have a steady population size governed by the carrying capacity (K) of the ecosystem. In plant pathology, r-strategist hosts have *epidemic* diseases, and K-strategist hosts have *endemic* diseases. This distinction applies equally to insect infestations. However, some K-strategist hosts have epidemic diseases of ephemeral tissue, such as the leaves of a deciduous tree.

Robinson (1976) made a distinction between the *continuous* and the *discontinuous* pathosystem. A discontinuous pathosystem has spatial discontinuity due to genetic heterogeneity in the populations of both the host and the parasite; and it has sequential discontinuity due to a regular dead season such as winter or a tropical dry season. A heteroecious aphid pathosystem, with its winter and summer species of host, is typically discontinuous. In a continuous pathosystem, on the other hand, there is a spatial continuity due to genetic uniformity in both the host and the parasite populations; and there is sequential continuity due to continuously favorable growing conditions. It can be argued that a vertical subsystem (gene-for-gene relationship) can only evolve with epidemic parasites that are r-strategists in a discontinuous pathosystem, although it need not necessarily do so. Conversely, endemic parasites that are K-strategists in a continuous pathosystem can only be controlled by the horizontal subsystem.

Robinson (1979) suggested that some of the principles of animal ethology could be applied to plants, although he was unaware at that time that Teryokhin (1972) had already made the same suggestion in a wider botanical context. Of particular interest in this context is Maynard Smith and Price's (1973) concept of the evolutionarily stable strategy (ESS), which they defined as follows: "Roughly, an ESS is a strategy such that, if most of the members of a population adopt it, there is no 'mutant' strategy which will give a higher reproduction fitness." This concept was formulated at the systems level of the individual and it concerned contests between pairs of animals disputing an advantage such as mates or territory. Later, Maynard Smith (1976) made it clear that he restricted the use of the term ESS to contests between individuals and he commented that the ESS was concerned solely with individual advantage. This restriction is perhaps undesirable in view of the choice of words in this three-word term, where the literal meaning clearly has far wider biological implications than just those of animal ethology at the systems level of the individual. Accordingly, the term ESS is used here in its wide sense of covering the higher systems levels of all biological systems.

An ESS is now defined as the best possible strategy of inherited behavior patterns, at a specified systems level. It has been arrived at by the competitive elimination of all other, inferior strategies. On this basis, it can be argued that every inherited behavior strategy must be an ESS; otherwise it could never have become genetically encoded; nor could it have survived to the present. It follows that the behavior of every wild plant pathosystem, and its subsystems, must be an ESS. Discernment of an ESS is thus the best test for the validity of any conceptual model of a wild pathosystem. Demonstration of a superior ESS will invalidate a postulated ESS; and, if no ESS can be discerned, the model is suspect.

11.3 THE WILD PATHOSYSTEM MODEL

For ease of discussion, a hypothetical aphid pathosystem is envisaged. The aphid is heteroecious; its winter host is a deciduous tree species, and its summer host is an annual herb with a population of one million individuals. It need hardly be added that the model is a deliberate over-simplification that will inevitably require modification or elaboration when applied to an actual aphid pathosystem. Equally, all population numbers are conjectured to the nearest order of magnitude and are intended solely for purposes of illustration.

The model pathosystem has sequential discontinuity due to the alternating host species and the dead season of winter. It also has spatial discontinuity due to genetic heterogeneity in both the aphid and the summer host populations. In particular, there is genetic heterogeneity with respect to vertical resistances and parasitic abilities. The aphid is polycyclic, that is, there are many aphid reproduction cycles in each epidemic cycle; and the epidemic cycle is of one year duration, beginning in the spring.

11.3.1 The Primary Exodemic

The epidemis begins with the hatching of the overwintered aphid eggs, refoliation in the winter host, and seed germination in the summer host. Quite soon, the nymphs become mature fundatrices, which produce winged spring migrants. These fly to the emerging summer host seedlings, which, being aphid-free, can only be alloinfected. This phase of the epidemic is the *primary exodemic* and is controlled by the vertical subsystem. It will become apparent that the vertical subsystem is an entirely discrete subsystem with its own evolutionary function and its own ESS.

Each individual spring migrant thus leaves its winter host and alloinfects an individual summer host. Each of these aphids has a vertical parasitic ability, which either does or does not match the vertical resistance of the summer host individual. If it does not match, the aphid cannot feed; in some pathosystems, it may even die. If merely prevented from feeding, it will search out another summer host individual in order to attempt another, different alloinfestion, which again may not match. This process could be repeated until the aphid dies of starvation or exhaustion. Alternatively, any one of the alloinfections could be a matching infection and, in common usage, the vertical resistance of that host individual would then "break down."

Some matching will always occur. Indeed, it must occur, for otherwise the pathosystem would not exist. It is assumed that there are 12 pairs of matching genes in the vertical subsystem. The ESS will ensure that these genes achieve the maximum effect, that is, an effect that has the maximum survival value, economy, resilience, flexibility, homeostasis and, above all, stability, in the control of the subsystem.

11.3.2 ESS of the Exodemic

It has been argued elsewhere (Robinson, 1980a) that such an ESS is obtained if all individuals in both populations possess $n/2$ vertical genes, where n is the total number of pairs of matching genes; and when all such $n/2$ pathotypes and pathodemes occur with equal frequency. The probability of any alloinfection being a matching infection is then constant throughout the pathosystem and is $1/e$, where e is the binomial mode derived from Pascal's triangle. In the present model, therefore, every individual has 6 vertical genes, and the binomial mode is 924; that is, there are 924 different, 6-gene vertical resistances and parasitic abilities. Given equal frequencies of all these vertical pathodemes and pathotypes, the probability of an alloinfection being a matching infection is constant at 1/924. For each of discussion, this probability is taken to the nearest order of magnitude and is 10^{-3}.

It may be assumed that the spring migrant population of aphids consists of 10^6 individuals, and that each individual can alloinfect only once. There are 10^3 vertical pathotypes within this population, occurring with equal frequencies. Each vertical pathotype thus consists of 10^3 individuals, randomly distributed throughout the parasite population. Similarly, there are 10^3 vertical pathodemes, each consisting of 10^3 individuals randomly distributed throughout the summer host population, which totals 10^6 individuals.

Given a probability of matching alloinfection of 10^{-3}, only 10^3 host individuals will be matched in the primary exodemic, assuming that, on average, each host individual is alloinfected only once. Again, for ease of discussion, it may be assumed that each matched host represents a different vertical pathodeme; and, as a consequence, each of the 10^3 vertical pathotypes becomes established on one host individual.

Whatever the actual rates of alloinfection may prove to be in practice, it is clear that the functioning of the vertical subsystem is totally dependent on spatial discontinuity due to genetic heterogeneity in both populations. This conclusion can be demonstrated by a reductio ad absurdum. It might be supposed that the host individual with all 12 vertical resistance genes would have the greatest survival advantage; the host population would then gradually become homogeneous with respect to this one vertical resistance. However, the same is true of the parasite population, which would become similarly homogeneous with respect to the one vertical parasitic ability possessing all 12 vertical parasitic ability genes. At this point, however, the vertical subsystem would cease to exist because every alloinfection would be a matching infection.

It is here that the discerning of an ESS becomes a valuable test of conceptual accuracy. It is self-evident that the gene-for-gene relationship constitutes an inherited behavior strategy that has evolved independently in many different host-parasite associations. It is clear also that the vertical resistance genes can only have evolved in relation to the vertical parasitic ability genes, and vice versa; the host and the parasite genes constitute one system. For such a system to become genetically encoded simultaneously in the two different species of host and parasite, it must be an ESS. The strategy must have evolutionary survival value; and it can only do so if it is stable over periods of geological time. Equally, it is clear that the ESS can be discerned only in terms of the interaction of the two populations of host and parasite, treated as one system; and that it is dependent on heterogeneity in both populations. Provided that this heterogeneity is maintained, the strategy has obvious survival value and stability. The heterogeneity is maintained in both populations by genetic homeostasis, which Lerner (1954) defined as "the tendency of a Mendelian population as a whole to retain its genetic composition arrived at by previous evolutionary history."

11.3.3 The Esodemic

The esodemic involves host individuals and begins with a matching alloinfection; the vertical resistance does not operate and the infection is successful. The matching, spring migrant aphid then commenced parthenogenetic production of genetically identical, wingless individuals (virginoparae), which autoinfect the one host individual. Because all parts of the one host individuals have the same vertical resistance, and all the virginoparae on that host have the same vertical parasitic ability, it follows that all autoinfection is matching infection. The vertical subsystem cannot influence the esodemic, which can only be controlled by the horizontal subsystem. To postulate that horizontal resistance does not exist is to postulate an absolute susceptibility. Indeed, it can be argued that every epidemic involves at least some matching infection; and, consequently, that every host must possess at least some horizontal resistance to all of its parasites.

During the esodemic, the aphid population has no need for genetic heterogeneity and, in common with most fungal parasites, it can afford the risk of asexual but very rapid reproduction. This makes it a typical r-strategist. Equally, there is no immediate need for migration and wingless individuals are an obvious economy. Eventually, however, crowding (or some other stimulus) promotes the production of winged migrants, which then alloinfect other summer host individuals; this is the secondary exodemic.

11.3.4 The Secondary Exodemic

It may be assumed that, as a result of the esodemic on 10^3 host individuals, the migrating aphid population in the secondary exodemic contains 10^7 individuals; that is, there is a tenfold increase over the primary exodemic. As a result, each summer host is now alloinfected by 10 different aphid individuals. Because the 10^3 vertical pathotypes are still equally represented in the migrating population, the probability of matching alloinfection is still 10^{-3}, and with 10^7 aphids, 10^4 host individuals are matched in the secondary exodemic. On average also, all vertical pathotypes remain equally represented in the new esodemics. Eventually, a further population of migrants is produced and this leads to the tertiary exodemic. It may be assumed that there are 10^8 migrants in the tertiary exodemic and 10^5 host individuals are matched. The consequent esodemics produce 10^9 migrants in the quarternary exodemic in which all 10^6 host individuals are matched. This succession of exodemic and esodemic is summarized in Table I.

11.3.5 Random Mating

The matching alloinfections are random and, by the end of the quarternary exodemic, it is unlikely that all vertical pathotypes still occur with equal frequencies. However, all vertical genes are likely to occur with equal frequencies in the aphid population. When the summer ends, both the parasite and the summer host populations undergo random mating. As a result of genetic homeostasis, it may be assumed that the overwintering populations of both eggs and seeds possess equal frequencies of all 924 six-gene vertical parasitic abilities and resistances. In the following spring, a new epidemic cycle is repeated under identical conditions of genetic heterogeneity.

11.3.6 The "Recovery" of Vertical Resistance

The esodemic ends with the death of all summer host tissue except the seed. The overall vertical resistance of the summer host population then "recovers" in the converse sense of its "breakdown." In the following spring, each new seedling will be aphid-free and it can only be alloinfected; all vertical resistances will be as effective as in the primary exodemic of the previous epidemic cycle. The esodemic thus begins with the breakdown of vertical resistance and ends with its recovery. This is the principal effect of a sequential discontinuity in the pathosystem and is an essential feature of the ESS of the vertical subsystem. It also explains why a vertical subsystem cannot evolve in a continuous pathosystem, where the vertical resistance would break down but could never recover. The cycle of breakdown and recovery also explains why temporary resistance due to the gene-for-gene relationship should evolve in a tree host in which the individual has a lifespan measured in centruies; such resistance evolves in leaf tissue of deciduous trees only; and such tissue has sequential discontinuity, which permits an annual recovery of the vertical resistance of each tree.

11.3.7 The Overall Effect of the Vertical Subsystem

From Table I, it will be seen that the vertical subsystem has three major effects on the epidemic as a whole. Also, provided that spatial and sequential discontinuity is maintained, these effects are stable over periods of geological time.

First, there is a sequential effect. In the primary exodemic, only 0.1% of host individuals are matched; in the secondary exodemic, 1% are matched; in the tertiary exodemic, 10% are matched. Only in the quarterary exodemic, which in this model is the last exodemic of the epidemic cycle, is 100% matching achieved.

There is a major effect on the migrating aphid population. In any exodemic, only 0.1% of migrants achieve a matching alloinfection. Only one in every thousand migrating individuals is able to infect and reproduce. The overall effect of the vertical subsystem is thus to reduce the population growth rate (r) of an r-strategist parasite. Only the increasing number of esodemics permits effective increases in the aphid population.

There is a corresponding effect on the amount of parasite damage caused to the summer host population. Host individuals that are matched in the primary exodemic suffer maximum (100%) damage; in the secondary, 75% damage; in the tertiary, 50% damage; and in the quarternary, only 25% damage. However,

TABLE I. Succession of Exodemic and Esodemic Populations in a Hypothetical Aphid Pathosystem

Exodemic	Size of Migrating Aphid Population	Number of Host Individuals Matched
Primary	10^6	10^3
Secondary	10^7	10^4
Tertiary	10^8	10^5
Quarternary	10^9	10^6

these figures are false in that they assume linear increases in the parasite population. The lag phase of the sigmoid growth curve will mean that host individuals that are matched in the quarternary exodemic will suffer only negligible damage; and they are 90% of the host population.

11.3.8 The ESS of the Horizontal Subsystem

An ESS for the horizontal subsystem is easily discerned and involves a balanced rate of reproduction in both the host and the parasite, relative to each other. Horizontal resistance is a polygenically controlled, quantitative variable; its level in the host population is governed by selection pressure exerted by the parasite. If the resistance is too low, it increases during the course of a few host generations because of the positive selection pressure of excessive parasitism. Conversely, excessive resistance is an unnecessary survival value, which declines due to genetic drift in the face of negative selection pressure. Similar selection pressures operate on the aphid population. Too high a horizontal parasitic ability will reduce the size of the future summer host population. There must be an absolute limit to the parasitic ability of the aphid because, if the summer host became extinct, the aphid would also become extinct. Conversely, too low a parasitic ability would threaten survival of the aphid. The point of balance is where the positive and negative selection pressures are equal, both for horizontal resistance in the host population, and for horizontal parasitic ability in the aphid population. This is an ESS based on feedback in quantitative genetics and is comparable with the parasitic systems studied by Pementel (1968).

11.3.9 The Relation between the Vertical and the Horizontal Subsystems

Robinson (1979, 1980a) has argued that wild plant pathosystems can be classified according to their degree of discontinuity in a number of variables; and that, in any one pathosystem, the degree of discontinuity is approximately equal, both in all variables and in the host and the parasite. He argued further that the survival value of the vertical subsystem was directly proportional to the degree of discontinuity. Conversely, the survival value of the horizontal subsystem is inversely proportional to the degree of discontinuity and is at its greatest in the continuous pathosystem.

The hypothetical model described here has a very high degree of discontinuity; the importance of the vertical subsystem is consequently obvious, while that of the horizontal subsystem is much less obvious. This should not permit an underestimation of the overall importance of the horizontal subsystem in plant pathosystems. For example, if the summer host in the hypothetical model had been a deciduous tree, the esodemic would involve a very large population of genetically identical leaves and the role of the horizontal subsystem in controlling the esodemic becomes more apparent.

11.3.10 Aphid-Transmitted Viruses

It is possible to envisage many different kinds of epidemics of aphid-transmitted viruses and, given our almost complete ignorance of wild pathosystems, speculation is largely fruitless. For example, the most simple postulation would be a virus that is carried by both the winter and the summer hosts; the distribution of the virus in the summer host population would then be identical to that of the aphid. However, there might be a gene-for-gene relationship between the summer host and the virus; and this vertical subsystem might operate entirely independently of the vertical subsystem of the summer host and the aphid. Alternatively, the virus reservoir might be in a different species of summer host of a polyphagous aphid, with or without its own vertical subsystems. There are many possibilities.

One general comment is that, from our knowledge of crop pathosystems, it seems that vertical resistance to viruses is fairly rare (Robinson, 1976).

11.3.11 The Winter Host

In the fall, winged gynoparae and, later, winged males migrate to the winter host. These migrations involve alloinfection but it is most unlikely that the winter host possesses vertical resistance. Vertical resistance in the winter host would greatly restrict the random mating between different vertical pathotypes of the aphid and/or prevent feeding in the spring by nonmatching nymphs on each winter host individual. An equal frequency of all $n/2$ vertical pathotypes would then occur in the following spring only if every vertical pathodeme were equally represented in the winter host population. It is thus difficult to discern an ESS for a vertical subsystem in the winter host, which is probably protected by a horizontal subsystem only.

11.4 THE CROP PATHOSYSTEM

11.4.1 Domestication

The crop pathosystem differs from the wild pathosystem in that man has interfered; he has exerted artificial control on the system and, as a result, cultivars differ from wild plants, cultivation differs from the wild ecosystem, and the behavior patterns of parasites are modified accordingly. In particular, the ESS of the wild pathosystem has been destroyed; the crop pathosystem is no longer a stable system.

The key feature of the crop pathosystem is domestication; by artificial crossing and selection, both the quantity and the quality of the crop product have been increased far beyond such capacities in a wild plant. But comparable attempts to domesticate resistance to crop parasites have been less successful, because it seems that the functions of the various subsystems in the wild pathosystem have not been understood. It is now argued that resistance can be domesticated in various ways; and that it is at least theoretically possible to produce new, artificial strategies tailored to the requirements of the crop pathosystem. For purposes of discussion, the same hypothetical, discontinuous, aphid pathosystem (described above) will be used throughout. It may be assumed that the wild summer host species (which might, perhaps, be a species of the Leguiminosae or the Cruciferae) has been domesticated. It is now a seed-propagated, annual crop, sown each spring in numbers comparable to those of the wild pathosystem already discussed. The loss of systems balance is examined first; this is followed by a discussion of the possibilities of restoring that balance.

11.4.2 Loss of Balance in the Vertical Subsystem

The summer host is now a cultivar. It is, above all, a pure line and, for simplicity, it may be assumed that all the cultivars within the one geographical pathosystem belong to the same vertical pathodeme (i.e., all have the same vertical resistance). Man has imposed homogeneity on the summer host population and the aphid responds by becoming similarly homogeneous. All the autumn migrants (both males and gynoparae) belong to the one, matching vertical pathotype. Initially, the autumn migrants may be heterozygous and sexual recombination will lead to a Mendelian segregation of nonmatching pathotypes. However, after a few epidemic cycles, the aphid populations will be both homogeneous and homozygous. In practice, some degree of heterogeneity will be retained because (1) the total crop is more likely to consist of several different vertical pathodemes and (2) some of the original, wild summer host population may still survive as weeds and wasteland inhabitants. Nevertheless, the frequency of the matching vertical pathotype(s) in the spring migrant population will be far higher than that of the wild pathosystem.

Given homogeneity in both the summer host population and the spring migrant population, the control of the primary exodemic by the vertical subsystem is lost. Virtually every alloinfection becomes a matching infection. The epidemiological effect of this is the equivalent of increasing the size of the spring migrant population; in the hypothetical model, this increase would be a thousandfold.

Nevertheless, one major (and new) effect of the crop pathosystem must be noted. If a new cultivar with a different vertical resistance is introduced, quite suddenly and on a large scale, the aphid population is foiled. The control of the primary exodemic is complete because every alloinfection is a nonmatching infection. The vertical resistance of the new cultivar appears perfect; it is an apparent immunity and it has even been called immunity. But it is obviously temporary resistance and will endue only for the few epidemic cycles required by the aphid population to adjust. When that happens, the vertical resistance "breaks down."

Another new cultivar, with another, different vertical resistance will be required; and the whole procedure is repeated. This has been called the "boom and bust" cycle of cultivar production.

This boom and bust situation has happened many times, in many species of crop and against even more species of parasite, because of a misconception. The conventional approach to breeding for resistance to crop parasites always involves a "good source" of resistance, which is often discovered in a related wild species of host. This resistance is then transferred to a suitable cultivar by pedigree breeding and back-crossing. Clearly, the process is greatly simplified if the good source of resistance involves a single mechanism that is genetically controlled by a single, Mendelian gene. The misconception arises because this entire breeding procedure is conducted during the exodemic. The "resistance" is demonstrated with nonparasitic (i.e., nonmatching) pathotypes of the parasite. The good source was not good at all; it only appeared to be good because there was a zero level of parasitism, which, however, was due to nonmatching. (There is confusion in the common usage because the terminology is based on the host; thus "resistance" and its "breakdown." The terminology could equally be based on the parasite; and then vertical resistance is not resistance; it is a lack of parasitic ability. In the pathosystem concept, however, the host and parasite are one system; the terminology should be based on the system rather than one or another of its components. It is important to recognize the ambiguity of the common usage terms and to refer only to "matching" and "nonmatching.")

The control of the secondary, tertiary, etc., exodemics by the vertical subsystem is also lost. Every summer migrant is able to make a matching alloinfection on the first host it finds. When the vertical subsystem does not operate, the crop pathosystem must endure a summer migrant population, which (in the present model) is 1000 times greater than in the wild pathosystem. In effect, the epidemiological phenomenon of the esodemic has been raised from the systems level of the individual to that of the population; every plant in the crop is the epidemiological equivalent of one leaf in one tree. The resulting exponential increase in aphids is a very prominent population "explosion," which would never occur in the wild pathosystem. This is one reason why the bust of the boom and busy cycle is so damaging.

If there were a wild pathosystem, homeostasis would gradually restore the situation because, in the following epidemics, the summer host would be largely absent, the aphid populations would decline accordingly, and the summer host population would gradually recover its numbers and heterogeneity. (In theory, anyway; in practice, it is doubtful if a wild pathosystem could survive such an extreme loss of stability in the face of competition from other species in the ecosystem.) However, this is not a wild pathosystem; the summer host is planted by farmers who, no doubt, will sow the newest cultivar with the latest vertical resistance. If no new resistance is available, the farmer can employ insecticides, and here there is a very close analogy with host resistance. Some insecticides are within the capacity for microevolutionary change of the aphid; their effectiveness is likely to break down and they correspond to vertical resistance. Other insecticides are beyond the capacity for microevolutionary change of the aphid; their effectiveness is permanent and they correspond to horizontal resistance.

11.4.3 Loss of Balance in the Horizontal Subsystem

A comparable loss of balance occurs in the control of the esodemic by the horizontal subsystem. Horizontal resistance is a polygenically controlled, variable survival value that responds to positive and negative selection pressures. The conventional breeding for vertical resistance is conducted exclusively during the exodemic. But positive selection pressure for horizontal resistance can only occur during the esodemic, that is, in the presence of a matching vertical pathotype after the vertical resistance has broken down. It follows that the level of horizontal resistance tends to decline, due to genetic drift in the face of negative selection pressure, as a result of breeding for vertical resistance. This crucially important phenomenon was discovered by Vanderplank (1963), who called it the *vertifolia effect* after a potato cultivar in which it was particularly prominent. The effect of conventional breeding methods is thus to destroy not only the ESS of the vertical subsystem but the ESS of the horizontal subsystem also. When the vertical resistance of a modern cultivar breaks down, its level of horizontal resistance is low. This is another reason why the bust of the boom and bust cycle is so damaging.

It may be assumed, therefore, that most modern cultivars possess less horizontal resistance than their wild progenitors, whereas, in fact, they need considerably more. The wild progenitor has enough horizontal resistance to control the esodemic and to ensure its evolutionary survival, but these are not significant criteria in the crop pathosystem. What matters is that neither the yield nor the quality of the crop product is damanged by parasites. In practice, this is likely to mean that the agricultural requirements for horizontal resistance are very high. Nevertheless, with new approaches to breeding for resistance (see Section 11.4.4), the prospect of achieving this requirement is at least within the bounds of possibility.

11.4.4 The Domestication of Vertical Resistance

Whatever future investigation may reveal, it is quite clear that the ESS of the vertical subsystem in a wild plant pathosystem is very different from the manner in which vertical resistance is employed in agriculture. The boom and bust situation of agriculture is patently not the "best possible" evolutionarily stable strategy that would become genetically encoded in many different host-parasite associations. It follows that many of the features of the crop pathosystem must be regarded as artifacts of agriculture; and most of them are disadvantageous (Robinson, 1980a). One artifact of the vertical subsystem, however, is advantageous and can be exploited. This is Vanderplank's (1968) concept of *strength* of vertical resistance and his second gene-for-gene hypothesis, which concerns gene quality as opposed to gene identity.

Strength refers to the rarity of the matching vertical pathotype, both before and after the breakdown of vertical resistance. If the matching vertical pathotype is rare, the vertical resistance endures for a longer period. If it becomes rare again quickly, after the vertical resistance has been abandoned, that vertical resistance can be reemployed after a shorter interval. Such a vertical resistance is described as strong. A weak vertical resistance breaks down quickly and is never reemployed because the matching vertical pathotype is common and remains common. By analogy, the insecticide DDT confers a weak protection against houseflies, and a relatively strong protection against malarial mosquitoes.

In the ESS of the wild, vertical subsystem, all vertical resistances are almost certainly weak; and, even more necessary, they must be equally weak, otherwise heterogeneity could not be maintained. During the breeding process, vertical resistance genes are transferred from a wild host, which may even be a different species, to a cultivar that will be grown in a very different environment. The strength of some of those resistances then increases. The mechanism of this change is probably a genetically linked reduction in the epidemiological competence of the matching vertical pathotype which must now parasitize a new host in a new environment. Vertical resistances that remain weak will be lost during the breeding process; they may even escape observation. However, any resistance that endures to the point of being employed in a popular cultivar must be assumed to have considerably greater strength than it did in the wild pathosystem. A few vertical resistances have proved to be exceptionally strong in agriculture and, clearly, they are very useful. Even more important, it appears that new combinations of vertical resistance genes can lead to further increases in strength, occasionally to the point where the matching pathotype lacks epidemiological competence completely and the resistance is effectively permanent (Vanderplank, 1975).

Strong vertical resistances permit an entirely new, artificial, manmade strategy that Robinson (1980a) suggested should be called the CSS (crop stable strategy). It involves employing patterns of different, strong vertical resistances. The patterns may be in space, time, or both and are designed essentially to exploit the boom and to avoid the bust. However, the CSS also involves the domestication of horizontal resistance.

11.4.5 The Domestication of Horizontal Resistance

The level of horizontal resistance in the wild pathosystem is enough (but only enough) to ensure the evolutionary survival of the host, and to provide an adequate control of the esodemic. That level probably corresponds to the natural levels of sucrose in the wild progenitors of sugarcane and sugar beet. Just as these sucrose levels were domesticated to levels far higher than those of the progenitors, so the levels of horizontal resistance can be comparably domesticated.

The limit of this domestication can only be determined by experiment; this limit refers both to the maximum obtainable levels of resistance to any one species of parasite, and to the numbers and kinds of parasite. It is at least theoretically possible to obtain resistance that is permanent, complete, and comprehensive (Robinson, 1979, 1980a), but concrete evidence is still largely lacking because this approach is so new.

The best evidence comes from crops that are derived from a wild pathosystem that is continuous and in which vertical resistance consequently does not occur. Thus sugarcane in Hawaii has horizontal resistance that is both complete and comprehensive (against all locally important parasites) in many cultivars that produce the highest sucrose yields in the world. Working with sweet potato, Jones *et al.* (1976) achieved good levels of horizontal resistance to six species of insect, one nematode, and one disease while retaining good marketing characteristics. With cassava, Hahn (1978, 1980) has produced many lines with good yield and quality as well as resistance to mosaic virus and bacterial blight. By selecting within heterogeneous crops of arabica coffee in Ethiopia, N. A. van der Graff (personal communication, 1979) obtained sufficient resistance to control the very damaging coffee berry disease, as well as good resistance to four other diseases and a leaf miner, while retaining good yield and cup quality. The International Programme for Horizontal Resistance (IPHR) of the Food and Agriculture Organisation of the United Nations is now testing the possibilities of domesticating horizontal resistance in a number of crops and countries (Robinson and Chiarappa, 1977). some of the best evidence comes from the behavior of the maizes of tropical Africa following the introduction of *Puccinia polysora* (Robinson, 1976).

The techniques used in breeding for horizontal resistance are beyond the scope of this chapter and have already been described (Robinson, 1973, 1976, 1979, 1980a, b). The most crucial feature is that all screening must be conducted during the esodemic; all vertical resistance must be either absent or inoperative. It is also necessary to employ population breeding methods, and this is a considerable departure from tradition in autogamous crops such as wheat, rice, and various grain legumes.

11.5 REFERENCES

Flor, H. H. (1942). Inheritance of pathogenicity in *Melampsora lini. Phytopathology* **32**: 653-659.

Gould, S. J. (1977). "Ever Since Darwin." Norton, New York and London.

Habgood, R. M. (1970). Designation of physiologic races of plant pathogens. *Nature* **227**: 1268-1269.

Hahn, S. K. (1978). Breeding of cassava for resistance to mosaic disease and bacterial blight. *In*: "Diseases of Tropical Food Crops," (H. Maraite and J. A. Meyer, eds.). Proc. Int. Symp., *Univ. Louvain*, Lovain, Belgium.

Hahn, S. K. (1980). Breeding cassava for resistance to bacterial blight. *Pans* **24**: 480-485.

Jones, A., Dukes, P. D. and Cuthbert, F. P., Jr. (1976). Mass selection in sweet potato;

breeding for resistance to insects and diseases and for horticultural characteristics. *J. Am. Soc. Hort. Sci.* **101**: 701-704.

MacArthur, R. H. and Wilson, E. O. (1967). "The Theory of Island Biogeography." Princeton University Press, New Jersey.

Maynard Smith, J. (1976). Evolution and the theory of games. *Am. Sci.* **64**: 41-45.

Maynard Smith, J. and Price, G. R. (1973). The logic of annual conflict. *Nature* **246**: 15-18.

Person, C. O. (1959). Gene-for-gene relationships in host parasite systems. *Can. J. Bot.* **37**: 1101-1130.

Pimentel, D. (1968). Population regulation and genetic feedback. *Science* **159**: 1432-1437.

Robinson, R. A. (1969). Disease resistance terminology. *Rev. Plant Pathol.* **48**: 593-606.

Robinson, R. A. (1973). Horizontal resistance. *Rev. Plant Pathol.* **52**: 483-501.

Robinson, R. A. (1976). "Plant Pathosystems." Springer, Berlin, Heidelberg, and New York.

Robinson, R. A. (1979). Permanent and impermanent resistance to crop parasites. *Z. Pflanzenzucht.* **83**: 1-39.

Robinson, R. A. (1980a). New concepts in breeding for disease resistance. *Annu. Rev. Phytopathol.* (in press).

Robinson, R. A. (1980b). The pathosystem concept. *In*: "Breeding Plants Resistant to Insects," (F. G. Maxwell and P. R. Jennings, eds.), pp. 00-00. John Wiley, New York.

Robinson, R. A. and Chiarappa, L. (1977). The international programme on horizontal resistance. *FAO Plant Protection Bull.* **25**: 197-200.

Teryokhin, E. S. (1972). On the possibility of using the concept "behaviour" in studying the evolution of plants.

Vanderplank, J. E. (1963). "Plant Diseases: Epidemics and Control." Academic Press, New York.

Vanderplank, J. E. (1968). "Disease Resistance in Plants." Academic Press, New York.

Vanderplank, J. E. (1975). "Principles of Plant Infection." Academic Press, New York.

Chapter 12

CONTROL OF VECTOR-BORNE MYCOPLASMAS

Karl Maramorosch

Waksman Institute of Microbiology
Rutgers–The State University
New Brunswick, New Jersey, USA

ISBN 0-12-326440-5

12.1 INTRODUCTION

Since 1967 a large number of plant diseases, earlier classified as virus diseases, has been recognized as caused by microorganisms (Doi *et al.*, 1967; Ishiie *et al.*, 1967; Nasu *et al.*, 1967; Maramorosch *et al.*, 1968; Maramorosch *et al.*, 1970). Later some of the causative microorganisms were found to have cell walls and these were described as rickettsia-like organisms (RLO) or, more correctly, as bacteria, belonging to the class Schizomycetes. The remainder, characterized by the presence of cell membranes and the lack of cell walls, was properly recognized as members of the class Mollicutes. Before any of these mycoplasma-like organisms, MLOs, were cultured in cell-free media, they were given the trivial name plant mycoplasmas, which we shall use in this chapter. During the past decade several MLOs have been isolated and cultured. All of these cultured MLOs have been placed in the newly created family Spiroplasmataceae. The trivial name spiroplasma has been coined for these microorganisms (Davis and Worley, 1973).

Is there a scientific basis for the assumption that the steadily increasing number of spiroplasmas, isolated from plants and insects and grown artificially, differ basically from the as yet uncultured, more fastidious MLOs? Until 1973 the MLOs of citrus stubborn disease and until 1975 the MLO of corn stunt disease resisted cultivation attempts. Only after they were successfully cultured were they removed from the MLOs and placed in the newly created family Spiroplasmataceae. A few authors, who have failed in their attempts to culture more fastidious MLOs, have proposed to divide plant mycoplasmas into two distinct groups–cultivable spiroplasmas and uncultivable MLOs. Such an artificial division, based on the degree of fastidiousness and the lack of proper media, does not seem justified. Therefore, in the present chapter, all plant mycoplasmas, irrespective of their status of cultivation, will be considered as a single group. Those interested in the current status of the controversy are referred to recent reviews dealing with spiroplasmas of plants and animals (Maramorosch, 1981; Davis *et al.*, 1981a).

In nature, as well as in the laboratory, transmission of plant mycoplasmas to plants requires grafting or specific insect vectors. The latter belong to leafhoppers (Cicadellidae), planthoppers (Fulgoridae) and psyllids (Psyllidae). At this point in time no other groups of insects have been incriminated as vectors of MLOs.

Plant mycoplasmas (MLOs) possess a unit-type membrane and lack a cell wall. They contain RNA in their ribosomes and DNA in the form of a coil in the nuclear region. The presence of both types of nucleic acid clearly distinguishes them from viruses that, by definition, contain only one type of nucleic acid. In thin section electron micrographs of infected plant tissues and insect organs MLOs resemble members of the Acholeplasmataceae and Mycoplasmataceae (Hirumi and Maramorosch, 1973; Hemmati, 1979; Hemmati and McLean, 1980) (Fig. 1). The cultured spiroplasmas require cholesterol in growth media and it is reasonable to assume that they utilize plant and insect sterols when

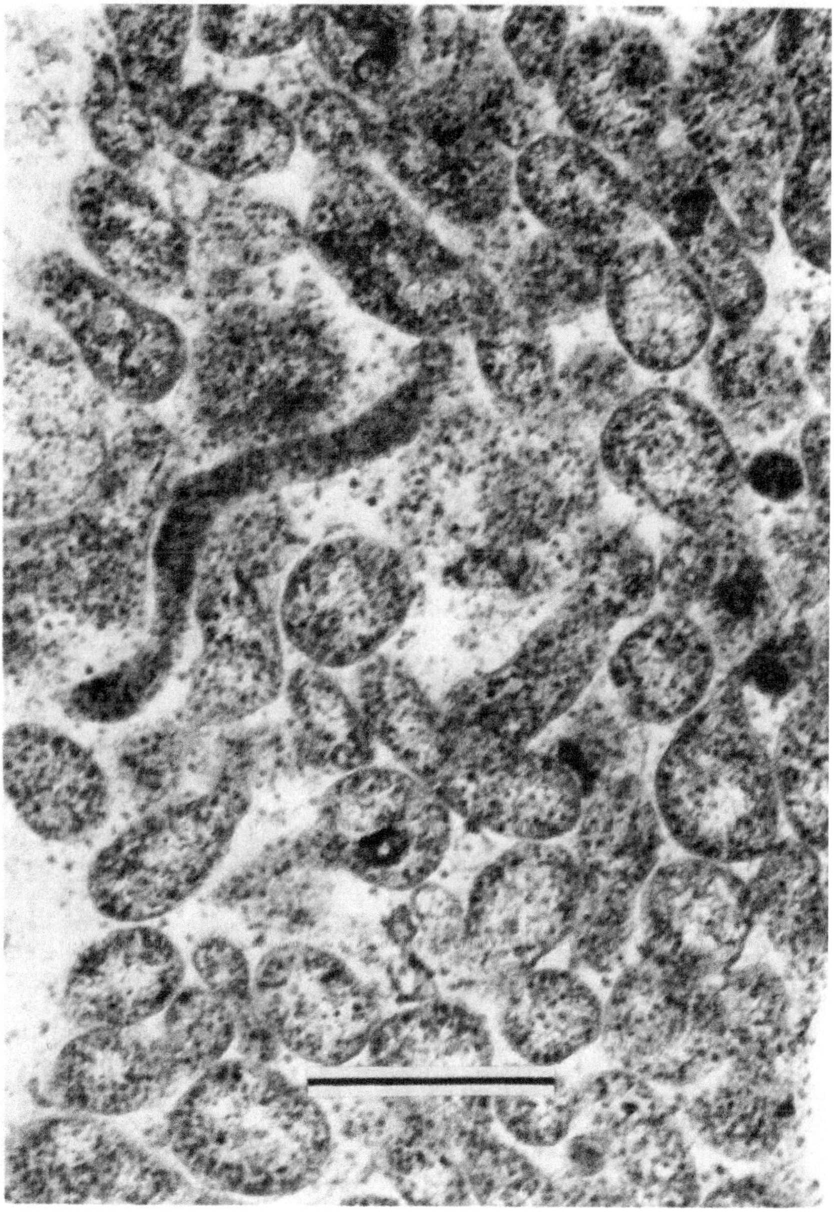

FIG. 12-1. Section of sieve element from an aster yellows-infected *Nicotiana rustica* plant. Round MLOs predominate but one typical spiroplasma is seen. The original electron micrograph, selected from over 2000 sections, has been prepared by H. Hirumi in 1971 when spiroplasmas were still unknown. Bar equals 0.5 μ.

growing in their alternate plant and insect host environment. On agar media spiroplasmas form "fried-egg" type colonies (Fig. 2). Scanning electron micrographs of agar cultures (Maramorosch and Phillips, 1981) do not show spiral forms, but consist of rounded bodies (Fig. 3). Spiral forms can be detected in broth cultures and, in certain instances, in crude plant extracts using dark field or phase contrast microscopy. Negatively stained preparations reveal characteristic spiral forms (Fig. 4). Unfortunately, negative staining applied to crude plant extracts results in artifacts that, although resembling negatively stained mycoplasmas, are indistinguishable from artifacts observed in extracts from healthy material (Wolanski and Maramorosch, 1970). Spiroplasmas can be distinguished by serological methods (Davis and Lee, 1981) and by polyacrylamide gel electrophoresis (Padhi and McIntosh, 1976; Padhi *et al.*, 1977).

Plant mycoplasmas cause numerous economically important diseases affecting, among others, rice, potato, maize, carrot, spinach, lettuce, eggplant, onion (Maramorosch *et al.*, 1970), and numerous species of trees and shrubs (Maramorosch and Raychaudhuri, 1981) all over the world. The diseases are not new, but until 1967 they have been confused with virus diseases. Many have been described as "yellows-type," "proliferation" or "witches' broom" diseases. Among insect diseases, the androcidal disease of fruitflies, in which the male offspring is eliminated by spiroplasma infection (Poulson and Sakaguchi, 1961), a spiroplasma that does not kill male flies (Ota *et al.*, 1979), and a fatal disease of honeybees (Clark, 1977) are of special interest. Although it is customary to divide spiroplasmas according to hosts they affect, this separation is artificial, as can be seen from the fact that plant-pathogenic spiroplasmas can have a deleterious effect on arthropod vectors. This has been described for the corn stunt spiroplasma and some of its leafhopper vectors (Granados and Mehan, 1975; Davis *et al.*, 1981a; Davis and Lee, 1981). There is a hypothesis, as yet untested, that a form of human dementia, the Creutzfeldt-Jacob disease, is linked with spiroplasmas (Bastian *et al.*, 1981). Although most spiroplasmas have arthropod vectors, some, found in flowers, cause no known plant disease and have no known vectors (Davis, 1978, 1981; Davis *et al.*, 1981b). Plant mycoplasmas have their own viruses (Cole, 1977). Some of these viruses morphologically resemble tailed bacteriophages, while others resemble rod-shaped plant viruses.

Control methods of plant mycoplasma diseases can be directed against the disease agents or against the specific vectors (Maramorosch, 1963). A few of these methods have been developed and successfully applied long before the microbial disease agents have been recognized, while others have been tried only after 1967. Breeding for resistance, control of vector populations by chemical and biological means, antimycoplasmal drugs, heat therapy, plant surgery and screens to eliminate vectors are among the already tried control approaches. Future approaches, some on the drawing board and others not yet researched, include the use of mycoplasma viruses, mycoplasma strain interference for cross-protection, use of antimetabolites, alteration of feeding habits of vectors by alien mycoplasmas, and induction of mycoplasma mutants for cross-protection or for vector destruction.

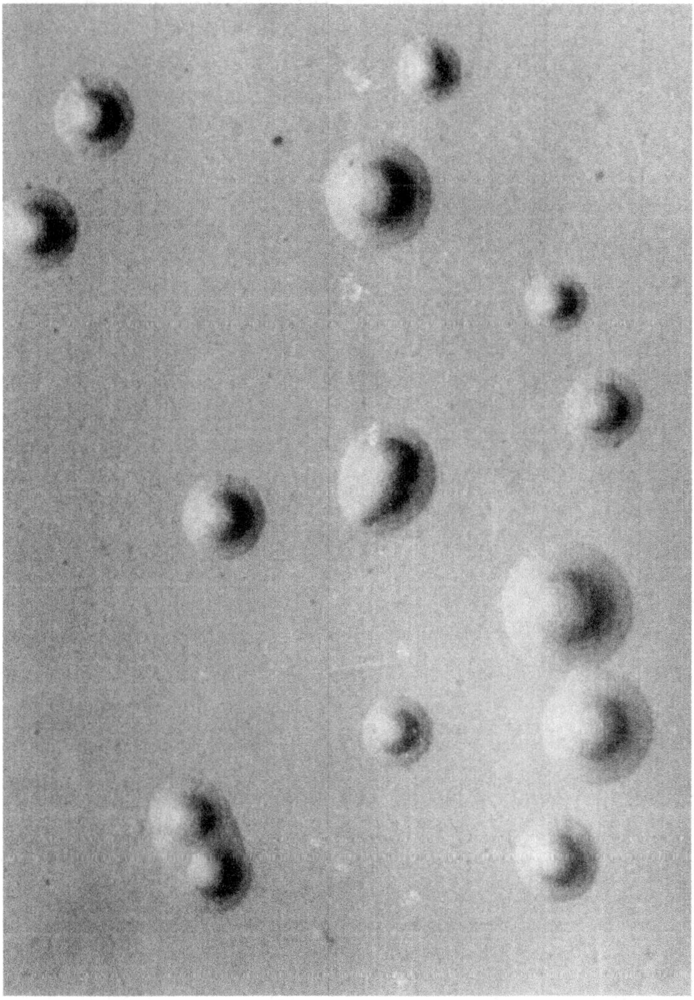

FIG. 12-2. Colonies of aster yellows spiroplasma, isolated from diseased lettuce (*Lactuca sativa*), on agar plate, 13 days after inoculation.

12.2 BREEDING FOR RESISTANCE TO MYCOPLASMAS

One of the most efficient, traditional and proven approaches to disease control has always been breeding for resistance to disease agents. When plant mycoplasma diseases were still believed to be caused by viruses, attempts to find resistance and to incorporate it into plant populations have been attempted and there is no reason to believe that in the future this approach would be less successful with mycoplasmas than with other plant pathogens. Breeding for resistance to disease agents is not only the most successful, but also the least expensive and safest way to protect crops. This has been demonstrated for numerous plant diseases, including mycoplasma diseases, but experimental testing for resistance to MLOs is complicated because of the necessity of exposing plants to insect vectors. Until now, all attempts to infect plants mechanically with MLOs have failed and breeders have to rely on vectors or grafting.

The coconut lethal yellowing disease provides a good example of controlling a mycoplasma disease by the use of resistance to the pathogen. The mycoplasma etiology of this disease was recognized when Plavsic-Banjac *et al.* (1972) demonstrated the presence of mycoplasmas in diseased phloem of palms from Jamaica. On that island, palms are now being destroyed at the rate of one million a year, and in southern Florida, where palms are grown as ornamentals, more than 100,000 have been destroyed in recent years. In Jamaica, one coconut palm variety known under the name Malayan Dwarf, has shown high resistance to natural infection. In fact, only one in 10,000 palms succumbs to the disease in Jamaica. Therefore, replacement of other coconut palms with the Malayan Dwarf has been recommended in Florida. In Jamaica, where the palms are grown for copra production, the Malayan Dwarf was not an adequate substitute but a large breeding program, crossing Malayan Dwarf palms with other coconut palms that produce high quality copra, is now under way. Some of the crosses have shown high resistance. In a few instances, where crosses became diseased, instead of being killed within the usual period of four months they survived for several years, and some even recovered completely.

In addition to coconut palms, 28 other species of palms, including the oil palm and the date palm, are susceptible to lethal yellowing. The disease occurs in America and Africa, and the possibility of its accidental introduction into the coconut growing areas of Asia has to be considered at this time (Maramorosch and Hunt, 1981). Therefore, breeding for resistance is of great importance in this instance, especially since the natural mode of transmission is still uncertain (Howard and Thomas, 1980; Tsai and Thomas, 1981).

China asters (*Callistephus chinensis*) are very susceptible to the aster yellows spiroplasma (Maramorosch, 1981), while members of the genus *Aster* are completely resistant to this MLO infection. Until now, crossing of the genera *Callistephus* and *Aster* has not been achieved, and the resistance of *Aster* to the pathogen could not be incorporated into China asters. Recent evidence suggests that the aster yellows spiroplasma is a strain of the spiroplasma that causes citrus stubborn disease (Christiansen *et al.*, 1980). The aster yellows disease affects plants in more than 40 families, including important crops, such as potato, carrot, lettuce, spinach and onion. Therefore, attempts to breed for resistance to this spiroplasma are of considerable interest.

Considerable resistance to corn stunt disease was detected nearly three decades ago in certain Mexican maize varieties, from Oaxaca and from Chiapas. An effort is now under way to incorporate the resistance into commercial hybrid corn varieties (Ortega, 1974; Osler, 1977).

In the state of Karnataka in South India a search is under way to find sandal (*Santalum album*) trees resistant to the sandal spike disease. Methods have been developed to propagate sandal through plant tissue culture and if and when a resistant tree is found, rapid propagation could be achieved and resistant material provided for reforestation.

12.3 CONTROL OF VECTOR POPULATIONS

Readers interested in the basic principles of insect population suppression and management and the present status of the fundamentals of insect control are referred to the authoritative and thorough presentation by Knipling (1980). Control procedures of vector populations take into consideration the number of insects present in the ecosystem, basic principles of conventional insecticides, insect attractants, chemosterilants, and various cytogenetic techniques. Special consideration has to be given to the circumstances under which various control methods might be used in integrated control (Burges and Hussey, 1971).

In the present volume, Harpaz (1981) lists several indirect approaches to control plant viruses, including the common genetic manipulation aimed at producing plant varieties that are resistant to infection, as well as the much less common production of plants resistant to vectors. In another chapter, Cohen (1981) discusses the control of whitefly vectors by the use of color munches. Some of these methods could also find application in the control of MLO vectors.

12.3.1 Chemical Control

Chemical control of leafhopper and planthopper vectors of mycoplasmas has been attempted in temperate as well as in tropical climates. In most instances the results were less than satisfactory and the cost very high. During the past three decades chemical pesticides have been used widely and often indiscriminately for the control of agriculturally important insects, but applications to control mycoplasma vectors were usually limited to insects that had a single generation per year. The best example of such an application of chemical insecticides and of successful control of a mycoplasma disease is the application of persistent chemical insecticides to control *Scleroracus vaccinii* (Van Duzee), the vector of cranberry false blossom mycoplasma in the United States. The vector usually has only one generation per year, cranberry plants are the main host plant of the vector, and the insects are not efficient transmitters. A single application of a persistent insecticide provides adequate control of cranberry false blossom disease for up to two years. The finding of residual and potentially harmful insecticides in cranberries grown in the western United States a few years ago resulted in the banning of certain insecticides and substitution by other, less harmful compounds. Public awareness of, and resistance to, chemical insecticides have played an important part in the search for alternative control methods.

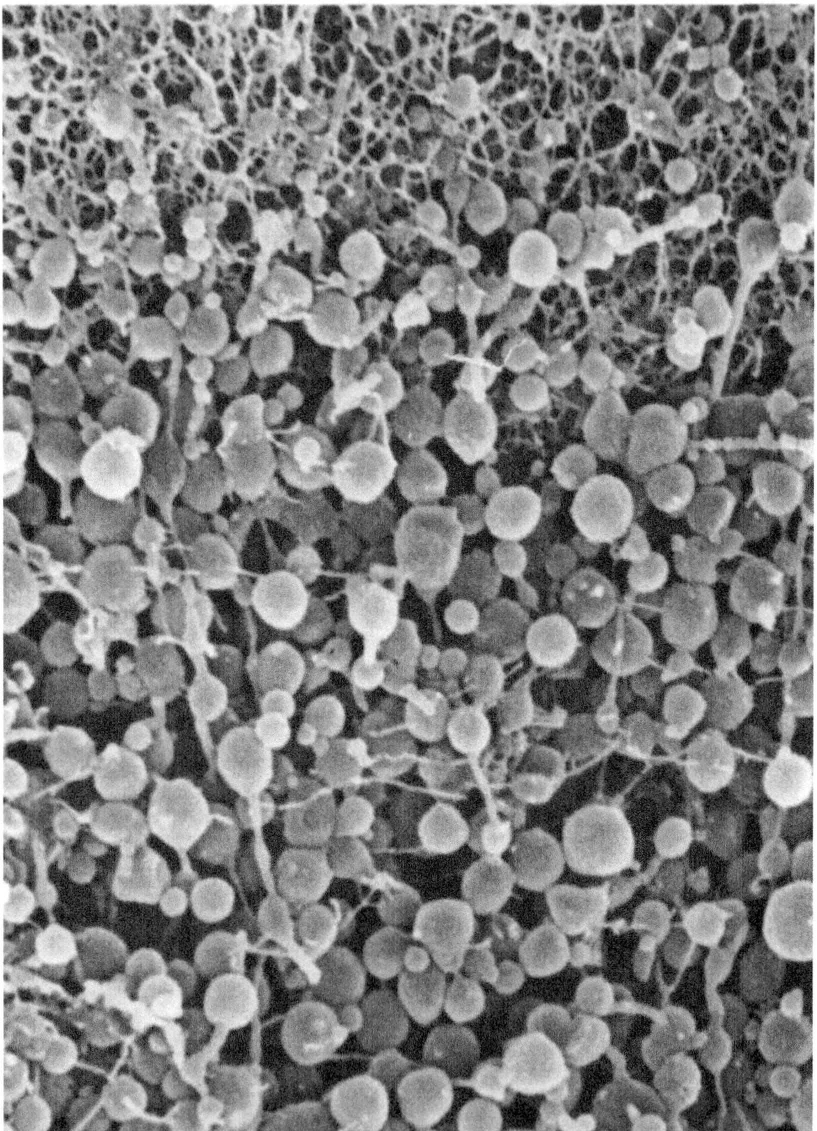

FIG. 12-3. Scanning electron micrograph of aster yellows spiroplasma colony grown on agar. Original micrograph by D. M. Phillips. Magnification 5000×.

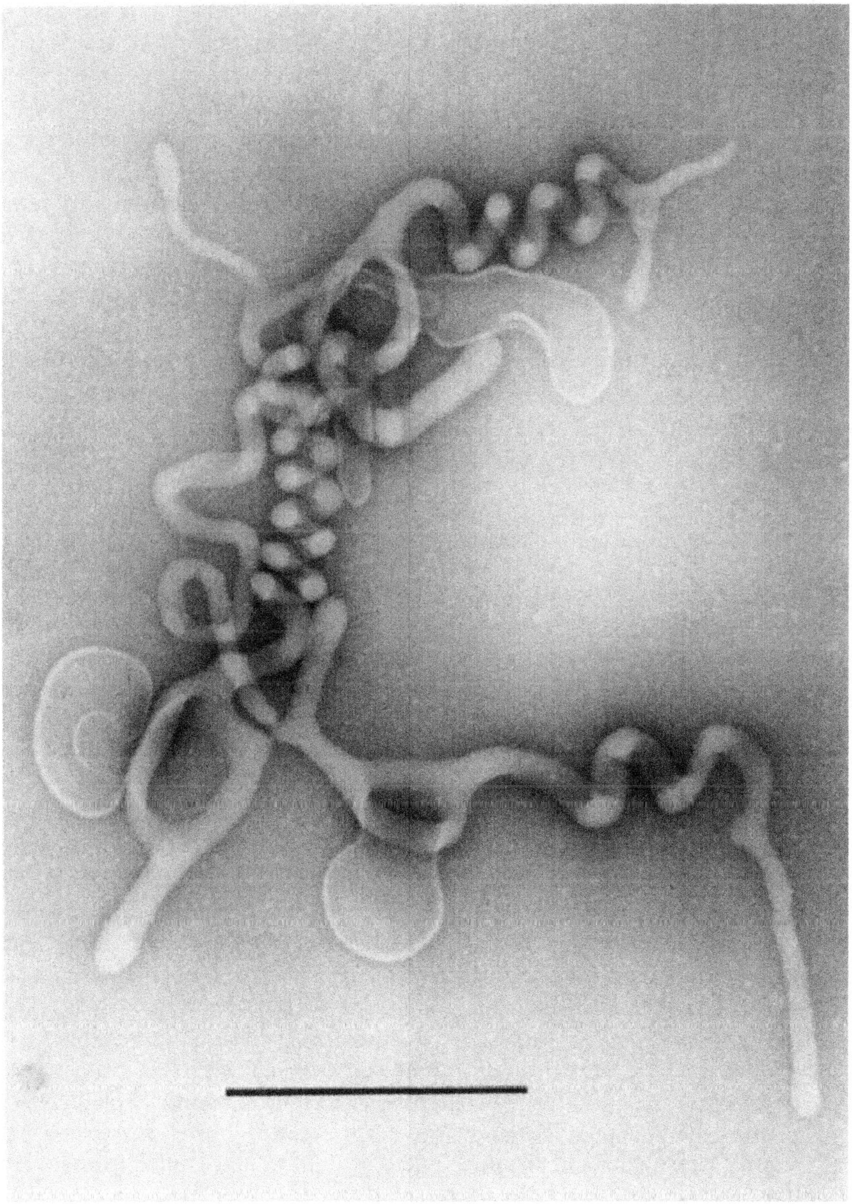

FIG. 12-4. Negatively stained spiroplasma, isolated from lettuce, with typical helical morphology. Note attachment to blebs. Stained with 6% ammonium molybdate. Bar equals 1 μ.

The second example of successful, but not necessarily desirable chemical control of mycoplasma vectors has been in the tropics. There is considerable pressure in the tropics to use pesticides on rice, which is the staple food of over 60% of the world's population. Although most of the 130 million ha under rice is wet paddy, upland rice accounts for 60% of the production in West Africa, 75% in South America, and 20% in Asia. In parts of Asia a systemic insecticide, Carbofuran (Heindrichs, 1979), has provided adequate control of planthoppers and of leafhoppers, including *Nephotettix cincticeps* (Uhler) and *N. virescens* (Distant), the vectors of the rice yellow dwarf mycoplasma. The spectacular success was unexpected, because systemic insecticides kill insects after they have had a chance to feed on a treated plant—usually after they already transmitted the plant-pathogenic mycoplasma. Furthermore, in the tropics, where the leafhoppers have several generations a year and where they are constantly present in rice paddies and grasses, prevention of the spread of a vector-borne pathogen presented a formidable problem. Despite the striking success (Heindrichs, 1979), the use of the compound has been declared unsafe in several countries, including Japan, where rice is the staple food of the population and possible residues in consumable rice had to be avoided.

12.3.2 Nonchemical and Biological Control of Vectors

The term biological control is usually used to describe the action of pathogens, parasites, and predators, in contrast to chemical control, which pertains to the use of chemicals. In recent years the use of persistent and potentially dangerous pesticides has contributed to wide-scale pollution and has intensified the search for safer alternatives. While considerable progress has been achieved in the biological control of insect pests and, to a certain degree, the control of vectors of medical importance, very little has been done so far to control vectors of plant mycoplasma agents. Even less has been achieved in the direct control of plant mycoplasmas within infected plants. A directory of "Biological Control Projects in Plant Pathology," published in 1978 (Charudattan, 1978) lists among 224 projects only one that deals with biological control of plant mycoplasmas. This single project had to be abandoned because the U.S. Department of Agriculture did not agree to provide financial support for its continuation within the framework of the competitive grant system.

The environment is of paramount importance in the establishment of baseline data and biological vector control measures require careful consideration of climatic factors. The need for new approaches to control vectors of plant mycoplasmas has become more obvious because of the lack of proper chemotherapeutic compounds that could be used safely on crops and in the environment, because of public concern about the use of chemical vector control, and last, but not least, because of the development of resistance to chemical pesticides and their high cost. It is difficult to predict whether biological control measures will eventually replace completely chemical insecticides, but they promise to be very useful within the framework of an integrated vector control program.

I shall describe several approaches that have already been tested as well as possible future methods to control plant mycoplasma diseases. Practical aspects of microbial control have been outlined in an earlier multiauthor treatise (Burges and Hussey, 1971). This extensive source includes the control of insect pests and mites, but not insect vectors of viruses or mycoplasmas.

12.3.3 Viral Insecticides

Baculoviruses are among the biological control agents that, theoretically, could be used alone, or in combination with selected, small quantities of chemical insecticides in integrated control (Miltenburger, 1980a,b). These viruses have been given high priority because they were believed to be species specific, cause no pollution, and induce no insect resistance. Some of these obvious advantages are not well-documented however (Padhi and Maramorosch, 1980). It can be expected that eventually resistance shall develop against viruses, just as it did develop against chemicals, even though at present such resistance has been reported infrequently. Viral insecticides work slower than chemical pesticides in destroying pests. The species specificity has to be tested in each instance, because several baculoviruses are now known to infect more than one insect species. As far as mycoplasma vectors are concerned, no baculovirus has been found as yet that could be used as "living insecticide" against planthoppers or leafhoppers. Theoretically, viral insecticides have the advantage over chemicals in that they occur naturally and are apparently innocuous to all but their host species. Until recently it was generally believed that baculoviruses cannot replicate in vertebrate cells (Anonymous, 1973), but the opposite has now been proven (McIntosh and Shamy, 1980). The possibility that insect viruses may cause long-term effects in man, such as cancer or birth defects are very difficult to test (Wade, 1973). The ecological impact of viral insecticides has not been considered or evaluated to the same degree as the impact of chemical insecticides. For the treatment of vectors in a monoculture such as rice, the usual approach would be to develop a viral insecticide that would kill as many leafhoppers as possible to obtain complete control and eliminate losses. This approach requires, in turn, repeated applications of large quantities of viral insecticides and such an "unnatural" release of virus would seem prone to encounter problems in the long run (Maramorosch, 1979). Safety considerations of baculovirus insecticides will require long-term studies of possible effects on the environment during prolonged exposure to large baculovirus doses (Kurstak *et al.*, 1978).

Since no viral insecticides of the nuclear polyhedrosis (NPV) or granulosis (GV) type are available for vectors of plant mycoplasmas, it would seem worthwhile to search for such viruses by collecting planthopper and leafhopper vectors that died in nature. Theoretically, baculoviruses are particularly suitable for vector control, being relatively safe in use, efficient, and easy to produce and to store. The use of other types of insect viruses (Maramorosch, 1977) might involve greater difficulties from the standpoint of specificity as well as safety (Kurstak *et al.*, 1978).

12.3.4 Mycoplasmas Deleterious to Vectors

Years ago insects were considered to be only passive carriers of mycoplasmas, not affected by the pathogens carried to other hosts. More recently studies on mycoplasma-vector interactions revealed very complex relationships between the microorganisms and the arthropods, and, in some instances, direct effects on the metabolism of vectors, as well as of nonvectors, that acquired the mycoplasmas (Maramorosch, 1980b). Mycoplasmas multiply in their transmitters as well as in their host plants and some have been shown to play a beneficial role in the biology of insects by altering the insect in such a way that its reproductive capacity is increased, that the diseased plants can serve as food hosts for insects that die on healthy plants of the same species, or that the mycoplasma-carrying insects can survive on healthy but strange plant species that previously were not acceptable or suitable as nourishment (Maramorosch, 1958b).

Mycoplasmas with alternative arthropod and plant hosts constitute a bridge between the typical animal pathogens and typical plant pathogens. The multiplication cycle of aster yellows spiroplasma, alternating in nature between plants and leafhopper vectors, could have arisen through evolution by very simple nutritive requirements or by a well-organized adaptation of all metabolic activities. Our present knowledge of biologically transmitted plant mycoplasmas shows that, with certain exceptions, they have a fairly large host range among plant species but a narrow one among the auchennorrhynchous vectors. This favors the assumption that their adaptation to the invertebrate host is well organized and, from an evolutionary point of view, probably also of much longer existence (Maramorosch, 1954). It seems conceivable that these mycoplasmas were originally pathogens of insects, and due to the long association between the host and parasite, ideally balanced in their almost symbiotic-like relationship. This could explain the tolerance of the highly virulent plant mycoplasmas in respect to the "susceptible" arthropod vectors.

The deleterious effects of plant-pathogenic spiroplasmas on insect vectors ranges from cytopathic changes of certain organs to decreased life span and impaired fecundity. The aster yellows spiroplasma causes cytopathic changes in the fat body tissues of the leafhopper vector, *Macrosteles fascifrons* Stål (Littau and Maramorosch, 1956). These changes can be detected in males, but not in females, and they become pronounced only after a few weeks, and not during the beginning of the transmission period (Littau and Maramorosch, 1960). The cytopathic changes were detected in some, but not all infected males. Nuclei of the fat body were stellate instead of round, and the cytoplasm appeared sparse. These findings indicated that arthropod vectors might be harmed directly by mycoplasmas that multiply in their bodies and the observations stimulated further research on direct, deleterious effects of mycoplasmas on vectors.

The first mycoplasma demonstrated to be injurious to an insect vector was that causing Western X-disease of stone fruits and a yellows disease in celery. This mycoplasma has been shown to cause a significant reduction in the life span of one of its vectors, *Colladonus montanus* (Van Duzee) (Jensen, 1958, 1959,

1962, 1963). Infected *C. montanus* also produce fewer eggs than do noninfective controls (Maramorosch and Jensen, 1963). The mean longevity of normal leafhoppers averaged 51 days, while the mean longevity of mycoplasma transmitting insects was 20 days only. The premature deaths of mycoplasma-carrying insects appeared to be due to the Western X-disease mycoplasma. The effect on fecundity was very pronounced, as infected leafhoppers laid only about one-third as many eggs as healthy controls.

The corn stunt mycoplasma is transmitted in nature by several leafhopper species, the most efficient being *Dalbulus maidis* (DeL. & W.) and *D. elimatus* (DeL. & W.) (Wolanski and Maramorosch, 1979). The first is commonly present in Mesoamerica from sea level to 3,500 feet elevations, while the latter occurs at higher elevations. During 20 years of experiments we detected no harmful effects of the Rio Grande corn stunt spiroplasma on *D. maidis*. This vector species is one of the most efficient mycoplasma transmitters known. Practically every insect that feeds on a corn stunt-diseased plant is rendered into a transmitter, provided it lives long enough for the completion of the protracted, 2-3 week incubation period of the pathogen in the body of the vector. The incubation period of the corn stunt spiroplasma in the highland vector *D. elimatus* is somewhat longer and the vector is directly harmed by one strain of the corn stunt spiroplasma (Granados and Meehan, 1975). The longevity and fecundity of transmitting leafhoppers in that instance is significantly impaired, compared to noninfected controls. Claims of reduced longevity caused by the Rio Grande corn stunt spiroplasma on *D. maidis* (Williamson and Whitcomb, 1974) have been based on small samples and the data are unconvincing.

The corn stunt spiroplasma is easily grown in spiroplasma media (Chen and Liao, 1975). Therefore, standard bacteriological methods could be devised to induce, chemically or physically, mutations and selected spiroplasmas with desired characteristics (Maramorosch, 1980b). Some mutants can be expected to be more harmful to the vectors than the original isolate, and through such manipulation and selection one could expect to develop strains that would kill leafhopper vectors before these vectors get an opportunity to transmit the spiroplasma to susceptible plants. On the other hand, mutants that cause very mild plant disease could also be obtained. This approach to the control of vectors is still highly speculative but the huge importance of the agricultural spheres warrants the basic research effort.

12.3.5 Integrated Control

It is unlikely that biological control measures could replace entirely chemical insecticides, but they promise to be useful eventually within the framework of an integrated vector control program. In a broader sense, integrated control can be seen as the combination of several types of control, including more than one type of nonchemical control, as well as the combination of nonchemical and chemical insecticides. The general idea is to make use of the synergistic effect of different methods to reduce the vector population. As pointed out by Harpaz, elsewhere in this volume, the integrated control of vectors depends on the judicious use of selective pesticides together with intensive development and application of biological, cultural and other control measures. This approach is currently regarded as the most reasonable, most promising, and most feasible alternative to the exclusively chemical destruction of vectors. While considerable success has already been achieved with integrated control methods in reducing the population of certain aphid vectors, integrated control against leafhopper and planthopper vectors is merely on the drawing board at this time.

Integrated control of vectors requires a thorough knowledge of the vector biology as well as of the ecology. Attention has to be focused at the total ecosystem rather than its component parts (Kurstak *et al.*, 1978).

Integrated control of vectors of plant-pathogenic mycoplasmas promises to result in the drastic reduction or elimination of the vectors, or in the efficient interruption of the transmission cycle. The ultimate goal of integrated control is the prevention of transmission, which, in turn, requires that each newly arriving vector would have to be killed before it had a chance to feed or even to probe. Interruption or even suppression of transmission is a formidable task but in the case of persistent, biologically transmitted mycoplasmas it is a realistic goal.

In 1977, at the 7th session of the FAO Panel of experts on integrated pest control, priority areas have been specified for certain crops, such as millet, grain legumes, roots, tubers, and vegetables. At the same time, priority areas have been singled out for human and animal diseases such as malaria, schistosomiasis, filariasis and trypanosomiasis. It has become obvious that chemical control measures alone no longer can be used, either because of resistance problems, or because of environmental problems. Biological control by itself is not sufficiently effective in most instances. Therefore, the combined action of biological agents and small quantities of carefully selected chemicals is now generally recommended (Anonymous, 1977).

12.3.6 Breeding for Resistance to Vectors

Breeding plants for resistance to vectors, one of several possible approaches to the general concept of breeding plants for resistance to insect pests, has been reviewed recently (Maramorosch, 1980a). In a few instances, breeding for resistance to aphid vectors of viruses has been very successful, but the success has been accidental, and not the result of a planned experimental breeding, because virus control has not been the aim of the breeding program. Aphids are not known to carry plant mycoplasma agents and the only established groups of vectors of plant mycoplasmas are, at this point in time, leafhoppers, planthoppers, and psyllids. Future attention will have to be focused at control approaches that specifically deal with these types of vectors.

In the future, consideration must be given to aspects of plant resistance to vectors of mycoplasmas, including the various types and optimal levels of resistance, as well as genetic, environmental and behavioral factors affecting the stability and expression of resistance. Attention will have to be paid to diverse interactions between vectors and mycoplasmas, as well as between vectors and diseased plants. It is well known that diseased plants attract certain vectors, such as aphids, because of color changes from green to yellow. Attraction of leafhoppers and planthoppers to diseased plants, due to chemical and morphological alterations of diseased plants, have been observed but not studied in detail. Phytochemical and nutritional changes might benefit vectors nutritionally or act as attractants. Morphological changes, including bushy growth, leaf curling, proliferations and other abnormalities also may benefit vectors. These complexities were responsible in the past for the omission of vector resistance from most older breeding programs, but hopefully future programs will be aimed not only at insect pests, but also at vector resistance (Maramorosch, 1980a). As of now, vector resistance has been obtained in a few instances only, as a secondary result of breeding programs aimed at an insect pest that, coincidentally, also happened to carry a plant mycoplasma.

12.4 MYCOPLASMA ANTIBIOTICS

Tetracycline antibiotics have been used successfully to inhibit mycoplasmas and combat human and animal mycoplasma diseases. The first tetracycline antibiotic, aureomycin, was discovered by B. M. Duggar, not a medical researcher, but a plant pathologist, who at the age of 72, after compulsory retirement from the University of Wisconsin at Madison, became Research Director at Lederle Laboratories in Pearl River, New York, where he embarked on a large screening program for new antibiotics. Duggar did not foresee that plant pathologists will some day profit from his discovery. Rocky Mountain spotted fever, a rickettsial disease, and human "atypical virus pneumonia," caused by *Mycoplasma pneumoniae*, as well as chlamydial diseases were among the first human ailments successfully treated with aureomycin. Since the report by Ishiie *et al.* (1967), it became apparent that most, if not all, MLO diseases of plants can be affected by water-soluble tetracycline compounds. Temporary disease remission, and parallel disappearance of MLOs can be used to diagnose the MLO nature of a disease. Extensive tests, carried out in Japan, established that plants were not cured by tetracycline antibiotics permanently. Unfortunately, the tests that were carried out at 13 experiment stations by nearly 50 senior scientists and 100 technical assistants, were seldom cited outside Japan, because the publication appeared in Japanese without an English abstract (Anonymous, 1969). The tests demonstrated conclusively that tetracyclines merely provide temporary remission, lasting for days, weeks, or even months, but that in every instance disease recurred after awhile. This was not surprising, because tetracyclines have been known to act as mycoplasmastatic, but not mycoplasmacidal, compounds.

While the work in Japan was concluded and published, investigators in several other countries tried unsuccessfully to obtain permanent cures of MLO diseases by treating plants with tetracycline antibiotics. Failure to test the actual uptake and activity of the drugs resulted in erroneous conclusions that plant mycoplasmas differ strikingly from animal mycoplasmas in their response to tetracycline chemotherapy. Eventually it was found that plant mycoplasmas respond to antibiotics in a similar manner as do animal-infecting mycoplasmas and acholeplasmas. Tetracycline uptake by plants was studied first by Frederick *et al.* (1971), using *Bacillus cereus* to test the presence and retention of the antibiotic in plant tissues. Similar tests were subsequently carried out by Sinha and Petersen (1972). These early tests had to depend on *in vivo* systems. After 1973 the cultured *Spiroplasma citri* and other spiroplasmas provided accurate testing and evaluation systems of antibiotic spectra of different spiroplasmas (Kondo and Maramorosch, 1977; Liao and Chen, 1981). Today there is no evidence to suggest that MLOs associated with plant diseases differ significantly in their antibiotic spectrum from other mycoplasmas.

In practice, certain diseases of plants have been treated with some degree of success with tetracyclines. Sugarcane white leaf disease has been treated with promising results in Asia. Pear decline in commercial orchards in the United States, and coconut palms in Dade County, Florida, affected with lethal yellowing disease, have been treated with tetracyclines to delay the progress of the respective diseases. In Karnataka State, India, sandal trees in forests where spike disease is rampant, have also been treated in a similar manner (Maramorosch and Raychaudhuri, 1981), and in Korea, witches' broom of Jujube has been controlled (La *et al.*, 1976). There are obvious disadvantages and dangers of using tetracycline antibiotics, but, surprisingly, these seem to have been overlooked. The antibiotic treatment might lead to the production or selection of strains of microorganisms that are completely resistant to tetracyclines, as has been the case with mycoplasmas contaminating mammalian tissue cultures. More important, perhaps, is the danger that large-scale use of the antibiotics needed to treat human diseases might result in the unintentional consumption of these antibiotics by those who eat pears from treated orchards, or drink coconut water in Florida. No large-scale tests have been published and there is no adequate information available in the literature concerning the possible retention of tetracyclines in coconuts or pears. Perhaps the ornamental nature of coconut palms in Florida has obscured their edible features when large-scale use of antibiotics was permitted a few years ago. Even if the mentioned objections should prove completely theoretical, the use of tetracyclines for treating plant mycoplasma diseases seems inadvisable, since no permanent cure can be achieved. The Florida program appears to be a complete failure. There is an urgent need for further research on different groups of chemicals, that might eliminate MLOs without affecting plants. This could be achieved by altering the pH, or the osmotic pressure in the phloem. More than three decades ago, when Stoddard (1942) reported successful treatment of "virus infected" peach trees with sulfa drugs, his report was received with downright scepticism, but his discovery was valid, since the X-disease of peaches which he studied, is caused by a plant mycoplasma. Stoddard reported that bud wood from peach trees with X-disease was cured by soaking in water solutions of quinhydrone, urea and sodium thiosulphate (1942). Chemotherapy is urgently needed for the production of MLO-free clones as well as to combat MLO diseases of trees.

In the case of cultivable spiroplasmas, testing for antispiroplasmal compounds can be carried out in the laboratory (Goldstein *et al.*, 1976) in order to discover chemicals that could be applied to agricultural crops. At present, chemotherapy is much too costly and neither rice, sugarcane, or potatoes affected by MLO diseases can be treated in the field with tetracycline drugs, even though these drugs inhibit the growth of MLOs. It is also important to remember that tetracyclines work only when introduced via roots or injected into the stems of trees, but not when sprayed onto leaves or poured on the soil.

12.5 HEAT THERAPY

Heat therapy for plant diseases has been used successfully for the treatment of certain, but not all, MLO diseases. Kunkel (1936b) discovered that peach yellows-affected trees and bud stocks can be completely cured by immersing them for appropriate periods in hot water, and growing trees could be cured by keeping them continuously for two or three weeks at 35°C. Kunkel also demonstrated that *Vinca rosea* and *Nicotiana rustica* could be freed completely from aster yellows and some other MLOs by heat therapy (1941, 1943, 1945, 1952), and insect vectors can be freed from the same MLOs by subjecting them to elevated temperatures (Kunkel, 1937). Cured plants and insects could be reinfected, because no immunity to MLOs was induced by the previous infection.

The thermal inactivation of peach MLO was found to be 29°C in living plants, and that of aster yellows 32°C. *In vivo* inactivation of corn stunt mycoplasma has not been achieved.

Similarly to certain viruses, MLOs seem unable to enter apical meristems and propagation of plants from MLO-free meristem or callus tissues appears to be a promising approach to produce disease-free plants. The combination of heat treatment and plant tissue culture could be used for practical elimination of MLOs from clonal varieties.

Heat treatment can also be used to obtain mild strains of MLOs. This has been found by Kunkel (1937, 1941) who obtained an attenuated strain of the aster yellows spiroplasma by heat treatment.

12.6 SURGERY

In 1904, Baur demonstrated that surgery can be applied to cure a plant virus disease (Baur, 1904). Surgery has been tried but rarely to cure MLO diseases, because it has been assumed generally that the systemic MLO infection precluded the use of this method. The success of surgical cure depends on the slow movement of MLOs within an infected plant, or the incomplete invasion of the plant by the pathogen. The only instance of successful surgical therapy has been the cure of papaya trees affected by bunchy top disease (Maramorosch *et al.*, 1970). When the "bunchy top" is cut off and the decapitated tree subsequently protected from reinfection, by preventing the access of the leafhopper vector *Empoasca papayae*, healthy leaves and fruits are formed.

The slow movement of MLOs might not be as rare as sometimes believed, especially in woody plants, and it might be worthwhile to study further this aspect, as well as the removal of diseased branches by surgery. In the witches' broom disease of willow, caused by MLOs, the progress of the disease is checked naturally by the frost destruction of diseased branches during the winter (Holmes *et al.*, 1972).

12.7 SCREENS

Screening plants can provide complete elimination of insect vectors and thus effectively interrupt the alternating cycle of the MLOs by preventing plant mycoplasma infection. Screening is now used on a small scale by seed growers, and in experiments, where screened greenhouses and screened cages are standard features.

Screens have been used for many years to prevent the access of sixspotted leafhoppers to susceptible aster plants, and thus to prevent infection with the aster yellows spiroplasma (Kunkel, 1929). Although this method can only be applied to plants grown on small plots, it has proven economical and effective in the production of China aster (*C. chinensis*) seed by commercial seed growers. Copper mesh, saran monofilament, or nylon screens have been used routinely to protect greenhouse plants from vector-borne MLOs.

12.8 MYCOPLASMA VIRUSES FOR MYCOPLASMA CONTROL

In 1973, Cole *et al.* reported the first occurrence of viruses in cultured *S. citri*. Subsequently, two bacteriophage-type, tailed viruses, and one rod-shaped virus have been reported associated with the citrus stubborn MLO (Cole, 1977). More recently, indicator strains for one of the viruses have been discovered (Cole *et al.*, 1977; Maniloff *et al.*, 1977). This permitted quantitative studies and definitive infectivity tests of the spiroplasma viruses. Calavan and Gumpf (1974) described viruses in their *S. citri* cultures. A rod-shaped particle, resembling the filamentous virus found in 1973 by Cole *et al.*, has been detected in the aster yellows spiroplasma isolate (Kondo *et al.*, 1977; Maramorosch and Kondo, 1978). In the cultured corn stunt spiroplasma, virus-like particles have also been detected.

The first notion that MLOs might be associated with viruses came from electron micrographs of thin sections of infected plants and insect vectors (Giannotti, personal communication, 1970; Allen, 1972). Subsequently, such observations have been made in several laboratories where aster yellows, stolbur, corn stunt and other MLO-caused diseases were being studied (Anonymous, 1974).

It was known for many years that aster leafhoppers, *M. fascifrons*, confined to older leaves of systemically infected plants, are unable to acquire the aster yellows spiroplasma (Maramorosch, 1964). In 1972, electron microscopy revealed that the older leaves of systemically infected plants contain deteriorating MLOs, often associated with rod-shaped virus particles (Hirumi and Maramorosch, 1972). Since the aster yellows spiroplasma culture was found to contain similar virus particles (Fig. 5) and the particles seemed to increase in number in successive subcultures, their role was of particular interest. We have found that some of the subcultures were destroyed and the destruction caused by the rupture of the spiroplasma membranes (Fig. 6). Apparently spiroplasma's own virus can provide a means for the spontaneous elimination of the pathogen. Could such spontaneous elimination be harnessed so as to cure living, diseased plants? Future artificial manipulation of the interaction between spiroplasma viruses and spiroplasmas might provide means for viral control of certain mycoplasma diseases.

Virus particles have been reported associated with corn stunt spiroplasma in salivary glands of *D. maidis* (Granados, 1969). Similar findings pertain also to viruses observed in salivary glands of experimentally infected *Euscelis plebejus* that carried *S. citri*. The presence of such viruses might be crucial in the mechanism by which spiroplasmas cause plant disease. It is conceivable that pathogenicity as well as antigenic variations among spiroplasmas depend on their infection by specific viruses. These viruses might be required for the production of spiroplasma toxins that, in turn, elicit the disease syndrome. If viruses are required for toxin production, the removal of such viruses would provide a possible means of control. The role of viruses of spiroplasmas in respect to virulence and toxin production is conjectural at present and the speculation might seem far-fetched, but this subject deserves special attention (Liss and Cole, 1981).

12.9 CROSS PROTECTION (STRAIN INTERFERENCE)

Interference between related strains of spiroplasmas has been discovered by Kunkel (1955, 1957). The discovery of this "immunization" led to further studies of the very complex host parasite-vector interrelationships. Kunkel used the eastern strain and celery infecting strain of aster yellows, the latter derived from California. Kunkel was able to show that *M. fascifrons* which acquired first the eastern strain of spiroplasma, and in which the spiroplasma had completed its incubation, was unable to transmit the California strain, after it fed on plants infected with that strain. The same phenomena occurred when the procedure was reversed. This "cross-immunization" or interference suggested that the spiroplasma first acquired by a vector multiplied in certain tissues and organs of the insect in such a manner as to occupy all the available multiplication sites, leaving no opportunity for the second spiroplasma to reproduce itself. This interpretation was further strengthened by Kunkel's finding that some insects which were permitted access to both strains of virus through alternating feeding during a 48-hr period, sometimes transmitted both strains of spiroplasma. Probably the simultaneous invasion of sites of multiplication permitted both strains to develop and to be transmitted subsequently.

A somewhat different type of "immunization" or interference by spiroplasma strains was described by Maramorosch (1958a). When *D. maidis* leafhoppers were confined to maize plants infected with the Rio Grande strain of corn stunt spiroplasma, and a few days later were transferred to maize plants infected with the Mesa Central strain (Maramorosch, 1955), the insects transmitted the Rio Grande strain exclusively, that is, the strain they acquired first. When, however, the acquisition of the two strains was reversed, and the insects first fed on plants with Mesa Central spiroplasma and later on plants with Rio Grande, they transmitted first the strain acquired first, but after a while they transmitted the Rio Grande strain. The transition period between the pure transmission of the first-acquired and that of the later-acquired strain was characterized by sporadic cases of double transmission. This interesting case of unilateral immunization may have been the result of a difference in the virulence or rate of multiplication of the two corn stunt spiroplasma strains, or the difference in the susceptibility of *D. maidis* to the respective spiroplasma strains. The insect immunization by related spiroplasma strains indicates an action of the respective spiroplasma in certain cells of the leafhopper vector, but the mechanism of this phenomenon is not yet understood.

Unilateral interference was also reported by Freitag (1967, 1969) to occur in the *M. fascifrons* leafhopper, infected with three California-derived strains of the aster yellows spiroplasma. The strains were named severe, dwarf, and Tulelake. Individual leafhoppers fed for two weeks on dwarf and then two weeks on severe strain transmitted only the dwarf strain. However, leafhoppers fed two weeks on Tulelake and then two weeks on dwarf all transmitted only Tulelake for a time but, later, 16 of 24 individuals stopped transmitting Tulelake and subsequently transmitted only the dwarf strain.

These examples demonstrate how closely related strains of spiroplasmas may compete with each other and prevent or delay multiplication and transmission of challenging strains.

For control purposes, it would seem conceivable to infect artificially insect vectors with a mild, almost nonpathogenic strain of spiroplasma, to prevent the acquisition and transmission by such insects of the virulent strain. The outcome of such experiments cannot be predicted, because interference might be unilateral or incomplete.

Bilateral or unilateral interference is known to occur only between closely related strains of spiroplasmas, but not between unrelated spiroplasmas nor between spiroplasmas and other plant pathogens (Kunkel, 1936a). Thus, the described interference between Rio Grande and Mesa Central corn stunt provides additional evidence that the two disease syndromes are caused by related strains of spiroplasmas. The media used for the cultivation and isolation of Rio Grande spiroplasmas are not suitable for Mesa Central pathogens and the abundance of spiral forms in crude extracts from Rio Grande diseased plants contrasts sharply with the absence of spiral forms in the crude sap of Mesa Central-diseased maize.

Based on the above differences, Nault and Bradfute (1979) hypothesized that Rio Grande is caused by corn stunt spiroplasma, but Mesa Central is caused by MLOs that differ from spiroplasmas. These MLOs were given the designation of "MBSM", "maize bushy stunt mycoplasma." The hypothesis overlooks the fact that spirals are not the unique characteristic feature of spiroplasmas, as demonstrated by Townsend *et al.* (1977), who discovered nonhelical *S. citri* that otherwise were indistinguishable from the helical forms. The hypothesis also overlooks the Mesa Central-Rio Grande interference, a phenomenon known to occur only with closely related strains.

12.10 TOXIN ANTIMETABOLITES

The syndrome of yellows diseases, including chlorosis, wilting, growth abnormalities, such as stunting or axillary proliferation, necrosis, breaking of dormancy, proliferation of roots, and sterility could be the result of toxins. This hypothesis was tested by Daniels (1979) and at least two toxins incriminated in *S. citri*. Unfortunately, chemical characterization of the toxins has not been completed because of their instability and the difficulties encountered in their purification.

At present, no similar work has been performed with other spiroplasmas. It can be assumed that in other plant diseases, caused by MLOs, toxins are also responsible for plant chlorosis, stunting, breaking of dormancy, production of axillary shoots, seed sterility, and death of plants. When other MLO toxins will be isolated and the mechanism of toxin action worked out, it might become feasible to use antimetabolites that would overcome the action of toxins and either destroy them in diseased plants, or counteract their specific activity. Such approaches would only seem warranted in trees, such as pears with pear decline, elms with phloem necrosis, mulberry with dwarf disease, coconut palms with lethal yellowing, or sandal trees affected by spike disease, where the individual plants are of high value.

12.11 ALTERING VECTOR FEEDING BY ALIEN MYCOPLASMAS

In 1958 an observation was made that the accidental acquisition of the aster yellows spiroplasma by *D. maidis* leafhoppers altered the feeding habits of the insects and permitted their survival on plants that previously were not suitable as food (Maramorosch, 1958b). The acquisition of corn stunt spiroplasma and other MLOs by nonvector species has not been given much attention until now. The availability of cultured spiroplasmas and the detailed knowledge of feeding habits of numerous vector species of MLOs now permits a study of possible effects of MLOs on the feeding habits of nonvector species. Whether acquisition of alien MLOs by potential vectors can be used to divert their feeding from susceptible plants is not known at present, but if this could be achieved, it would provide an additional means of preventing the spread and transmission of MLOs (Orenski, 1964).

12.12 INDUCED MYCOPLASMA MUTATIONS FOR CROSS-PROTECTION AND VECTOR CONTROL

Artificially attenuated strains of spiroplasmas could be produced in the same manner in which induced mutations of bacteria are produced. Attenuated strains could be used as possible agents for immunization. The phenomenon of interference between strains of MLOs resembles viral interference. The systemic infection by a mild strain can sometimes prevent superinfection by a more virulent strain. Cross-protection could be used in instances where a disease is impossible to eradicate by other means, and where losses caused by the severe strains of MLOs are such that they outweigh those from widespread use of the milder strains. Technical difficulties to identify nonvirulent strains and to propagate them could be solved with present technology. The efficiency of such new biological control methods will have to be tested to evaluate their usefulness.

One can easily see how cross-protection with mild strains would be of practical use in trees, for instance, in coconut palms endangered by lethal yellowing disease. Before such tests could be made, however, the MLO of lethal yellowing would have to be cultured and means for its introduction into palms would have to become available. Neither of these prerequisites exists at this time and the vectors of lethal yellowing MLOs are still uncertain (Howard and Thomas, 1980; Tsai and Thomas, 1981). Even more difficult is the future application of mild strains of MLOs to such plants as onions, spinach, lettuce or carrots. In the absence of mechanical means of plant inoculation, vectors must be relied on to inoculate plants with mild strains, developed artificially. In a limited way, transmission by grafting or by dodder could be carried out. The use of insect vectors requires a greenhouse-type insectary for the inoculation of seedlings, that would be planted subsequently in strips along the fields days or weeks in advance of the planting of the commercial crop. Leafhoppers and planthoppers are often attracted to diseased plants and prefer them to healthy ones. These potential insect vectors would be able to acquire the mild strain of MLO from planted strips of diseased plants and would no longer be able to acquire and transmit the severe, naturally occurring strain. Such approaches might seem too costly or too difficult in large, commercial farming. They are actually used with mild strains of viruses in several countries at present.

The deleterious effect of certain MLOs on insect vectors has been mentioned (3.4). Induced mutant strains of cultured spiroplasmas could be selected that would affect insect vectors deleteriously, by shortening their life span or by destroying their embryos. This approach to biological control has not yet been attempted.

12.13 CONCLUSIONS

The control of MLO diseases can be achieved in various ways. Breeding for resistance to disease and breeding for resistance to insect vectors are among the most promising approaches. The control of leafhopper or planthopper vectors that transmit MLOs requires the drastic reduction or the complete elimination of the vectors, or the interruption of the transmission cycle. In all instances, the insect vectors constitute alternate hosts of the plant-pathogenic MLOs. Transmission could be prevented if each newly arriving vector would be killed before it had a chance to feed on a susceptible plant (Maramorosch, 1980c). The control of MLO dissemination by conventional pesticides has been effective in a limited number of cases and alternative approaches are urgently needed. In the seed production of certain plants, as in experimental tests, use of screens can provide the needed interruption of the disease cycle, effectively preventing insects from feeding on plants. Transitory repellency by sprays, to prevent feeding of potential vectors, has merely a temporary effect. Plant breeders could incorporate breeding for resistance to specific MLO vectors so as to achieve permanent resistance to vectors (Maramorosch, 1980a). Such breeding programs present exciting and challenging research problems and they are urgently needed. With proper collaboration among breeders, plant pathologists, and entomologists, and with increased international collaboration, this goal can be achieved.

The development of new approaches to biological control of MLO diseases, through the development of attenuated strains that could be used for plant immunization and of strains that would affect vectors deleteriously, through the use of alien MLOs that would alter the feeding preference of vectors, seems realistic and worthwhile. A search for chemicals other than tetracycline antibiotics can yield compounds that would effectively control MLOs in infected trees, or prevent their establishment in plants. The insidious dangers of using chemical pesticides, including systemic compounds, has been reviewed recently by Sullivan (1980). As pointed out by him, enormous amounts of funds are involved in the production, administration and sale of pesticides and the chemical industry has a vested interest in chemical pest control. Unfortunately, much of modern pest and vector control is dangerous and often irresponsible. In the next 15 years it can be expected that the world population will reach 6.5 billion. Therefore, food production will remain a major concern. Instead of using chemical insecticides alone to control vectors of MLOs, integrated control will have to be used. This will involve the combined use of biological control, such as natural insect predators, genetic manipulation, plant breeding for resistance to vectors, use of insect pheromones, microbial and viral pathogens, vector sterilization, cross-protection by strain interference, and selective use of small amounts of chemical insecticides (Knipling, 1979). In a limited number of instances, the proven, older methods of control, such as heat therapy, surgery, or screens will still find their applications.

In the next two decades, agricultural production not only in developed, but

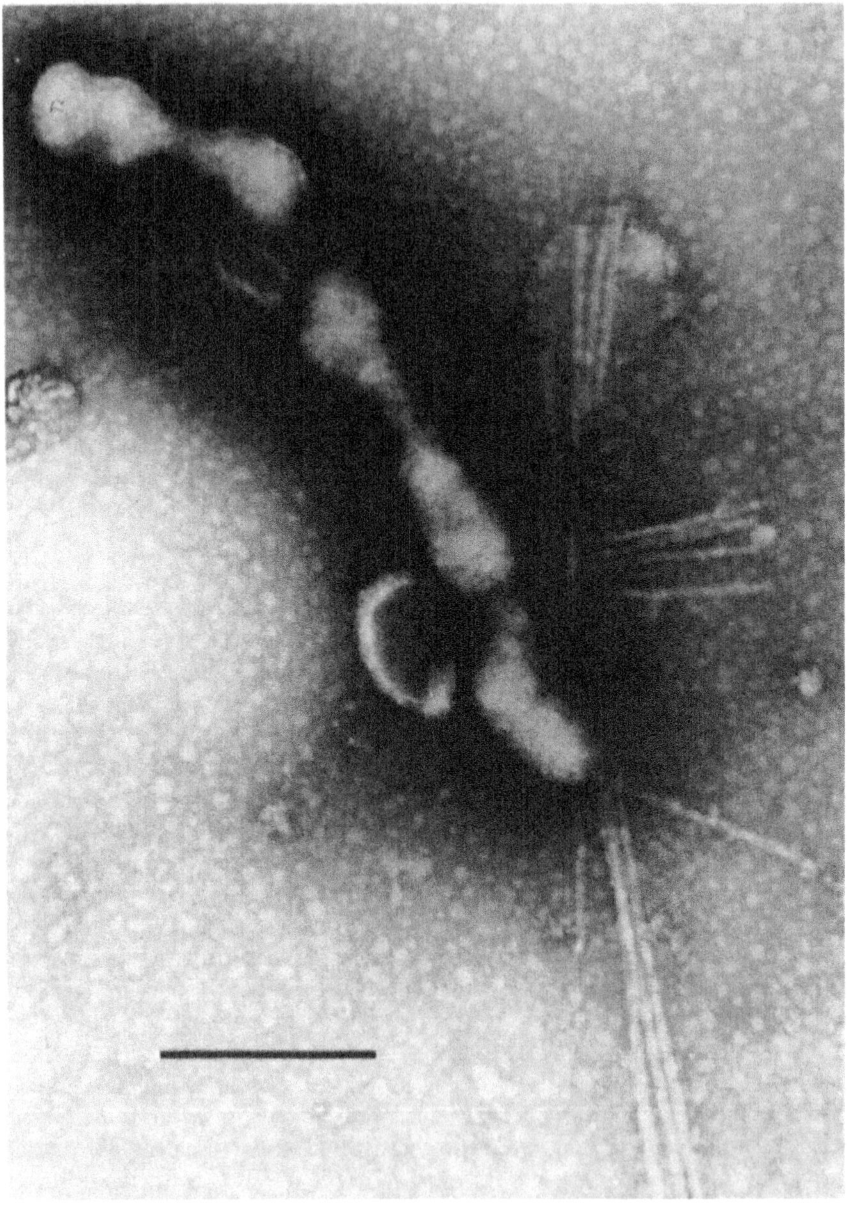

FIG. 12-5. Cultured aster yellows spiroplasma from diseased lettuce, associated with rod-shaped virus particles. Negative staining. Bar equals 0.2 μ.

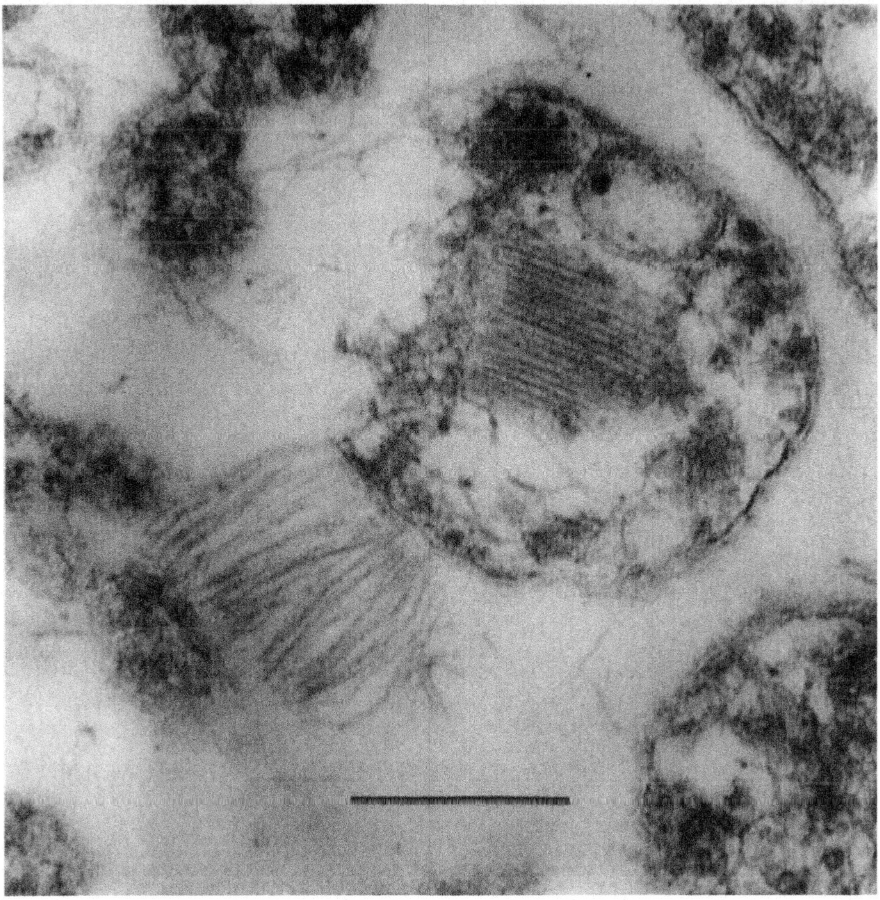

FIG. 12-6. Thin section of cultured aster yellows spiroplasma. Note ruptured spiroplasma membrane and rod-shaped virus particles inside and outside the spiroplasma. Bar equals 0.2 μ.

also, even more so, in developing countries, will have to be increased considerably to prevent a serious food deficit. Between 80 and 90% of the population in developing countries depends on agriculture directly or indirectly. Small farmers in developing countries require inexpensive, effective, and efficient methods to control diseases of plants. Farm products are not only the source of food for the world's population, but also the source of many industrial products. There are no easy and simple solutions to the problem of food supply but there are several research approaches that can be pursued now, so as to provide ways to eliminate or reduce losses caused by plant diseases, especially diseases caused by plant mycoplasmas. Integrated pest control promises to prevent economic losses, while taking into consideration the ecological constraints and preserving the environment. The ultimate goal in developing new approaches to the control of MLO-caused plant diseases is to achieve optimum plant production at the lowest cost, so as to benefit the farmers and the society.

New strategies against plant mycoplasma diseases and vectors of MLOs should be directed not only at epidemic outbreaks, when successful control would appear least likely to succeed. Attention should be given to vector populations during interepidemic periods, when insect populations are low and foci of disease more restricted and when the opportunity for successful management would be greater.

12.14 ACKNOWLEDGMENTS

Work on spiroplasmas, carried out at the Waksman Institute of Microbiology, Rutgers–The State University, was partially supported by grants from The Rockefeller Foundation (GA AS 7721), The National Science Foundation (BMS-74-13608), National Institutes of Health (PHS-AI-14509), by Waksman-Merck and Busch Postdoctoral Fellowships, and by a Charles and Johanna Busch award.

12.15 REFERENCES

Allen, T. C. (1972). Bacilliform particles within asters infected with a western strain of aster yellows. *Virology* **47**:491-493.

Anonymous. (1969). Tetracycline Research Group. Effects of tetracycline antibiotics on various plants with presumptive mycoplasmic diseases. 80 p. Tokyo, Japan (in Japanese).

Anonymous. (1973). World Health Organization. The use of viruses for the control of insect pests and disease vectors. WHO Tech. Rep. Ser. 531, Geneva.

Anonymous (1974). Studies on the pathogen of mulberry dwarf disease. I. Viruslike particles and mycoplasmalike bodies associated with mulberry yellow dwarf disease. *Sci. Sinica* **17**:421-427.

Anonymous. (1977). Integrated pest control. Report of 7th Session of FAO Panel of experts, Rome, 21-28 April 1977. FAO, Rome. 45 p.

Bastian, F. O., Hart, M. N., and Cancilla, P. A. (1981). Additional evidence of spiroplasma in Creutzfeldt-Jacob disease. *Lancet* No. 8221, **1**, 660.

Baur, E. (1904). Zur Aetiologie der infektiösen Panaschierung. *Ber. Deut. Bot. Gaz.* **22**:453-460.

Burges, H. D., and Hussey, N. W., eds. (1971), "Microbial Control of Insects and Mites." Academic Press, 861 p.

Calavan, E. C., and Gumpf, D. J. (1974). Studies on citrus stubborn disease. *Colloq. INSERM* **33**:181-185.

Charudattan, R. (1978). Biological control projects in plant pathology: A directory. Misc. Publications, Univ. of Florida, Gainesville.

Chen, T. A., and Liao, C. H. (1975). Corn stunt spiroplasma: isolation, cultivation, and proof of pathogenicity. *Science* **188**:1015-1017.

Christiansen, C., Freundt, E. A., and Maramorosch, K. (1980). Identity of cactus and lettuce spiroplasmas with *Spiroplasma citri* as determined by DNA-DNA hybridization. *Curr. Microbiol.* **4**:353-356.

Clark, T. B. (1977). *Spiroplasma* sp., a new pathogen in honey bees. *J. Invertebr. Pathol.* **29**: 112-113.

Cohen, S. (1981). Control of whitefly vectors of viruses by color mulches. *In* "Pathogens, Vectors, and Plant Diseases: Approaches to Control" (K. F. Harris and K. Maramorosch, eds.). Academic Press, New York.

Cole, R. M. (1977). Spiroplasmaviruses. *In* "The Atlas of Insect and Plant Viruses" (K. Maramorosch, ed.), pp. 451-457. Academic Press.

Cole, R. M., Tully, J. G., Popkin, T. J., and Bové, J. M. (1973). Morphology, ultrastructure, and bacteriophage infection of the helical mycoplasma-like organism (*Spiroplasma citri* gen. nov. sp. nov.) cultured from "stubborn disease" of citrus. *J. Bacteriol.* **115**:367-386.

Cole, R. M., Mitchell, W. O., and Garon, C. (1977). Sprioplasmavirus citri 3: Propagation, purification, proteins, and nucleic acid. *Science* **198**:1262-1263.

Daniels, M. J. (1979). Mechanisms of spiroplasma pathogenicity. *In* "The Mycoplasmas III" (R. F. Whitcomb and R. J. Tully, eds.), pp. 209-227. Academic Press.

Davis, R. E. (1978). Spiroplasma isolated from flowers of the tulip tree (*Liriodendron tulipifera* L.). *Can. J. Microbiol.* **24**:954-959.

Davis, R. E. (1981). The enigma of the flower spiroplasmas. *In* "Mycoplasma Diseases of Trees and Shrubs" (K. Maramorosch and S. P. Raychaudhuri, eds.), Academic Press, in press.

Davis, R. E., and Lee, I. M. (1981). Spiroplasmas: comparative properties and emerging taxonomic concepts; a proposal. *Rev. Inf. Dis.* In press.

Davis, R. E., and Worley, J. F. (1973). Spiroplasma: motile, helical microorganism associated with corn stunt disease. *Phytopathology* **63**:403-408.

Davis, R. E., Chen, T. A., and Worley, J. F. (1981a). Corn stunt spiroplasma. *In* "Virus and Viruslike Diseases of Maize in the United States" (D. T. Gordon, J. K. Knoke, and G. E. Scott, eds.), Vol. 247, pp. 43-53. Southern Coop. Series Bulletin.

Davis, R. E., Lee, I. M., and Worley, J. F. (1981b). *Spiroplasma floricola,* a new species isolated from surfaces of flowers of the tulip tree, *Liriodendron tulipifera* L. *Intl. J. System. Bacteriol.* In press.

Doi, Y., Terenaka, M., Yora, K., and Asuyama, H. (1967). Mycoplasma—or PLT group-like microorganisms found in the phloem elements of plants infected with mulberry dwarf, potato witches' broom, aster yellows, or paulownia witches' broom. (In Japanese.) *Ann. Phytopathol. Soc. Japan* **33**:259-266.

Frederick, R. J., Klein, M., and Maramorosch, K. (1971) Acquisition and retention of tetracycline-HC1 by plants. *Plant Dis. Rep.* **5**:223-226.

Freitag, J. H. (1967). Interaction between strains of aster yellows virus in six-spotted leafhopper *Macrosteles fascifrons. Phytopathology* **57**:535-537.

Freitag, J. H. (1969). Interactions of plant viruses and virus strains in their insect vectors. *In* "Viruses, Vectors, and Vegetation" (K. Maramorosch, ed.), pp. 303-325. Wiley (Interscience), New York.

Granados, R. R. (1969). Electron microscopy of plants and insect vectors infected with the corn stunt disease agent. *Contrib. Boyce Thompson Inst.* **24**:173-187.

Granados, R. R., and Meehan, D. J. (1975). Pathogenicity of the corn stunt agent to an insect vector, *Dalbulus elimatus. J. Invert. Pathol.* **26**:313-320.

Goldstein, N. I., McIntosh, A. H., Fisher, P. B., Maramorosch, K. and Schaffner, C. P. (1976). *In vitro* anti-mycoplasmal activity of amphotericin B methyl ester. *J. Antibiotics* **29**:656-661.

Harpaz, I. (1981). Nonpesticidal control of vector-borne viruses. *In* "Pathogens, Vectors, and Plant Diseases: Approaches to Control" (K. F. Harris and K. Maramorosch, eds.). Academic Press.

Heinrichs, E. A. (1979). Control of leafhopper and planthopper vectors of rice viruses. *In* "Leafhopper vectors and Plant Disease Agents" (K. Maramorosch and K. F. Harris, eds.), pp. 529-560. Academic Press.

Hemmati, K. (1979). Localization and pathogenicity of dwarf aster yellows mycoplasma-like agent in the midgut of the leafhopper vector *Macrosteles fascifrons. Iranian J. Plant Pathol.* **15**:45-51.

Hemmati, K., and McLean, D. L. (1980). Ultrastructure and morphological characteristics of mycoplasma-like organisms associated with Tulelake aster yellows. *Phytopathol. Z.* **99**: 146-154.

Hirumi, H., and Maramorosch, K. (1972). Natural degeneration of mycoplasma-like bodies in an aster yellows infected host plant. *Phytopathol. Z.* **75**:9-26.

Hirumi, H., and Maramorosch, K. (1973). Ultrastructure of the aster yellows agents: Mycoplasma-like bodies in sieve tube elements of *Nicotiana rustica. Ann. N. Y. Acad. Sci.* **225**:201-222.

Holmes, F. O., Hirumi, H., and Maramorosch, K. (1972). Witches' broom of willow: Salix yellows. *Phytopathology* **62**:826-828.

Howard, F. W., and Thomas, D. L. (1980). Transmission of palm lethal decline to *Veitchia merrillii* by a planthopper *Myndus crudus. J. Econ. Entomol.* **73**:715-717.

Ishiie, T., Doi, Y., Yora, K., and Asuyama, H. (1967). Suppressive effects of antibiotics of tetracycline group on symptom development of the mulberry dwarf disease. (In Japanese.) *Ann. Phytopathol. Soc. Japan* **33**:267-275.

Jensen, D. D. (1958). Reduction in longevity of leafhoppers carrying peach yellow leaf roll virus. *Phytopathology* **48**:394.

Jensen, D. D. (1959). A plant virus lethal to its insect vector. *Virology* **8**:164-175.

Jensen, D. D. (1962). Pathogenicity of Western-X-disease virus of stone fruits to its leafhopper vector, *Colladonus montanus* (Van Duzee). *Proc. 11th Intl. Congress Entomology Vienna,* **1960**:790-791.

Jensen, D. D. (1963). Effects of plant-viruses on insects. *Ann. N. Y. Acad. Sc.* **105**:685-712.

Knipling, E. F. (1979). The basic principles of insect population suppression and management, Agriculture Handbook No. 512, p. 659, U. S. Department of Agriculture, Washington, D. C.

Kondo, F., and Maramorosch, K. (1977). Antibiotic spectrum of plant pathogenic spiroplasmas. Abstracts ASM Annual Meeting: G-19.

Kondo, F., Maramorosch, K., McIntosh, A. H., and Varney, E. H. (1977). Aster yellows spiroplasma: isolation and cultivation *in vitro. Proc. Amer. Phytopathol. Soc.* **4**:190-191.

Kunkel, L. O. (1929). Wire screen fences for the control of aster yellows. *Phytopathology* **19**:100.

Kunkel, L. O. (1936a). Immunological studies on the three peach diseases, yellows, rosette, and little peach. *Phytopathology* **26**:201-219.

Kunkel, L. O. (1936b). Heat treatments for the cure of yellows and other virus diseases of peach. *Phytopathology* **26**:809-830.

Kunkel, L. O. (1937). Effect of heat on ability of *Cicadula sexnotata* (Fall.) to transmit aster yellows. *Am. J. Botany* **24**:316-327.

Kunkel, L. O. (1941). Heat cure of aster yellows in periwinkles. *Amer. J. Botany* **28**:761-769.

Kunkel, L. O. (1943). Potato witches' broom transmission by dodder and cure by heat. *Proc. Amer. Phil. Soc.* **86**:470.

Kunkel, L. O. (1945). Studies on cranberry false blossom. *Phytopathology* **35**:805.

Kunkel, L. O. (1952). Transmission of alfalfa witches' broom to nonleguminous plants by dodder and cure in periwinkle by heat. *Phytopathology* **42**:27-31.

Kunkel, L. O. (1955). Cross protection between strains of yellows-type viruses. *Adv. Virus Res.* **3**:251-273.

Kunkel, L. O. (1957). Acquired immunity from infection by strains of aster yellows virus in the aster leafhopper. *Science* **126**:1233.

Kurstak, E., Tijssen, P., and Maramorosch, K. (1978). Safety considerations and development problems make an ecological approach of biocontrol by viral insecticides imperative. *In* "Viruses and Environment" (E. Kurstak and K. Maramorosch, eds.), pp. 571-592. Academic Press.

La, Y. J, Brown, W. M., and Moon, D. S. (1976). Control of witches' broom disease of Jujube with oxytetracycline injection. *Korean J. Plant Protection* **15**:107-110.

Liao, C. H. and Chen, T. A. (1981). *In vitro* susceptibility and resistance of two spiroplasmas to antibiotics. *Phtyopathology* **71**:442-445.

Liss, A., and Cole, R. M. (1981). Spiroplasmavirus group I: isolation, growth and propagation. *Curr. Microbiol.* **5**:357-362.

Littau, V. C., and Maramorosch, K. (1956). Cytological effects of aster yellows virus on its insect vector. *Virology* **2**:128-130.

Littau, V. C., and Maramorosch, K. (1960). A study of the cytological effects of aster yellows virus on its insect vector. *Virology* **10**:483-500.

Maniloff, J., Das, J., and Christensen, J. R. (1977). Viruses of mycoplasmas and spiroplasmas. *Adv. Virus Res.* **21**:343-380.

Maramorosch, K. (1954). Biological transmission of plant viruses by animal vectors. *Trans. N. Y. Acad. Sci.* **16**:189-195.

Maramorosch, K. (1955). The occurrence of two distinct types of corn stunt in Mexico. *Plant Dis. Reporter* **39**:896-898.

Maramorosch, K. (1958a). Cross-protection between two strains of corn stunt virus in an insect vector. *Virology* **6**:448-459.

Maramorosch, K. (1958b). Beneficial effect of virus diseased plants on non-vector insects. *Tijdschrift over Plantenziekten* **63**:383-391.

Maramorosch, K. (1963). Arthropod transmission of plant viruses. *Ann. Rev. Entomol.* 8: 369-414.

Maramorosch, K. (1964). Interrelationships between plant pathogenic viruses and insects. *Ann. N. Y. Acad. Sci.* **118**:363-370.

Maramorosch, K. (1977). "The Atlas of Insect and Plant Viruses" (K. Maramorosch, ed.), pp. 478. Academic Press, New York.

Maramorosch, K. (1979). Biological control of insect pests with viruses. *In* "Practical Tissue Culture Applications" (K. Maramorosch and H. Hirumi, eds.), pp. 387-398. Academic Press.

Maramorosch, K. (1980a). Insects and plant pathogens. *In* "Breeding Plants Resistant to Insects" (F. G. Maxwell and P. R. Jennings, eds.), pp. 137-155. Wiley & Sons.

Maramorosch, K. (1980b). Spiroplasmas of plants and animals: an overview. Golden Jubilee Commemoration Volume, *National Academy Sc. India*, pp. 43-62.

Maramorosch, K. (1980c). Spread of plant viruses and mycoplasmas through airborne vectors. *Ann. N. Y. Acad. Sci.* **353**:179-185.

Maramorosch, K. (1981). Spiroplasmas: agents of animal and plant diseases. *BioScience* **31**:374-380.

Maramorosch, K., and Hunt, P. (1981). Lethal yellowing disease of coconut and other palms. *In* "Mycoplasma Diseases of Trees and Shrubs" (K. Maramorosch and S. P. Raychaudhuri, eds.). Academic Press, New York.

Maramorosch, K., and Jensen, D. D. (1963). Harmful and beneficial effects of plant viruses in insects. *Ann. Rev. Microbiol.* **17**:495-530.

Maramorosch, K., and Kondo, F. (1978). Aster yellows spiroplasma: infectivity and association with a rod-shaped virus. *Zentr. Bakteriol. Parasitol. Infekt. u. Hygiene* **241**:196.

Maramorosch, K., and Phillips, D. M. (1981). Scanning electron microscopy of aster yellows and cactus spiroplasmas. *Phytopathol. Z.* In press.

Maramorosch, K., and Raychaudhuri, S. P., eds. (1981). "Mycoplasma Diseases of Trees and Shrubs." Academic Press, New York.

Maramorosch, K., Shikata, E., and Granados, R. R. (1968). Structures resembling mycoplasma in diseased plants and in insect vectors. *Trans. N. Y. Acad. Sci.* **30**:841-855.

Maramorosch, K., Granados, R. R., and Hirumi, H. (1970). Mycoplasma diseases of plants and insects. *Adv. Virus Res.* **16**:135-193.

McIntosh, A. H., and Shamy, R. (1980). Biological studies of a baculovirus in a mammalian cell line. *Intervirology* **13**:331-341.

Miltenburger, H. G., ed. (1980a). Safety aspects of baculoviruses as biological insecticides. Symposium Proceedings. Bundesministerium für Forschung und Technologie, Bonn. Germany, 301 p.

Miltenburger, H. G. (1980b). Viral pesticides: hazard evaluation for non-target organisms and safety testing, pp. 57-75. *In* "Environmental Protection and Biological Forms of Control of Pest Organisms" (B. Lundholm and M. Stackerud, eds.), Vol. 31. Ecol. Bull. Stockholm.

Nasu, S., Sugiura, M., Wakimoto, T., and Iida, T. T. (1967). On the pathogen of rice yellow dwarf virus. (In Japanese) *Ann. Phytopathol. Soc. Japan.* **33**:343-344.

Nault, L. R., and Bradfute, O. E. (1979). Corn stunt: involvement of a complex of leafhopper-borne pathogens, pp. 561-586. *In* "Leafhopper Vectors and Plant Disease Agents" (K. Maramorosch and K. F. Harris, eds.). Academic Press.

Orenski, S. W. (1964). Effects of a plant virus on survival, food acceptability, and digestive enzymes of corn leafhoppers. *Ann. N. Y. Acad. Sci.* **118**:374-383.

Ortega, O. (1974). Proc. World Wide Maize Improv. 70s and Role of CYMMIT, El Batan, Mexico.

Osler, R. D. (1977). Proc. Maize Virus Dis. Colloq. Workshop 1976, Ohio Agric. Res. and Develop. Center, Wooster, p. 132.

Ota, T., Kawabe, M., Oishi, K., and Poulson, D. F. (1979). Non-male killing spiroplasmas in *Drosophila hydei*. *J. Hered.* **70**:211-213.

Padhi, S. B., and Maramorosch, K. (1980). Effect of a commercial preparation of *Heliothis zea* baculovirus on silkworm, *Bombyx mori*. *ASM Annu. Abstr.* **Q-79**:207.

Padhi, S. B., and McIntosh, A. H. (1976). Comparison of two plant spiroplasmas by electrophoresis of their proteins. *In Vitro* **12**:334-335.

Padhi, S. B., McIntosh, A. H., and Maramorosch, K. (1977). Characterization and identification of spiroplasmas by polyacrylamide gel electrophoresis. *Phytopathol. Z.* **90**:268-272.

Plavsic-Banjac, B., Hunt, P., and Maramorosch, K. (1972). Mycoplasma-like bodies associated with lethal yellowing disease of coconut palms. *Phytopathology* **62**:298-299.

Poulson, D. F., and Sakaguchi, B. (1961). Nature of "sex-ratio" agent in *Drosophila*. *Science* **133**:1489-1490.

Sinha, R. C., and Petersen, E. E. A. (1972). Uptake and persistence of oxytetracycline in aster plants and vector leafhoppers in relation to inhibition of clover phyllody agent. *Phytopathology* **62**:377-383.

Stoddard, E. M. (1942). Inactivating *in vivo* the virus of X-disease of peach by chemotherapy. *Phytopathology* **32**:17.

Sullivan, D. J. (1980). Pesticide treadmill. America, Oct. 25, 1980, 247-249.

Townsend, R., Markham, P. G., Plaskitt, K. A., and Daniels, M. J. (1977). Isolation and characterization of a non-helical strain of *Spiroplasma citri*. *J. Gen. Microbiol.* **100**:15-21.

Tsai, J. H., and Thomas, D. L. (1981). Transmission of lethal yellowing mycoplasma by *Myndus crudus. In* "Mycoplasma Diseases of Trees and Shrubs" (K. Maramorosch and S. P. Raychaudhuri, eds.). Academic Press.

Wade, N. (1973). Insect viruses: a new class of pesticides. *Science* **181**:925-928.

Williamson, D. L., and Whitcomb, R. F. (1974). Helical, wall-free prokaryotes in *Drosophila,* leafhoppers and plants. *INSERM (Inst. Natl. Sante. Rech. Med.)* 283-290.

Wolanski, B., and Maramorosch, K. (1970). Negatively stained mycoplasmas: fact or artifact? *Virology* **42**:319-327.

Wolanski, B. S., and Maramorosch, K. (1979). Rayado fino virus and corn stunt spiroplasma, phloem restriction and transmission by *Dalbulus elimatus* and *D. maidis. Fitopatol. Brasiliera* **4**:47-54.

INDEX

D

H

I

J

K

Q

T

U

V

W

Printed in Dunstable, United Kingdom

91341566R00181